Continuum Mechanics

Continuum Mechanics

Prasun Kumar Nayak
*Department of Mathematics (UG & PG)
Midnapore College (Autonomous),
P.O.& Dist. Midnapore–721 101*

Mijanur Rahaman Seikh
*Department of Mathematics
Kazi Nazrul University,
Asansol–713 340
West Bengal, INDIA*

First published 2022
by CRC Press
4 Park Square, Milton Park, Abingdon, Oxon, OX14 4RN

and by CRC Press
6000 Broken Sound Parkway NW, Suite 300, Boca Raton, FL 33487-2742

© 2022 Manakin Press

CRC Press is an imprint of Informa UK Limited

The right of Prasun Kumar Nayak and Mijanur Rahaman Seikh to be identified as the authors of this work has been asserted in accordance with sections 77 and 78 of the Copyright, Designs and Patents Act 1988.

All rights reserved. No part of this book may be reprinted or reproduced or utilised in any form or by any electronic, mechanical, or other means, now known or hereafter invented, including photocopying and recording, or in any information storage or retrieval system, without permission in writing from the publishers.

For permission to photocopy or use material electronically from this work, access www.copyright.com or contact the Copyright Clearance Center, Inc. (CCC), 222 Rosewood Drive, Danvers, MA 01923, 978-750-8400. For works that are not available on CCC please contact mpkbookspermissions@tandf.co.uk

Trademark notice: Product or corporate names may be trademarks or registered trademarks, and are used only for identification and explanation without intent to infringe.

Print edition not for sale in South Asia (India, Sri Lanka, Nepal, Bangladesh, Pakistan or Bhutan).

British Library Cataloguing-in-Publication Data
A catalogue record for this book is available from the British Library

Library of Congress Cataloging-in-Publication Data
A catalog record has been requested

ISBN: 978-1-032-29047-8 (hbk)
ISBN: 978-1-003-29977-6 (ebk)

DOI: 10.1201/9781003299776

Contents

Preface	*vii*
1. Theory of Strain	**1**
1.1 Continuum and Continuum Particle	1
1.2 Continuum Mechanics	2
1.3 Configuration	2
1.4 Formulation of Models	3
1.5 Deformation	4
1.6 Methods of Descriptions	6
1.7 Displacement	9
1.8 Deformation Gradients, Finite Strain Tensors	13
1.9 Infinitesimal Strain Tensor	42
1.10 Strain Quadric	67
1.11 Principal Strain	70
1.12 Strain Equations of Compatibility for Infinitesimal Strains	89
1.13 Stretch and Rotation Tensors	95
1.14 Plane Strain and Mohr's Circles for Strain	100
Exercise	103
2. Theory of Motion	**111**
2.1 Motion	111
2.2 Material and Spatial Description	112
2.3 Material and Spatial Time Derivatives	113
2.4 Velocity and Acceleration	116
2.5 Flow	128
2.6 Boundary Surface	138
2.7 Material Derivative of the Element of Arc, Surface, and Volume	143
2.8 Kinematics of Line, Surface and Volume Integrals	157
2.9 Spin vector and Spin Tensor	159
2.10 Irrotational motion and Velocity potential	165
2.11 Objective Tensor	167

	Exercise	170

3. Theory of Stress — **179**
   3.1 Forces of continuum: Body and Surface Forces, Mass Density   179
   3.2 Euler-Cauchy Stress Principle   181
   3.3 Cauchy's Stress Formula: Stress Tensor   186
   3.4 Cauchy's Stress Quadric   192
   3.5 Transformation of Stress Components   196
   3.6 Principal Stress   197
   3.7 Stress Invariants   202
   3.8 Extremum of Stress Values   204
   3.9 Mohr's Circles for Stress   211
   3.10 Plane Stress   220
   Exercise   225

4. Fundamental Principles of Continuum Mechanics — **229**
   4.1 Principle of Conservation of Mass   229
   4.2 Balance of Linear Momentum   234
   4.3 Balance of Angular Momentum   235
   4.4 Thermodynamics of Continuous Media   237
   Exercise   241

5. Linear Elasticity — **243**
   5.1 Linearly Elastic Solid   243
   5.2 Strain Energy   245
   5.3 Elastic Symmetry   249
   5.4 Isotropic Elastic Body   253
   5.5 Strains in Terms of Stresses   257
   5.6 Fundamental Boundary Value Problem in Elastostatics   274
   5.7 Fundamental Boundary Value Problem in Elasto-dynamics   286
   Exercise   292
   Bibliography   294

Preface

This book has been designed to introduce the fundamental concepts of Continuum Mechanics. The basic concepts and also the advanced topics of Continuum Mechanics have been dealt within simple and logical language.

This book has been written according to the different areas of Science and Engineering. A unique feature of the book is that each chapter has been presented with different types of solved problems that are explained in a simple way. This book also contains a wide variety of exercises which are intended to be an important part of the text.

Our indebtedness remains to the authors of some advanced texts which have been consulted during the presentation of this book.

The authors have tried to put their best efforts for improving the subject matter. In spite of that, there might have crept some mistakes and misprints due to quick submission. Any kind of corrections, suggestions and critical evaluation for further improvement of this book will be highly appreciated and acknowledged.

Finally, we are thankful to Manakin Press for their keen interest and sincere efforts to bring out the book.

Authors

1

Theory of Strain

1.1 Continuum and Continuum Particle

Materials, such as solids, liquids and gases, are composed of molecules separated by "empty" space. On a microscopic scale, materials have cracks and discontinuities. However, certain physical phenomena can be modeled assuming the materials exist as a *continuum*. Continuum means the matter in the body is continuously distributed and fills the entire region of space it occupies with no empty space. This ensures that it possesses unique physical properties (such as unique density, unique displacement, unique velocity at every point of space) which can be expressed as continuous functions of position and time. A continuum is a body that can be continually sub-divided into infinitesimal elements with properties being those of the bulk material.

Thus, matter is idealized as a *continuum*, which has two properties:
   (*i*) it is infinitely divisible (one can subdivide some region of the solid as many times as he/she wish); and
   (*ii*) it is locally homogeneous—in other words, if you subdivide it sufficiently many times, all sub-divisions have identical properties (eg mass density).

A continuum can be thought of as an infinite set of vanishingly small particles, connected together.

The *material point* or the *particle* can be defined as a smallest piece of matter containing a large number of molecules within an infinitesimally small volume whose physical dimensions are so small that it may be regarded as concentrated at a point. Accordingly, in continuum we can associate a material point with each and every spatial point of the region of space occupied by a continuum body. One also speaks of a particle in a continuum, which means an infinitesimal volume of material.

1.2 Continuum Mechanics

The French mathematician Augustin-Louis Cauchy was the first to formulate such models in the 19th century, but research in the area continues today.

Continuum mechanics, a scientific discipline, is a branch of mechanics that deals with the analysis of the kinematics and the mechanical behavior of substances (materials modeled as a continuous mass rather than as discrete particles) under the influence of external agents that make changes in the state of medium. These changes may appear in the form of contact forces, such as chemical, electrical, mechanical or any other type of disturbances.

Such mechanics, a combination of mathematics and physical laws that approximate the large-scale behavior of matter that is subjected to mechanical loading, deals with physical properties of solids and fluids which are independent of any particular coordinate system in which they are observed. These physical properties are then represented by tensors, which are mathematical objects that have the required property of being independent of coordinate system. These tensors can be expressed in coordinate systems for computational convenience.

1.2.1 Major Areas of Continuum Mechanics

Continuum mechanics The study of the physics of continuous materials	Solid mechanics The study of the physics of continuous materials with a defined rest shape.	Elasticity Describes materials that return to their rest shape after applied stresses are removed.	
		Plasticity Describes materials that permanently deform after a sufficient applied stress.	Rheology The study of materials with both solid and fluid characteristics.
	Fluid mechanics The study of the physics of continuous materials which deform when subjected to a force.	Non-Newtonian fluids do not undergo strain rates proportional to the applied shear stress.	
		Newtonian fluids undergo strain rates proportional to the applied shear stress.	

1.3 Configuration

The complete specification of the position of all of the particles of B with respect to a fixed origin at some instant of time is said to define the *configuration* of the body at that instant.

The *configuration* of a solid is a region of space occupied (filled) by the solid. When we describe motion, we normally choose some convenient configuration of the solid to use as *reference* - this is often the initial, undeformed solid, but it can be any convenient region that could be occupied by the solid. The material changes its shape under the action of external loads, and at some time t occupies a new region which is called the *deformed* or *current* configuration of the solid.

We give special meaning to certain configurations of the body. In particular, we single out a *reference configuration* from which all displacements are reckoned. For the purpose it serves, the reference configuration need not be one the body ever actually occupies. Often, however, the *initial configuration*, that is, the one which the body occupies at time $t = 0$ is chosen as the reference configuration, and the ensuing deformations and motions related to it. The *current configuration* is that one which the body occupies at the current time t.

In developing the concepts of strain, we confine attention to two specific configurations without any regard for the sequence by which the second configuration is reached from the first. It is customary to call the first (reference) state the *undeformed configuration,* and the second state the *deformed configuration*. Additionally, time is not a factor in deriving the various strain tensors, so that both configurations are considered independent of time.

In fluid mechanics, the idea of specific configurations has very little meaning since fluids do not possess a natural geometry, and because of this it is the *velocity field* of a fluid that assumes the fundamental kinematic role.

1.4 Formulation of Models

Modeling an object as a continuum assumes that the substance of the object completely fills the space it occupies. Modeling objects in this way ignores the fact that matter is made of atoms, and so is not continuous; however, on length scales much greater than that of inter-atomic distances, such models are highly accurate. Fundamental physical laws such as the conservation of mass, the conservation of momentum, and the conservation of energy may be applied to such models to derive differential equations describing the behavior of such objects, and some information about the particular material studied is added through a constitutive relation.

A particular particle within the body in a particular configuration is characterized by a position vector where are the coordinate vectors in some frame of reference chosen for the problem (See Fig. 1.1).

This vector can be expressed as a function of the particle position in some *reference configuration*, for example, the configuration at the initial time, so that

$$x = \kappa(X, t) \tag{1.1}$$

We assume that the mappings (1.1) are single-valued and possess continuous partial derivatives with respect to their arguments for whatever order is desired, except possibly at some singular points, curves, and surfaces.

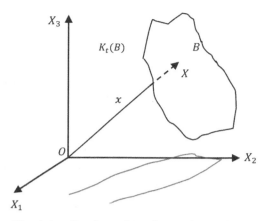

Fig. 1.1: Configuration of a continuum body

Moreover, each member of (1.1) is the unique inverse of the other in a neighborhood of the material point. This assumption is known as the *axiom of continuity*. It expresses the fact that the matter is indestructible, that is, no region of positive, finite volume of matter is deformed into a zero or infinite volume. Another implication of this axiom is that the matter is impenetrable, that is, the motion carries every region into a region, every surface into a surface, and every curve into a curve. One portion of matter never penetrates into another. In practice, there are cases in which this axiom is violated. For example, materials may break or may transmit shock and other types of discontinuities. Special attention must be given to these. The axiom of continuity is secured through a well- known theorem of calculus known as the implicit function theorem.

Thus the mapping (1.1) needs to have various properties so that the model makes physical sense such as–

(*i*) continuous in time, so that the body changes in a way which is realistic,
(*ii*) globally invertible at all times, so that the body cannot intersect itself,
(*iii*) orientation-preserving, as transformations which produce mirror reflections are not possible in nature.

For the mathematical formulation of the model, $\kappa(X,t)$ is also assumed to be twice continuously differentiable, so that differential equations describing the motion may be formulated.

1.5 Deformation

A Continuum body will change its configuration and orientation as well as its shape in general deformation. A deformation may be caused by external loads, body forces (such as gravity or electromagnetic forces), or temperature

1.5 Deformation

changes within the body. Our first task is to separate that part of the general deformation which causes a change in configuration and orientation of the body from that part which causes a change in the shape of the body.

(*a*) **Rigid-body deformation:** When the deformation is such that there are no changes in the relative positions of constituent material points of the continuum firmly bound together so that the length of any line joining any two material points does not change, then the deformation is a combination of translation and rotation about an axis at any point causing a change in configuration and orientation of the body only and is called rigid-body deformation and the body is called rigid body.

(*b*) **Strain deformation:** When the deformation is such that there are changes in the relative positions of constituent material points of the continuum body so that the length and orientation of any line joining any two points changes, then the deformation causes a change in the shape of the body only and is called strain deformation and the body is called deformable body. The existence of strain deformation depends on the occurrence of relative displacement of points in the medium with respect to each other.

Strain is a description of deformation in terms of *relative* displacement of particles in the body that excludes rigid-body motions. Different equivalent choices may be made for the expression of a strain field depending on whether it is defined with respect to the initial or the final configuration of the body and on whether the metric tensor or its dual is considered.

In a continuous body, a deformation field results from a stress field induced by applied forces or is due to changes in the temperature field inside the body. The relation between stresses and induced strains is expressed by constitutive equations, *e.g.*, Hooke's law for linear elastic materials. Deformations which are recovered after the stress field has been removed are called *elastic deformations*. In this case, the continuum completely recovers its original configuration. On the other hand, irreversible deformations remain even after stresses have been removed. One type of irreversible deformation is *plastic deformation*, which occurs in material bodies after stresses have attained a certain threshold value known as the *elastic limit* or yield stress, and are the result of slip, or dislocation mechanisms at the atomic level. Another type of irreversible deformation is *viscous deformation*, which is the irreversible part of viscoelastic deformation.

In the case of elastic deformations, the response function linking strain to the deforming stress is the compliance tensor of the material.

To begin with, we will describe all motions and deformations by expressing positions of points in both undeformed and deformed solids as components in a Cartesian reference frame (which is also taken to be an inertial frame).

Thus x_i will denote components of the position vector of a materal particle before deformation, and $X_i(x_k)$ will be components of its position vector after deformation.

Mathematically, we describe a deformation as a *one-one* mapping which transforms points from the reference configuration of a solid to the deformed configuration. Specifically, let ξ_i be three numbers specifying the position of some point in the undeformed solid (these could be the three components of position vector in a Cartesian coordinate system, or they could be a more general coordinate system, such as polar coordinates). As the solid deforms, each the values of the coordinates change to different numbers. We can write this in general form as

$$\eta_i = f_i(\xi_i, t) \tag{1.2}$$

This is called a *deformation mapping*. To be a physically admissible deformation

(*i*) The coordinates must specify positions in a Newtonian reference frame. This means that it must be possible to find some coordinate transformation, such that x_i are components in an orthogonal basis, which is taken to be 'stationary' in the sense of Newtonian dynamics.

(*ii*) The functions $f_i(\xi_k)$ must be *one-one* on the full set of real numbers; and f_i must be invertible

(*iii*) f_i must be continuous and continuously differentiable (we occasionally relax these two assumptions, but this has to be dealt with on a case-by-case basis)

(*iv*) The mapping must satisfy $det\left|\dfrac{\partial \eta_i}{\partial \xi_i}\right| > 0$.

1.6 Methods of Descriptions

Continuum mechanics models begin by assigning a region in three dimensional Euclidean spaces to the material body being modeled. The points within this region are called particles or material points. Different configurations or states of the body correspond to different regions in Euclidean space. The region corresponding to the body's configuration at time is labeled.

It is convenient to identify a reference configuration or initial condition which all subsequent configurations are referenced from. The reference configuration need not be one that the body will ever occupy. Often, the configuration at $t = 0$ is considered the reference configuration $K_0(B)$. The components of the position vector $X = (X_1, X_2, X_3)$ of a particle, taken with respect to the reference configuration, are called the material or reference coordinates. All physical properties associated with this material point will then be functions of X_1, X_2, X_3 and t.

When analyzing the deformation or motion of solids, or the flow of fluids, it is necessary to describe the sequence or evolution of configurations throughout time. There are two methods of analyzing the properties of the deformation or motion of a continuum:

(*i*) Lagrangian or Material method

(*ii*) Eulerian or Spatial method.

(i) **Lagrangian Description/Material Method**

Continuum mechanics theory makes use of a set of coordinates, known as material (or reference) coordinates. In the material or Lagrangian method, our object of study is material points of a continuum body. In this method, we identify individual material points and describe the motion of each individual material point of fixed identity for all time by following its motion through out its course. In this approach, individual material points are endowed with physical properties, which may be changed in two ways:

(*a*) they change as we pass from one material point to another and

(*b*) they change as time changes for a fixed material point.

In other words, these properties are considered as functions of time and of those data which identify the material points. These are normally denoted by uppercase variables *X, Y,* and *Z* (or *R,* Φ, and *Z*) and are used to label material particles. For such data we usually take the rectangular Cartesian coordinates X_1, X_2, X_3 of the position of a material point of the continuum body in its initial deformed state. We identify the given material point by (X_1, X_2, X_3). It is given a fixed identity by spacifying its initial position. The three symbols (X_1, X_2, X_3) are assigned once and for all time as the label or name of the material point considered, because it retains them throughout the whole investigation. All physical properties associated with this material point will then the functions of X_1, X_2, X_3 and time *t*. The primary quantity in this is the position of the material point in the deformed state of the body at subsequent time *t*. If (x_1, x_2, x_3) be the rectangular Cartesian coordinates of this position, them

$$x_i = x_i(X_1, X_2, X_3, t); \quad i = 1, 2, 3. \tag{1.3}$$

Equation (1.3) describes motion of the material point completely in material method giving the subsequent position at time *t*. The coordinates X_1, X_2, X_3 are independent coordinates called *material coordinates* or *Lagrangian coordinates*, whereas x_1, x_2, x_3 are dependent coordinates called *spatial coordinates*.

If a physical property of the body B such as its density ρ, or a kinematic property of its motion such as the velocity **v**, is expressed in terms of the material coordinates **X**, and the time t, we say that property is given by the *referential* or *material description*. When the referential configuration is taken as the actual configuration at time $t = 0$, this description is usually called the *Lagrangian description*. Any material particle is uniquely identified by its position in some given initial or reference configuration. As long as the solid stays in this configuration, material and spatial coordinates of every particle coincide and displacements are zero by definition.

Material coordinate variables X, Y, and Z must be used in coordinate-dependent expressions that refer to positions in the original geometry, for example, for material properties that are supposed to follow the material during deformation.

(ii) Eulerian description/Spatial Method

The Solid Mechanics interface, through its equations, describes the motion and deformation of solid objects in a 2- or 3-dimensional space, this physical space is known as the spatial frame and positions in the physical space are identified by lowercase spatial coordinate variables x, y, and z (or r, φ, and z in axisymmetric models).

In this spatial or Eulerian method, our object of study strictly speaking, is not moving material points but fixed spatial point. We identify the spatial points and describe the motion of the medium at each spatial point at different times without considering the whereabouts of individual material points. We focus our attention on a fixed spatial point in space occupied by different material points at different times and observe what changes of various properties are taking place at the spatial point. The spatial points, in a manner of speaking are endowed with physical properties and a material point is said to acquire these properties associated with the fixed spatial point when it pass through that spatial point. In this approach, physical properties change in two ways:

(a) when we pass from one spatial point to another point and

(b) with time at a fixed spatial point.

If a material point which was at the position (X_1, X_2, X_3) in the undeformed state at $t = 0$ happens to occupy the position (x_1, x_2, x_3) at subsequent time t, then coordinates x_1, x_2, x_3 identify the spatial point in the deformed state. The physical properties will be functions of the position (x_1, x_2, x_3) at time t. In particular

$$X_i = X_i(x_1, x_2, x_3, t), i = 1, 2, 3, \qquad (1.4)$$

which traces the material point occupying spatial position (x_1, x_2, x_3).

1.7. Displacement

Quantities that have a coordinate dependence in physical space, for example, a spatially varying electromagnetic field acting as a force on the solid, must be described using spatial coordinate variables x, y, and z.

Example 1.1: Let the motion equations be given in component form by the Lagrangian description $x_1 = X_1 e^t + X_3(e^t - 1)$, $x_2 = X_2 + X_3(e^t - e^{-t})$; $x_3 = X_3$. Determine the Eulerian description of this motion.

Solution: The given transformation equation can be written in the form $x = x(X, t)$. Notice first that for the given motion $x_1 = X_1$, $x_2 = X_2$; and $x_3 = X_3$, at $t = 0$, so that the initial configuration has been taken as the reference configuration. The Jacobian determinant is given by

$$J = \left|\frac{\partial x_i}{\partial X_k}\right| = \begin{vmatrix} \frac{\partial x_1}{\partial X_1} & \frac{\partial x_2}{\partial X_2} & \frac{\partial x_3}{\partial X_3} \\ \frac{\partial x_2}{\partial X_1} & \frac{\partial x_2}{\partial X_2} & \frac{\partial x_2}{\partial X_3} \\ \frac{\partial x_3}{\partial X_1} & \frac{\partial x_3}{\partial X_2} & \frac{\partial x_3}{\partial X_3} \end{vmatrix} = \begin{vmatrix} e^t & 0 & e^t - 1 \\ 0 & 1 & e^t - e^{-t} \\ 0 & 0 & 1 \end{vmatrix} = e^t \neq 0$$

Since the Jacobian determinant does not vanish, we can solve for the inverse equations $X = X(x, t)$. Because of the simplicity of these Lagrangian equations of the motion, we may substitute for into the first two equations and solve these directly to obtain the inverse equations

$$X_1 = x_1 e^{-t} + x_3(e^{-t} - 1), \; X_2 = x_2 + x_3(e^{-t} - e^t), \; X_3 = x_3$$

1.7 Displacement

A change in the configuration of a continuum body results in a *displacement*. The displacement of a body has two components: a *rigid-body displacement* and a *deformation*. A rigid-body displacement consists of a simultaneous translation and rotation of the body about an axis at any point causing a change in configuration and orientation without changing its shape or size. Deformation implies the change its configuration and orientation as well as in shape and/or size of the continuum body from an initial or undeformed configuration $K_0(B)$ to a current or deformed configuration $K_t(B)$ (Fig. 1.2).

If after a displacement of the continuum there is a relative displacement between particles, a deformation has occurred. On the other hand, if after displacement of the continuum the relative displacement between particles in the current configuration is zero, i.e. the distance between particles remains unchanged, then there is no deformation and a rigid-body displacement is said to have occurred.

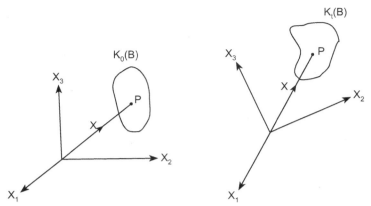

Fig. 1.2: Motion of a continuum body.

The displacement of a typical material point from its position (X_1, X_2, X_3) in the undeformed state at $t = 0$ to its position (x_1, x_2, x_3) in the deformed state at time t is defined by

$$u_i = x_i - X_i, \quad i = 1, 2, 3. \tag{1.5}$$

In the material description and are regarded as functions of X_1, X_2, X_3 and t so that displacement

$$u_i(X_1, X_2, X_3, t) = x_i(X_1, X_2, X_3, t) - X_i. \tag{1.6}$$

In the spatial description u_i and X_i are regarded as functions of x_1, x_2, x_3 and t so that displacement

$$u_i(x_1, x_2, x_3, t) = x_i - X_i(x_1, x_2, x_3, t). \tag{1.7}$$

In the first of this pair of equations we are describing the displacement that will occur to the particle that starts at **X**, and in the second equation we present the displacement that the particle now at **x** has undergone.

Example 1.2: Obtain the displacement field for the motion in both material and spatial descriptions.

$$x_1 = X_1 e^t + X_3(e^t - 1), \; x_2 = X_2 + X_3(e^t - e^{-t}); \; x_3 = X_3.$$

Solution: From the motion equations of motion, namely

$$x_1 = X_1 e^t + X_3(e^t - 1), \; x_2 = X_2 + X_3(e^t - e^{-t}); \; x_3 = X_3$$

we may compute the displacement field in material form directly as

1.7. Displacement

$$u_1 = x_1 - X_1 = X_1 e^t + X_3(e^t - 1) - X_1 = (X_1 + X_3)(e^t - 1)$$
$$u_2 = x_2 - X_2 = X_2 + X_3(e^t - e^{-t}) - X_2 = X_3(e^t - e^{-t})$$
$$u_3 = x_3 - X_3 = X_3 - X_3 = 0$$

By using the inverse equations, namely,

$$X_1 = x_1 e^{-t} + x_3(e^{-t} - 1), \; X_2 = x_2 + x_3(e^{-t} - e^t), \; X_3 = x_3$$

we obtain the spatial description of the displacement field in component form

$$u_1 = x_1 - X_1 = x_1 - x_1 e^{-t} - x_3(e^{-t} - 1) = (x_1 + x_3)(1 - e^{-t})$$
$$u_2 = x_2 - X_2 = x_2 - x_2 - x_3(e^{-t} - e^t) = x_3(e^t - e^{-t})$$
$$u_3 = x_3 - X_3 = x_3 - x_3 = 0$$

Example 1.3: With respect to superposed material axes X_1 and spatial axes x_1, the displacement field of a continuum body is given by $x_1 = X_1, x_2 = X_2 + AX_3, x_3 = X_3 + AX_2$ where is a constant. Determine the displacement vector components in both the material and spatial forms. Determine the displaced location of the material particles which originally comprise (for $A = 1/2$)

(a) the plane circular surface $X_1 = 0, X_2^2 + X_3^2 = 1/(1 - A^2)$
(b) the infinitesimal cube with edges along the coordinate axes of length $dX_i = dX$.

Solution: From the motion equations of motion, the displacement field in material form can be written as

$$u_1 = x_1 - X_1 = X_1 - X_1 = 0$$
$$u_2 = x_2 - X_2 = X_2 + AX_3 - X_2 = AX_3$$
$$u_3 = x_3 - X_3 = X_3 + AX_2 - X_3 = AX_2$$

The Jacobian determinant is given by

$$J = \left| \frac{\partial x_i}{\partial X_k} \right| = \begin{vmatrix} \frac{\partial x_1}{\partial X_1} & \frac{\partial x_1}{\partial X_2} & \frac{\partial x_1}{\partial X_3} \\ \frac{\partial x_2}{\partial X_1} & \frac{\partial x_2}{\partial X_2} & \frac{\partial x_2}{\partial X_3} \\ \frac{\partial x_3}{\partial X_1} & \frac{\partial x_3}{\partial X_2} & \frac{\partial x_3}{\partial X_3} \end{vmatrix} = \begin{vmatrix} 1 & 0 & 0 \\ 0 & 1 & A \\ 0 & A & 1 \end{vmatrix} = 1 - A^2 \neq 0.$$

Since the Jacobian determinant J does not vanish, we can solve for the inverse equations $X = x(x, t)$. Inverting the given displacement relations to obtain

$$X_1 = x_1, X_2 = (x_2 - Ax_3)/(1-A^2), X_3 = (x_3 - Ax_2)/(1-A^2)$$

The spatial description of the displacement field in component form are given by

$$u_1 = x_1 - X_1 = x_1 - x_1 = 0$$
$$u_2 = x_2 - X_2 = x_2 - (x_2 - Ax_3)/(1-A^2) = A(x_3 - Ax_2)/(1-A^2)$$
$$u_3 = x_3 - X_3 = x_3 - (x_3 - Ax_2)/(1-A^2) = A(x_2 - Ax_3)/(1-A^2)$$

From these results it is noted that the originally straight line of material particles expressed by $X_1 = 0$, $X_2 + X_3 = 1/(1+A)$ occupies the location $x_1 = 0$, $x_2 + x_3 = 1$ after displacement. Likewise the particle line $X_1 = 0$, $X_2 = X_3$ becomes after displacement $x_1 = 0$, $x_2 = x_3$.

(a) By the direct substitutions $X_2 = (x_2 - Ax_3)/(1-A^2)$, $X_3 = (x_3 - Ax_2)/1-A(1-A^2)$, the circular surface becomes the elliptical surface

$$(1+A^2)x_2^2 - 4Ax_2 x_3 + (1+A^2)x_3^2 = (1-A^2)$$

For $A = 1/2$, this is bounded by the ellipse $5x_2^2 - 8x_2 x_3 + 5x_3^2 = 3$ which when referred to its principal axes x_i^* (at 45° with x_i, $i = 2, 3$) has the equation $x_2^{*2} + 9x_3^{*2} = 3$. Fig. 1.3 below shows this displacement pattern.

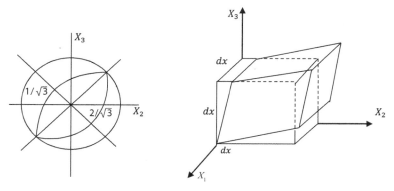

Fig. 1.3: Displacement Pattern

(b) From this problem, the displacements of the edges of the cube are readily calculated. For the edge $X_1 = X_1$, $X_2 = X_3 = 0$, $u_1 = u_2 = u_3 = 0$. For the edge $X_1 = X_2 = 0$, $X_3 = X_3$, $u_1 = u_3 = 0$, $u_2 = AX_3$ and the particles on this edge are displaced in the direction proportionally to their distance from the origin. For the edge $X_1 = X_3 = 0$, $X_2 = X_2$,

$u_1 = u_2 = 0, u_3 = AX_2$. The initial and displaced positions of the cube are shown in Fig. 1.3.

1.8 Deformation Gradients, Finite Strain Tensors

The deformation gradient is the fundamental measure of deformation in continuum mechanics. In deformation analysis we confine our attention to two stationary configurations and disregard any consideration for the particular sequence by which the final *deformed configuration* is reached from the initial *undeformed configuration*. Accordingly, the mapping function is not dependent upon time as a variable, so that Eq. (1.3) takes the form

$$x = x(X,t) \Rightarrow x_i = x_i(X_1, X_2, X_3, t); \quad i = 1,2,3$$

1.8.1 Lagrangian Finite Strain Tensor

The deformation gradient is a second order tensor which maps line elements in the reference configuration into line elements (consisting of the same material particles) in the current configuration. Consider a material line element P_0Q_0, joining a pair of neighboring points P_0, Q_0, of length oriented in the direction $N = (N_1, N_2, N_3)$ in the initial undeformed region B_0 at time $t = 0$. If P_0 has coordinates $X = (X_1, X_2, X_3)$ and Q_0 has coordinates $(X_1 + dX_1, X_2 + dX_2, X_3 + dX_3)$ with respect to an orthogonal set of coordinate axes fixed in space, then the magnitude squared of dX is

$$dL^2 = dX \cdot dX = dX_1^2 + dX_2^2 + dX_3^2 = dX_i \, dX_i = \delta_{ij} dX_i \, dX_j \quad (1.8)$$

and

$$N_i = \frac{X_i + dX_i - X_i}{dL} = \frac{dX_i}{dL} \quad (1.9)$$

where, δ_{ij} is a Kronecker delta. When the body undergoes deformation, the same material points which lie on P_0Q_0 at $t = 0$ will lie on a new line element PQ of length dl oriented in the direction in current region B at time t (Fig. 1.4). If P has coordinates $x = (x_1, x_2, x_3)$ and Q has coordinates $(x_1 + dx_1, x_2 + dx_2, x_3 + dx_3)$ with respect to an orthogonal set of coordinate axes fixed in space, then

$$dl^2 = dx \cdot dx = dx_1^2 + dx_2^2 + dx_3^2 = dx_k \, dx_k = \delta_{kl} \, dx_k \, dx_l \quad (1.10)$$

and

$$n_i = \frac{x_i + dx_i - x_i}{dl} = \frac{dx_i}{dl} \quad (1.11)$$

In the material method, this deformation is characterized by the equation

$$x_i = x_i(X_1, X_2, X_3, t). \quad (1.12)$$

Since $x_k + dx_k$ are coordinates of Q at the same time t, therefore

$$x_k + dx_k = x_k(X_1, X_2, X_3) + \frac{\partial x_k}{\partial X_j} dX_j$$

$$\Rightarrow \quad dx_k = \frac{\partial x_k}{\partial X_j} dX_j = \frac{\partial x_k}{\partial X_i} dX_i = x_{k,i}\, dX_i \quad (1.13)$$

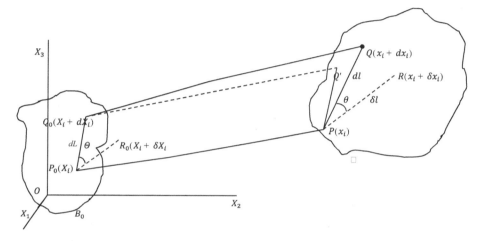

Fig. 1.4: Deformation gradient

The quantity $x_{k,i} = \dfrac{\partial x_k}{\partial X_i}$ is called *deformation gradient tensor* or simply the *deformation gradient* and is denoted by F_{ki}. Sometimes the dyadic notation F is used to express the deformation gradient as

$$F = \frac{\partial x}{\partial X} = Grad\ x \quad \Rightarrow \quad F_{ki} \equiv x_{k,i} = \frac{\partial x_k}{\partial X_i}$$

The capital G is used on 'Grad' to emphasize that this is a gradient with respect to the material coordinates, the *material gradient*. The tensor F characterizes the local deformation at X, and may depend explicitly upon X, in which case the deformation is termed *inhomogeneous*. If F is independent of X, the deformation is called *homogeneous*. In symbolic notation Eq. (1.13) appears in either of the forms

$$dx = F\ dX \Rightarrow dx = FdX\ \ dx_k = F_{ki}\ dX_i : \text{action of } F$$

where, as indicated by the second equation, the dot is often omitted for convenience. Using Eq.(1.13), from Eq.(1.10), we get

1.8 Deformation Gradients, Finite Strain Tensors

$$dl^2 = \delta_{kl} dx_k\, dx_l = \delta_{kl} \frac{\partial x_k}{\partial X_i} dX_i \frac{\partial x_l}{\partial X_j} dX_j$$

$$= \frac{\partial x_k}{\partial X_i} \cdot \frac{\partial x_{kl}}{\partial X_j} dX_i\, dX_j\,; \text{as } \delta_{kl} = 1, \text{ when } k = l \quad (1.14)$$

In describing motions and deformations, several measures of deformation are commonly used. First, let us consider that one based upon the change during the deformation in the magnitude squared of the distance between the particles originally at P_0 and Q_0. The difference $dl^2 - dL^2$ for two neighboring particles of a continuum is used as the *measure of deformation* that occurs in the neighborhood of the particles between the initial and final configurations. If this difference is identically zero for all neighboring particles of a continuum, a rigid displacement is said to occur. A measure of change of length of a line element is given by

$$dl^2 - dL^2 = \frac{\partial x_k}{\partial X_i} \cdot \frac{\partial x_k}{\partial X_j} dX_i\, dX_j - \delta_{ij} dX_i\, dX_j$$

$$= \left[\frac{\partial x_k}{\partial X_i} \cdot \frac{\partial x_k}{\partial X_j} - \delta_{ij} \right] dX_i\, dX_j = 2 r_{ij}\, dX_i\, dX_j \quad (1.15)$$

where,

$$r_{ij} = \frac{1}{2} \left[\frac{\partial x_k}{\partial X_i} \cdot \frac{\partial x_k}{\partial X_j} - \delta_{ij} \right] = \frac{1}{2}[C_{ij} - \delta_{ij}] \quad (1.16)$$

in which we have a symmetric tensor $C_{ij} = \frac{\partial x_k}{\partial X_i} \cdot \frac{\partial x_k}{\partial X_j}$, i.e., $C = F^T \cdot F$ known as the Green's deformation tensor. From this we immediately define the Lagrangian finite strain tensor r_{ij} as

$$2 r_{ij} = C_{ij} - \delta_{ij} \text{ or } 2R = C - I$$

where the factor of two is introduced for convenience in later calculations. Finally, we can write,

$$dl^2 - dL^2 = 2 r_{ij}\, dX_i\, dX_j = dX \cdot 2E \cdot dX$$

Therefore, we can write

$$\frac{dl^2 - dL^2}{dL^2} = 2 r_{ij} \frac{dX_i}{dL} \cdot \frac{dX_j}{dL} = 2 r_{ij} N_i N_j, \quad (1.17)$$

which is a scalar, but product $N_i N_j$ of two vector components is known to be a *tensor of order two*. The deformation of a body is completely described by the displacement vector. The Lagrangian finite strain tensor expressed by

Eq. (1.16) is given in terms of the appropriate deformation gradients. These same tensors may also be developed in terms of *displacement gradients*. For this purpose we begin by writing Eq. (1.16) in its time-independent form consistent with deformation analysis. In component notation, the material description is

$$u_i = x_i - X_i.$$

If $u_i + du_i$ be the displacement of the material point from its position Q_0 to Q, then

$$u_i + du_i = (x_i + dx_i) - (X_i + dX_i)$$

or, $\quad (x_i - X_i) + du_i = (x_i - X_i) + (dx_i - dX_i)$

or, $\quad du_i = dx_i - dX_i$, i.e., $dx_k = du_k + dX_k$.

Differentiating with respect to X_i, we get

$$\frac{\partial x_k}{\partial X_i} = x_{k,i} = \frac{\partial u_k}{\partial X_i} + \frac{\partial X_k}{\partial X_i} = \frac{\partial u_k}{\partial X_i} + \delta_{ik} = u_{k,i} + \delta_{ik}.$$

Similarly, differentiating with respect to X_j, we get

$$x_{k,j} = \frac{\partial u_k}{\partial X_j} \frac{\partial X_k}{\partial X_j} = \frac{\partial u_k}{\partial X_j} + \delta_{jk} = u_{k,j} + \delta_{jk}.$$

Therefore, from Eq.(1.16), the expression for in terms of the displacement of a material point from its position to is given by

$$r_{ij} = \frac{1}{2}\left[\frac{\partial x_k}{\partial X_i} \cdot \frac{\partial x_k}{\partial X_j} - \delta_{ij}\right] = \frac{1}{2}\left[x_{k,i} \cdot x_{k,j} - \delta_{ij}\right] = \frac{1}{2}\left[(u_{k,i} + \delta_{ik}) \cdot (u_{k,j} + \delta_{jk}) - \delta_{ij}\right]$$

$$= \frac{1}{2}\left[\left(\frac{\partial u_k}{\partial X_i} + \delta_{ik}\right) \cdot \left(\frac{\partial u_k}{\partial X_j} + \delta_{jk}\right) - \delta_{ij}\right]$$

$$= \frac{1}{2}\left[\frac{\partial u_k}{\partial X_i} \cdot \frac{\partial u_k}{\partial X_j} + \frac{\partial u_k}{\partial X_i} \cdot \delta_{jk} + \frac{\partial u_k}{\partial X_j} \cdot \delta_{ik} + \delta_{ik} \cdot \delta_{jk} - \delta_{ij}\right]$$

$$= \frac{1}{2}\left[\frac{\partial u_i}{\partial X_j} + \frac{\partial u_j}{\partial X_i} + \frac{\partial u_k}{\partial X_i} \cdot \frac{\partial u_k}{\partial X_j}\right] [\text{Since } \delta_{ik} \cdot \delta_{jk} = \delta_{ij}]$$

$$= \frac{1}{2}\left[u_{i,j} + u_{j,i} + u_{k,i} \cdot u_{k,j}\right]. \quad (1.18)$$

Change in the angle between two line elements in material method: Here we consider change in angle between two material line elements P_0Q_0 and P_0R_0 at P_0 inclined at an angle θ whereand P_0Q_0 is of length dL oriented

1.8 Deformation Gradients, Finite Strain Tensors

in the direction (N_1, N_2, N_3) and P_0R_0 is of length δL oriented in the direction (M_1, M_2, M_3) in the region B_0. If Q_0 has coordinates $(X_i + dX_i)$ and R_0 has coordinates $(X_i + \delta X_i)$, then

$$M_i = \frac{\delta X_i}{\delta L}, \; N_i = \frac{dX_i}{dL}; \; \cos\Theta = \frac{dX_i}{dL} \cdot \frac{\delta X_i}{\delta L} = N_i M_i \qquad (1.19)$$

The *unit relative displacement* is given by

$$\frac{du_i}{dL} = \frac{\partial u_i}{\partial X_j} \cdot \frac{\partial X_j}{\partial L} = \frac{\partial u_i}{\partial X_j} N_j$$

When the body undergoes deformation then the two line elements P_0Q_0 and P_0R_0 at P_0 will deform into two other line elements PQ and PR at P of length dl and δl, orientated in the direction (n_1, n_2, n_3) and (m_1, m_2, m_3) and inclined at an angle θ in the region B. If Q has coordinates $(x_i + dx_i)$ and R_0 has coordinates $(x_i + \delta x_i)$, then

$$m_i = \frac{\delta X_i}{\delta L}, \; n_i = \frac{dX_i}{dL}; \; \cos\theta = \frac{dx_i}{dl} \cdot \frac{\delta x_i}{\delta l} = n_i m_i$$

$$dx_k = \frac{\partial x_k}{\partial X_j} \cdot dX_j; \; \delta x_k = \frac{\partial x_k}{\partial X_j} \cdot \delta X_j. \qquad (1.20)$$

Therefore, we can write

$$\frac{\delta l^2 - \delta L^2}{\delta L^2} = 2r_{ij} \frac{\delta X_i}{\delta L} \cdot \frac{\delta X_j}{\delta L} = 2r_{ij} M_i M_j.$$

$$\frac{dl^2 - dL^2}{dL^2} = 2r_{ij} \frac{dX_i}{dL} \cdot \frac{dX_j}{dL} = 2r_{ij} N_i M_j. \qquad (1.21)$$

Again,

$$dx_i \delta x_i - dX_i \delta X_i = dx_k \delta x_k - dX_i \delta X_i$$

$$= \frac{\partial x_k}{\partial X_i} \cdot \frac{\partial x_k}{\partial X_j} \cdot dX_i \delta X_j \delta_{ij} - dX_i \delta X_j$$

$$= 2r_{ij} \, dX_i \delta X_j, \text{ where } r_{ij} = \frac{1}{2}\left[\frac{\partial u_i}{\partial X_j} + \frac{\partial u_j}{\partial X_i} + \frac{\partial u_k}{\partial X_i} \cdot \frac{\partial u_k}{\partial X_j}\right]$$

or, $\quad \dfrac{dx_i}{dL} \cdot \dfrac{\delta x_i}{\delta L} - \dfrac{dX_i}{dL} \cdot \dfrac{\delta X_i}{\delta L} = 2r_{ij} \dfrac{dX_i}{dL} \cdot \dfrac{\delta X_j}{\delta L}$

or, $\quad \dfrac{dx_i}{dl} \cdot \dfrac{\delta x_i}{\delta l} \cdot \dfrac{dl}{dL} \cdot \dfrac{\delta l}{\delta L} - \dfrac{dX_i}{dL} \cdot \dfrac{\delta X_i}{\delta L} = 2r_{ij} \dfrac{dX_i}{dL} \cdot \dfrac{\delta X_j}{\delta L}$

or, $\quad \dfrac{dl}{dL} \cdot \dfrac{\delta l}{\delta L} \cos\theta - \cos\Theta = 2r_{ij}N_i M_j \qquad (1.22)$

Eqns. (1.16), (1.21) and (1.22) show that, if $r_{ij} = 0$, (no strain) then $dl = dL$, $\delta l = \delta L$, $\theta = \Theta$. Thus, when $r_{ij} = 0$, length of a line element and angle between two line elements remain unchanged during deformation and the body undergone only rigid body deformation. Thus in the absence of strain, only a rigid body displacement can occur.

(*i*) Thus the necessary and sufficient condition for rigid body deformation at each point is $r_{ij} = 0$. In other words r_{ij} cause a change in the length and orientation, of a line element when body is deformed.

(*ii*) Since r_{ij} cause strain deformation, it may be taken as exact measure of strain produced during deformation.

(*iii*) The knowledge of r_{ij} at a point enables us to determine the change in length of a line element and change in angle between two line elements.

(*iv*) The symmetric tensor $C_{ij} = \dfrac{\partial x_k}{\partial X_i} \cdot \dfrac{\partial x_k}{\partial X_j}$ is known as the Green's deformation tensor.

(*v*) From Eq. (1.22), we see that, the term $2r_{ij}N_i N_j$ is a scalar, but product $N_i N_j$ of two vector components is known to be a tensor of order two. Hence by quotient law of tensors r_{ij} is a second order tensor, known as Lagrangian finite strain tensor. Now

$$r_{ji} = \frac{1}{2}\left[\frac{\partial u_j}{\partial X_i} + \frac{\partial u_i}{\partial X_j} + \frac{\partial u_k}{\partial X_j} \cdot \frac{\partial u_k}{\partial X_i}\right] = \frac{1}{2}\left[\frac{\partial u_i}{\partial X_j} + \frac{\partial u_j}{\partial X_i} + \frac{\partial u_k}{\partial X_i} \cdot \frac{\partial u_k}{\partial X_j}\right] = r_{ij}$$

so that r_{ij} is symmetric.

(*vi*) The relations given in Eq. (1.18)

$$r_{ij} = \frac{1}{2}\left[u_{i,j} + u_{j,i} + u_{k,i} \cdot u_{k,j}\right]$$

represent nonlinear strain-displacement relations independent of external forces and of constitution of the material. They are concerned with the geometry of the deformation without consideration of the influence of the forces which leads from undeformed position and are known as 'Kinematical relations.'

Example 1.4: Given the displacement field $x_1 = X_1 + 2X_3, x_2 = 2; X_2 - 2X_3; x_3 = X_3 - 2X_1 + 2X_2$. Determine the deformation gradient, Green's deformation tensor and Lagrangian finite strain tensor.

1.8 Deformation Gradients, Finite Strain Tensors

Solution: The deformation gradient F has the matrix form

$$(F_{ki}) = \begin{pmatrix} \dfrac{\partial x_1}{\partial X_1} & \dfrac{\partial x_1}{\partial X_2} & \dfrac{\partial x_1}{\partial X_3} \\ \dfrac{\partial x_2}{\partial X_1} & \dfrac{\partial x_2}{\partial X_2} & \dfrac{\partial x_2}{\partial X_3} \\ \dfrac{\partial x_3}{\partial X_1} & \dfrac{\partial x_3}{\partial X_2} & \dfrac{\partial x_3}{\partial X_3} \end{pmatrix} = \begin{pmatrix} 1 & 0 & 2 \\ 0 & 1 & -2 \\ -2 & 2 & 1 \end{pmatrix}.$$

The Green's deformation tensor $C = F^T \cdot F$ has the matrix form

$$(C_{ij}) = \begin{pmatrix} 1 & 0 & 2 \\ 0 & 1 & -2 \\ -2 & 2 & 1 \end{pmatrix}^T \begin{pmatrix} 1 & 0 & 2 \\ 0 & 1 & -2 \\ -2 & 2 & 1 \end{pmatrix} = \begin{pmatrix} 5 & -4 & 0 \\ -4 & 5 & 0 \\ 0 & 0 & 9 \end{pmatrix}$$

The displacements components $u_i = x_i - X_i$ of a material point are given by

$$u_1 = x_1 - X_1 = X_1 + 2X_3 - X_1 = 2X_3$$
$$u_2 = x_2 - X_2 = X_2 - 2X_3 - X_2 = -2X_3$$
$$u_3 = x_3 - X_3 = X_3 - 2X_1 + 2X_2 - X_3 = -2X_1 + 2X_2$$

Now the second order Lagrangian finite, strain tensors r_{ij} are given by Eq. (1.18). Therefore,

$$r_{11} = \frac{1}{2}\left[\frac{\partial u_1}{\partial X_1} + \frac{\partial u_1}{\partial X_1} + \frac{\partial u_k}{\partial X_1} \cdot \frac{\partial u_k}{\partial X_1}\right] = \frac{\partial u_1}{\partial X_1} + \frac{1}{2}\left[\left(\frac{\partial u_1}{\partial X_1}\right)^2 + \left(\frac{\partial u_2}{\partial X_1}\right)^2 + \left(\frac{\partial u_3}{\partial X_1}\right)^2\right]$$

$$= 0 + \frac{1}{2}\left[(0)^2 + (0)^2 + (-2)^2\right] = 2$$

$$r_{22} = \frac{1}{2}\left[\frac{\partial u_2}{\partial X_2} + \frac{\partial u_2}{\partial X_2} + \frac{\partial u_k}{\partial X_2} \cdot \frac{\partial u_k}{\partial X_2}\right] = \frac{\partial u_2}{\partial X_2} + \frac{1}{2}\left[\left(\frac{\partial u_1}{\partial X_2}\right)^2 + \left(\frac{\partial u_2}{\partial X_2}\right)^2 + \left(\frac{\partial u_3}{\partial X_2}\right)^2\right]$$

$$= 0 + \frac{1}{2}\left[(0)^2 + (0)^2 + (2)^2\right] = 2$$

$$r_{33} = \frac{1}{2}\left[\frac{\partial u_3}{\partial X_3} + \frac{\partial u_3}{\partial X_3} + \frac{\partial u_k}{\partial X_3} \cdot \frac{\partial u_k}{\partial X_3}\right] = \frac{\partial u_3}{\partial X_3} + \frac{1}{2}\left[\left(\frac{\partial u_1}{\partial X_3}\right)^2 + \left(\frac{\partial u_2}{\partial X_3}\right)^2 + \left(\frac{\partial u_3}{\partial X_3}\right)^2\right]$$

$$= 0 + \frac{1}{2}\left[(2)^2 + (-2)^2 + (0)^2\right] = 4$$

$$r_{12} = \frac{1}{2}\left[\frac{\partial u_1}{\partial X_2} + \frac{\partial u_2}{\partial X_1} + \frac{\partial u_k}{\partial X_1} \cdot \frac{\partial u_k}{\partial X_2}\right]$$

$$= \frac{1}{2}\left[\frac{\partial u_1}{\partial X_2} + \frac{\partial u_2}{\partial X_1} + \frac{\partial u_1}{\partial X_1} \cdot \frac{\partial u_1}{\partial X_2} + \frac{\partial u_2}{\partial X_1} \cdot \frac{\partial u_2}{\partial X_2} + \frac{\partial u_3}{\partial X_1} \cdot \frac{\partial u_3}{\partial X_2}\right]$$

$$= \frac{1}{2}\left[0 + 0 + 0 \cdot 0 + 0.0 + (-2) \cdot 2\right] = -2 = r_{21}$$

$$r_{13} = \frac{1}{2}\left[\frac{\partial u_1}{\partial X_3} + \frac{\partial u_3}{\partial X_1} + \frac{\partial u_k}{\partial X_1} \cdot \frac{\partial u_k}{\partial X_3}\right]$$

$$= \frac{1}{2}\left[\frac{\partial u_1}{\partial X_3} + \frac{\partial u_3}{\partial X_1} + \frac{\partial u_1}{\partial X_1} \cdot \frac{\partial u_1}{\partial X_3} + \frac{\partial u_2}{\partial X_1} \cdot \frac{\partial u_2}{\partial X_3} + \frac{\partial u_3}{\partial X_1} \cdot \frac{\partial u_3}{\partial X_3}\right]$$

$$= \frac{1}{2}\left[2 + (-2) + 0.2 + 0.(-2) + (-2) \cdot 0\right] = 0 = r_{31}$$

$$r_{23} = \frac{1}{2}\left[\frac{\partial u_2}{\partial X_3} + \frac{\partial u_3}{\partial X_2} + \frac{\partial u_k}{\partial X_2} \cdot \frac{\partial u_k}{\partial X_3}\right]$$

$$= \frac{1}{2}\left[\frac{\partial u_2}{\partial X_3} + \frac{\partial u_3}{\partial X_2} + \frac{\partial u_1}{\partial X_2} \cdot \frac{\partial u_1}{\partial X_3} + \frac{\partial u_2}{\partial X_2} \cdot \frac{\partial u_2}{\partial X_3} + \frac{\partial u_3}{\partial X_2} \cdot \frac{\partial u_3}{\partial X_3}\right]$$

$$= \frac{1}{2}\left[(-2) + 2 + 0.2 + 0 \cdot (-2) + 2.0\right] = 0 = r_{32}$$

In matrix notation, the second order Lagrangian finite, strain tensors r_{ij} are given by

$$(r_{ij}) = \begin{pmatrix} 2 & -2 & 0 \\ -2 & 2 & 0 \\ 0 & 0 & 4 \end{pmatrix} = \frac{1}{2}\begin{pmatrix} 5 & -4 & 0 \\ -4 & 5 & 0 \\ 0 & 0 & 9 \end{pmatrix} - \frac{1}{2}\begin{pmatrix} 1 & 0 & 0 \\ 0 & 1 & 0 \\ 0 & 0 & 1 \end{pmatrix} = \frac{1}{2}C - \frac{1}{2}I$$

Example 1.5: For the displacement field $u_1 = X_1^2 X_2$, $u_2 = X_2 - X_3^2$, $u_3 = X_2^2 X_3$, determine the unit relative displacement vector at $P(1, 2, -1)$ with respect to $Q(4, 2, 3)$

Answer: The direction cosines (N_1, N_2, N_3) of the line joining the points $P(1, 2, -1)$ and $Q(4, 2, 3)$ are given by

$$\frac{N_1}{4-1} = \frac{N_2}{2-2} = \frac{N_3}{3+1} = \frac{1}{\sqrt{3^2 + 0^2 + 4^2}} = \frac{1}{5}$$

$$\Rightarrow N_1 = \frac{3}{5}, N_2 = 0, N_3 = \frac{4}{5}.$$

1.8 Deformation Gradients, Finite Strain Tensors

The unit relative displacement is given by

$$\frac{du_1}{dL} = \frac{\partial u_1}{\partial X_j} N_j = \frac{\partial u_1}{\partial X_1} N_1 + \frac{\partial u_1}{\partial X_2} N_2 + \frac{\partial u_1}{\partial X_3} N_3$$

$$= 2X_1 X_2 N_1 + X_1^2 N_2 + 0.N_3 = 2.1.2.\frac{3}{5} + 1^2.0 + 0.\frac{4}{5} = \frac{12}{5}$$

$$\frac{du_2}{dL} = \frac{\partial u_2}{\partial X_j} N_j = \frac{\partial u_2}{\partial X_1} N_1 + \frac{\partial u_2}{\partial X_2} N_2 + \frac{\partial u_2}{\partial X_3} N_3$$

$$= 0.N_1 + 1 \cdot N_2 + (-2X_3).N_3 = 0 \cdot \frac{3}{5} + 1.0 + 2 \cdot \frac{4}{5} = \frac{8}{5}$$

$$\frac{du_3}{dL} = \frac{\partial u_3}{\partial X_j} N_j = \frac{\partial u_3}{\partial X_1} N_1 + \frac{\partial u_3}{\partial X_2} N_2 + \frac{\partial u_3}{\partial X_3} N_3$$

$$= 0.N_1 + 2X_2 X_3 \cdot N_2 + X_2^2 \cdot N_3 = 0 \cdot \frac{3}{5} + 2.2 \cdot (-1) \cdot 0 + 2^2 \cdot \frac{4}{5} = \frac{16}{5}$$

1.8.2 Eulerian Finite Strain Tensor

In the spatial method of description of deformation (x_1, x_2, x_3) are regarded as independent variables and the equations characterizing the deformation can be written as

$$X_k = X_k(x_1, x_2, x_3, t); k = 1, 2, 3. \quad (1.23)$$

where, X_1, X_2, X_3 represents the material coordinates of a material particle. Since $X_k + dX_k$ are coordinates of Q_0 at the same time t, therefore

$$dX_k = \frac{\partial X_k}{\partial x_i} dx_i = \frac{\partial X_k}{\partial x_j} dx_j = X_{k,i} \, dx_j \quad (1.24)$$

The quantity $X_{k,i} = \frac{\partial X_k}{\partial x_j}$ is called *deformation gradient tensor* or simply the *deformation gradient* and is denoted by F_{ki}^{-1}. In view of the smoothness conditions we have imposed on the mapping function, we know that $\mathbf{F}$ is invertible so that the inverse $\mathbf{F}^{-1}$ exists such that

$$d\mathbf{X} = \mathbf{F}^{-1} \cdot d\mathbf{x} \quad \Rightarrow \quad dX_k = \frac{\partial X_k}{\partial x_i} dx_i$$

Using Eq. (1.24), from Eq. (1.8), we get

$$dL^2 = dX_k \, dX_k = \frac{\partial X_k}{\partial x_i} \cdot \frac{\partial X_k}{\partial x_j} dx_i \, dx_j; \, dl^2 = d\mathbf{x} \cdot d\mathbf{x} = dx_i \, dx_i = \delta_{ij} dx_i \, dx_j \quad (1.25)$$

A measure of change of length of a line element

$$= dl^2 - dL^2 = \delta_{ij} dx_i\, dx_j - \frac{\partial X_k}{\partial x_i} \cdot \frac{\partial X_k}{\partial x_j} dx_i\, dx_j$$

$$= \left[\delta_{ij} - \frac{\partial X_k}{\partial x_i} \cdot \frac{\partial X_k}{\partial x_j} \right] dx_i\, dx_j = 2\eta_{ij}\, dx_i\, dx_j \qquad (1.26)$$

where,

$$\eta_{ij} = \frac{1}{2}\left[\delta_{ij} - \frac{\partial X_k}{\partial x_i} \cdot \frac{\partial X_k}{\partial x_j} \right] = \frac{1}{2}[\delta_{ij} - c_{ij}] \qquad (1.27)$$

in which we have a symmetric tensor $c_{ij} = \frac{\partial X_k}{\partial x_i} \cdot \frac{\partial X_k}{\partial x_j}$, i.e., $\mathbf{c} = (\mathbf{F}^{-1})^T \cdot \mathbf{F}^{-1}$ known as the Cauchy's deformation tensor. From this we immediately define the Eulerian finite strain tensor η_{ij} as

$$2\eta_{ij} = \delta_{ij} - c_{ij} \text{ or } 2\eta = \mathbf{I} - \mathbf{c}$$

where the factor of two is introduced for convenience in later calculations. Finally, we can write,

$$dl^2 - dL^2 = 2\eta_{ij}\, dX_i\, dX_j = d\mathbf{x} \cdot 2\eta \cdot d\mathbf{x}$$

Using Eq. (1.11), we can write

$$\frac{dl^2 - dL^2}{dL^2} = 2\eta_{ij} \frac{dx_i}{dl} \cdot \frac{dx_j}{dl} = 2\eta_{ij} n_i\, n_j, \qquad (1.28)$$

which is a scalar, but product of two vector components is known to be *a tensor of order two*. The deformation of a body is completely described by the displacement vector. The Eulerian finite strain tensor expressed by Eq. (1.27) is given in terms of the appropriate deformation gradients. These same tensors may also be developed in terms of *displacement gradients*. For this purpose we begin by writing Eq. (1.27) in its time-independent form consistent with deformation analysis. In component notation, the material description is

$$u_i = x_i - X_i.$$

The deformation of a body is completely described by the displacement vector. It is possible to express η_{ij} in terms of the displacement u_i of a spatial point from its position P_0 to P, then

$$u_i = x_i - X_i, \text{ i.e., } X_k = x_k - u_k.$$

If $u_i + du_i$ be the displacement of the spatial point from its position Q_0 to Q, then

1.8 Deformation Gradients, Finite Strain Tensors

$$u_i + du_i = (x_i + dx_i) - (X_i + dX_i)$$

or,
$$(x_i - X_i) + du_i = (x_i - X_i) + (dx_i - dX_i)$$

or,
$$du_i = dx_i - dX_i, \text{ i.e., } dX_k = dx_k - du_k$$

Differentiating with respect to x_i, we get

$$X_{k,i} = \frac{\partial X_k}{\partial x_i} = \frac{\partial x_k}{\partial x_i} - \frac{\partial u_k}{\partial x_i} = \delta_{ki} - \frac{\partial u_k}{\partial x_i} = \delta_{ki} - u_{k,i}.$$

Similarly, differentiating with respect to x_j, we get

$$X_{k,j} = \frac{\partial x_k}{\partial x_j} - \frac{\partial u_k}{\partial x_j} = \delta_{kj} - \frac{\partial u_k}{\partial x_j} = \delta_{kj} - u_{k,j}.$$

Therefore, from Eq. (1.27), the expression for u_i in terms of the displacement of a material point from its position P_0 to P is given by

$$\eta_{ij} = \frac{1}{2}\left[\delta_{ij} - \frac{\partial X_k}{\partial x_i} \cdot \frac{\partial X_k}{\partial x_j}\right] = \frac{1}{2}\left[\delta_{ij} - X_{k,i} \cdot X_{k,j}\right]$$

$$= \frac{1}{2}\left[\delta_{ij} - (\delta_{ki} - u_{k,i}) \cdot (\delta_{kj} - u_{k,j})\right]$$

$$= \frac{1}{2}\left[\delta_{ij} - \left(\delta_{ki} - \frac{\partial u_k}{\partial x_i}\right) \cdot (\delta_{kj} - \frac{\partial u_k}{\partial x_j})\right]$$

$$= \frac{1}{2}\left[\delta_{ij} - \frac{\partial u_k}{\partial x_i} \cdot \frac{\partial u_k}{\partial x_j} + \frac{\partial u_k}{\partial x_i}.\delta_{kj} + \frac{\partial u_k}{\partial x_j} \cdot \delta_{ki} - \delta_{ki}.\delta_{kj}\right]$$

$$= \frac{1}{2}\left[\frac{\partial u_i}{\partial x_j} + \frac{\partial u_j}{\partial x_i} - \frac{\partial u_k}{\partial x_i} \cdot \frac{\partial u_k}{\partial x_j}\right]; \text{ as } \delta_{ki} \cdot \delta_{kj} = \delta_{ij}$$

$$= \frac{1}{2}\left[u_{i,j} + u_{j,i} - u_{k,i} \cdot u_{k,j}\right]. \tag{1.29}$$

Change in the angle between two line elements in spatial method: Here we consider change in angle between two material line elements P_0Q_0 and P_0Q_0 at P_0 inclined at an angle θ where P_0Q_0 and is of length dL oriented in the direction (N_1, N_2, N_3) and P_0Q_0 is of length δL oriented in the direction (M_1, M_2, M_3) in the region B_0. If Q_0 has coordinates $(X_i + dX_i)$ and R_0 has coordinates $(X_i + \delta X_i)$, then

$$M_i = \frac{\delta X_i}{\delta L}, N_i = \frac{dX_i}{dL}; \cos\Theta = \frac{dX_i}{dL} \cdot \frac{\delta X_i}{\delta L} = N_i M_i \tag{1.30}$$

When the body undergoes deformation then the two line elements P_0Q_0 and P_0R_0 at P_0 will deform into two other line elements PQ and PR at P of length dl and δl, orientated in the direction (n_1, n_2, n_3) and (m_1, m_2, m_3) and inclined at an angle θ in the region B. If Q has coordinates $(x_i + dx_i)$ and R_0 has coordinates $(x_i + \delta x_i)$, then

$$m_i = \frac{\delta X_i}{\delta L}, n_i = \frac{dX_i}{dL}; \cos\theta = \frac{dx_i}{dl} \cdot \frac{\delta x_i}{\delta l} = n_i m_i \qquad (1.31)$$

It follows from Eq. (1.24) that $\delta x_k = \frac{\partial x_k}{\partial x_i} \cdot \delta X_i = \frac{\partial x_k}{\partial X_j} \cdot \delta X_j$. Therefore,

$$\frac{\delta l^2 - \delta L^2}{\delta L^2} = 2\eta_{ij} \frac{\delta X_i}{\delta L} \cdot \frac{\delta X_j}{\delta L} = 2\eta_{ij} M_i M_j. \qquad (1.32)$$

$$\frac{dl^2 - dL^2}{dL^2} = 2\eta_{ij} \frac{dX_i}{dL} \cdot \frac{dX_j}{dL} = 2\eta_{ij} N_i N_j.$$

Again,

$$dx_i \delta x_i - dX_i \delta X_i = \delta_{ij} dx_i \delta x_j - dX_k \delta X_k = \delta_{ij} dx_i \delta x_j - \frac{\partial X_k}{\partial x_i} \cdot \frac{\partial X_k}{\partial x_j} \cdot dx_i \delta x_j$$

$$= 2\eta_{ij} dx_i dx_j, \text{ where } \eta_{ij} = \frac{1}{2}\left[\frac{\partial u_i}{\partial x_j} + \frac{\partial u_j}{\partial x_i} - \frac{\partial u_k}{\partial x_i} \cdot \frac{\partial u_k}{\partial x_j}\right]$$

or, $\frac{dx_i}{dl} \cdot \frac{\delta x_i}{\delta l} - \frac{dX_i}{dl} \cdot \frac{\delta X_i}{\delta l} = 2\eta_{ij} \frac{dx_i}{dl} \cdot \frac{\delta x_j}{\delta l}$

or, $\frac{dx_i}{dl} \cdot \frac{\delta x_i}{\delta l} - \frac{dX_i}{dL} \cdot \frac{\delta X_i}{\delta L} \cdot \frac{dL}{dl} \cdot \frac{\delta L}{\delta l} = 2\eta_{ij} n_i m_j$

or, $\cos\theta - \cos\Theta \frac{dL}{dl} \cdot \frac{\delta L}{\delta l} = 2\eta_{ij} n_i m_j \qquad (1.33)$

Eqns. (1.32) and (1.33) show that, if $\eta_{ij} = 0$, then $dL = dl, \delta L = \delta l, \theta = \Theta$. Thus, when $\eta_{ij} = 0$, length of a line element and angle between two line elements remain unchanged during deformation and the body undergone only rigid body deformation. Thus

(i) The necessary and sufficient condition for rigid body deformation at each point is $\eta_{ij} = 0$. In other words η_{ij} cause a change in the length change of a line element as well as change in the angle between two line elements.

(ii) Since η_{ij} cause strain deformation, it may be taken as a measure of strain deformation.

1.8 Deformation Gradients, Finite Strain Tensors

(iii) The knowledge of η_{ij} at a point enables us to determine the change in length of a line element and change in angle between two line elements.

(iv) The tensor $F_{ij} = \dfrac{\partial X_k}{\partial x_i} \cdot \dfrac{\partial X_k}{\partial x_j}$ is known as the Cauchy's deformation tensor.

(v) From Eq.(1.28), we see that, the term $2\eta_{ij} N_i N_j$ is a scalar, but product $N_i N_j$ of two vector components is known to be a tensor of order two. Hence by quotient law of tensors η_{ij} is a second order tensor, known as Eulerian finite strain tensor. Now

$$\frac{1}{2}\left[\frac{\partial u_j}{\partial x_i} + \frac{\partial u_i}{\partial x_j} - \frac{\partial u_k}{\partial x_j} \cdot \frac{\partial u_k}{\partial x_i}\right] = \frac{1}{2}\left[\frac{\partial u_i}{\partial x_j} + \frac{\partial u_j}{\partial x_i} - \frac{\partial u_k}{\partial x_i} \cdot \frac{\partial u_k}{\partial x_j}\right] = \eta_{ij}$$

so that the tensor η_{ji} is symmetric.

Example 1.6: Given the displacement field $x_1 = X_1 + 2X_3$, $x_2 = X_2 - 2X_3$; $x_3 = X_3 - 2X_1 + 2X_2$. Determine the deformation gradient, Cauchy's deformation tensors and Eulerian finite strain tensor.

Solution: Now eliminating, X_3 we get

$$x_1 + x_2 = X_1 + X_2; x_2 + 2x_3 = -4X_1 + 5X_2$$

$$X_2 = \frac{1}{9}[4(x_1 + x_2) + x_2 + +2x_3] = \frac{1}{9}(4x_1 + 5x_2 + 2x_3)$$

$$X_1 = x_1 + x_2 - X_2 = x_1 + x_2 - \frac{1}{9}(4x_1 + 5x_2 + +2x_3) = \frac{1}{9}(5x_1 + 4x_2 - 2x_3)$$

$$X_3 = x_3 + 2X_1 - 2X_2 = x_3 + \frac{2}{9}(5x_1 + 4x_2 - 2x_3) - \frac{2}{9}$$

$$(4x_1 + 5x_2 + +2x_3) = \frac{1}{9}(2x_1 - 2x_2 + x_3)$$

The deformation gradient F^{-1} has the matrix form

$$\left(F_{ki}^{-1}\right) = \begin{pmatrix} \dfrac{\partial X_1}{\partial x_1} & \dfrac{\partial X_1}{\partial x_2} & \dfrac{\partial X_1}{\partial x_3} \\ \dfrac{\partial X_2}{\partial x_1} & \dfrac{\partial X_2}{\partial x_2} & \dfrac{\partial X_2}{\partial x_3} \\ \dfrac{\partial X_3}{\partial x_1} & \dfrac{\partial X_3}{\partial x_2} & \dfrac{\partial X_3}{\partial x_3} \end{pmatrix} = \frac{1}{9}\begin{pmatrix} 5 & 4 & -2 \\ 4 & 5 & 2 \\ 2 & -2 & 1 \end{pmatrix}.$$

The Cauchy's deformation tensor $c = (F^{-1})^T . F^{-1}$ has the matrix form

$$(c_{ij}) = \frac{1}{9}\begin{pmatrix} 5 & 4 & -2 \\ 4 & 5 & 2 \\ 2 & -2 & 1 \end{pmatrix}^T \frac{1}{9}\begin{pmatrix} 5 & 4 & -2 \\ 4 & 5 & 2 \\ 2 & -2 & 1 \end{pmatrix} = \frac{1}{81}\begin{pmatrix} 45 & 36 & 0 \\ 36 & 45 & 0 \\ 0 & 0 & 9 \end{pmatrix}$$

The displacements components $u_i = x_i - X_i$ of a material point are given by

$$u_1 = x_1 - X_1 = x_1 - \frac{1}{9}(5x_1 + 4x_2 - 2x_3) = \frac{1}{9}(4x_1 - 4x_2 + 2x_3)$$

$$u_2 = x_2 - X_2 = x_2 - \frac{1}{9}(4x_1 + 5x_2 + +2x_3) = \frac{1}{9}(4x_2 - 4x_1 - 2x_3)$$

$$u_3 = x_3 - X_3 = x_3 - \frac{1}{9}(2x_1 - 2x_2 + x_3) = \frac{1}{9}(8x_3 - 2x_1 + 2x_2)$$

Now the second order Eulearian finite, strain tensors η_{ij} are given by Eq.(1.29). Therefore,

$$\eta_{11} = \frac{1}{2}\left[\frac{\partial u_1}{\partial x_1} + \frac{\partial u_1}{\partial x_1} - \frac{\partial u_k}{\partial x_1}\cdot\frac{\partial u_k}{\partial x_1}\right] = \frac{\partial u_1}{\partial x_1} - \frac{1}{2}\left[\left(\frac{\partial u_1}{\partial x_1}\right)^2 + \left(\frac{\partial u_2}{\partial x_1}\right)^2 + \left(\frac{\partial u_3}{\partial x_1}\right)^2\right]$$

$$= \frac{4}{9} - \frac{1}{2}\left[\left(\frac{4}{9}\right)^2 + \left(-\frac{4}{9}\right)^2 + \left(-\frac{2}{9}\right)^2\right] = \frac{2}{9}$$

$$\eta_{22} = \frac{1}{2}\left[\frac{\partial u_2}{\partial x_2} + \frac{\partial u_2}{\partial x_2} - \frac{\partial u_k}{\partial x_2}\cdot\frac{\partial u_k}{\partial x_2}\right] = \frac{\partial u_2}{\partial x_2} - \frac{1}{2}\left[\left(\frac{\partial u_1}{\partial x_2}\right)^2 + \left(\frac{\partial u_2}{\partial x_2}\right)^2 + \left(\frac{\partial u_3}{\partial x_2}\right)^2\right]$$

$$= \frac{4}{9} - \frac{1}{2}\left[\left(-\frac{4}{9}\right)^2 + \left(\frac{4}{9}\right)^2 + \left(\frac{2}{9}\right)^2\right] = \frac{2}{9}$$

$$\eta_{33} = \frac{1}{2}\left[\frac{\partial u_3}{\partial x_3} + \frac{\partial u_3}{\partial x_3} - \frac{\partial u_k}{\partial x_3}\cdot\frac{\partial u_k}{\partial x_3}\right] = \frac{\partial u_3}{\partial x_3} - \frac{1}{2}\left[\left(\frac{\partial u_1}{\partial x_3}\right)^2 + \left(\frac{\partial u_2}{\partial x_3}\right)^2 + \left(\frac{\partial u_3}{\partial x_3}\right)^2\right]$$

$$= \frac{8}{9} - \frac{1}{2}\left[\left(\frac{2}{9}\right)^2 + \left(-\frac{2}{9}\right)^2 + \left(\frac{8}{9}\right)^2\right] = \frac{4}{9}$$

1.8 Deformation Gradients, Finite Strain Tensors

$$\eta_{12} = \frac{1}{2}\left[\frac{\partial u_1}{\partial x_2} + \frac{\partial u_2}{\partial x_1} - \frac{\partial u_k}{\partial x_1}\cdot\frac{\partial u_k}{\partial x_2}\right] = \frac{1}{2}\left[\frac{\partial u_1}{\partial x_2} + \frac{\partial u_2}{\partial x_1} - \frac{\partial u_1}{\partial x_1}\cdot\frac{\partial u_1}{\partial x_2} - \frac{\partial u_2}{\partial x_1}\cdot\frac{\partial u_2}{\partial x_2} - \frac{\partial u_3}{\partial x_1}\cdot\frac{\partial u_3}{\partial x_2}\right]$$

$$= \frac{1}{2}\left[\left(-\frac{4}{9}\right) + \left(-\frac{4}{9}\right) - \frac{4}{9}\cdot\left(-\frac{4}{9}\right) - \left(-\frac{4}{9}\right)\cdot\frac{4}{9} - \left(-\frac{2}{9}\right)\cdot\frac{2}{9}\right] = -\frac{2}{9} = \eta_{21}$$

$$\eta_{13} = \frac{1}{2}\left[\frac{\partial u_1}{\partial x_3} + \frac{\partial u_3}{\partial x_1} - \frac{\partial u_k}{\partial x_1}\cdot\frac{\partial u_k}{\partial x_3}\right] = \frac{1}{2}\left[\frac{\partial u_1}{\partial x_3} + \frac{\partial u_3}{\partial x_1} - \frac{\partial u_1}{\partial x_1}\cdot\frac{\partial u_1}{\partial x_3} - \frac{\partial u_2}{\partial x_1}\cdot\frac{\partial u_2}{\partial x_3} - \frac{\partial u_3}{\partial x_1}\cdot\frac{\partial u_3}{\partial x_3}\right]$$

$$= \frac{1}{2}\left[\frac{2}{9} + \left(-\frac{2}{9}\right) - \frac{4}{9}\cdot\frac{2}{9} - \left(-\frac{4}{9}\right)\cdot\left(-\frac{2}{9}\right) - \left(-\frac{2}{9}\right)\cdot\frac{8}{9}\right] = 0 = \eta_{31}$$

$$\eta_{23} = \frac{1}{2}\left[\frac{\partial u_2}{\partial x_3} + \frac{\partial u_3}{\partial x_2} + \frac{\partial u_k}{\partial x_2}\cdot\frac{\partial u_k}{\partial x_3}\right] = \frac{1}{2}\left[\frac{\partial u_2}{\partial x_3} + \frac{\partial u_3}{\partial x_2} - \frac{\partial u_1}{\partial x_2}\cdot\frac{\partial u_1}{\partial x_3} - \frac{\partial u_2}{\partial x_2}\cdot\frac{\partial u_2}{\partial x_3} - \frac{\partial u_3}{\partial x_2}\cdot\frac{\partial u_3}{\partial x_3}\right]$$

$$= \frac{1}{2}\left[\left(-\frac{2}{9}\right) + \frac{2}{9} - \left(-\frac{4}{9}\right)\cdot\frac{2}{9} - \frac{4}{9}\cdot\left(-\frac{2}{9}\right) - \frac{2}{9}\cdot\frac{8}{9}\right] = 0 = \eta_{32}$$

In matrix notation, the second order Eulerian finite, strain tensors η_{ij} are given by

$$(\eta_{ij}) = \frac{1}{9}\begin{pmatrix} 2 & -2 & 0 \\ -2 & 2 & 0 \\ 0 & 0 & 4 \end{pmatrix} = \frac{1}{2}\begin{pmatrix} 1 & 0 & 0 \\ 0 & 1 & 0 \\ 0 & 0 & 1 \end{pmatrix} - \frac{1}{2}\frac{1}{81}\begin{pmatrix} 45 & 36 & 0 \\ 36 & 45 & 0 \\ 0 & 0 & 1 \end{pmatrix} \Rightarrow \eta = \frac{1}{2}\mathbf{I} - \frac{1}{2}\mathbf{C}$$

Example 1.7 Find the Lagrangian and Eulerian finite strain tensors for deformation $x_1 = a_1(X_1 + \alpha X_2), x_2 = a_2 X_2, x_3 = a_3 X_3$, where a_1, a_2, a_3 are constants.

Solution: The displacements component $u_i = x_i - X_i$ of a material point are given by

$$u_1 = x_1 - X_1 = a_1(X_1 + \alpha X_2) - X_1 = (a_1 - 1)X_1 + a_1\alpha X_2$$

$$u_2 = x_2 - X_2 = a_2 X_2 - X_2 = (a_2 - 1)X_2$$

$$u_3 = x_3 - X_3 = a_3 X_3 - X_3 = (a_3 - 1)X_3$$

Now the second order Lagrangian finite, strain tensors r_{ij} are given by Eq.(1.18). Therefore,

$$r_{11} = \frac{1}{2}\left[\frac{\partial u_1}{\partial X_1} + \frac{\partial u_1}{\partial X_1} + \frac{\partial u_k}{\partial X_1}\cdot\frac{\partial u_k}{\partial X_1}\right] = \frac{\partial u_1}{\partial X_1} + \frac{1}{2}\left[\left(\frac{\partial u_1}{\partial X_1}\right)^2 + \left(\frac{\partial u_2}{\partial X_1}\right)^2 + \left(\frac{\partial u_3}{\partial X_1}\right)^2\right]$$

$$= (a_1 - 1) + \frac{1}{2}\left[(a_1 - 1)^2 + (0)^2 + (0)^2\right] = \frac{1}{2}a_1^2 - \frac{1}{2}$$

$$r_{22} = \frac{1}{2}\left[\frac{\partial u_2}{\partial X_2} + \frac{\partial u_2}{\partial X_2} + \frac{\partial u_k}{\partial X_2}\cdot\frac{\partial u_k}{\partial X_2}\right] = \frac{\partial u_2}{\partial X_2} + \frac{1}{2}\left[\left(\frac{\partial u_1}{\partial X_2}\right)^2 + \left(\frac{\partial u_2}{\partial X_2}\right)^2 + \left(\frac{\partial u_3}{\partial X_2}\right)^2\right]$$

$$= (a_2 - 1) + \frac{1}{2}\left[(a_1\alpha)^2 + (a_2 - 1)^2 + (0)^2\right] = \frac{1}{2}a_1^2\alpha^2 + \frac{1}{2}a_2^2 - \frac{1}{2}$$

$$r_{33} = \frac{1}{2}\left[\frac{\partial u_3}{\partial X_3} + \frac{\partial u_3}{\partial X_3} + \frac{\partial u_k}{\partial X_3}\cdot\frac{\partial u_k}{\partial X_3}\right] = \frac{\partial u_3}{\partial X_3} + \frac{1}{2}\left[\left(\frac{\partial u_1}{\partial X_3}\right)^2 + \left(\frac{\partial u_2}{\partial X_3}\right)^2 + \left(\frac{\partial u_3}{\partial X_3}\right)^2\right]$$

$$= (a_3 - 1) + \frac{1}{2}\left[(0)^2 + (0)^2 + (a_3 - 1)^2\right] = \frac{1}{2}a_3^2 - \frac{1}{2}$$

$$r_{12} = \frac{1}{2}\left[\frac{\partial u_1}{\partial X_2} + \frac{\partial u_2}{\partial X_1} + \frac{\partial u_k}{\partial X_1}\cdot\frac{\partial u_k}{\partial X_2}\right]$$

$$= \frac{1}{2}\left[\frac{\partial u_1}{\partial X_2} + \frac{\partial u_2}{\partial X_1} + \frac{\partial u_1}{\partial X_1}\cdot\frac{\partial u_1}{\partial X_2} + \frac{\partial u_2}{\partial X_1}\cdot\frac{\partial u_2}{\partial X_2} + \frac{\partial u_3}{\partial X_1}\cdot\frac{\partial u_3}{\partial X_2}\right]$$

$$= \frac{1}{2}\left[a_1\alpha + 0 + (a_1 - 1).a_1\alpha + 0.(a_2 - 1) + 0.0\right] = \frac{1}{2}a_1^2\alpha = r_{21}$$

$$r_{13} = \frac{1}{2}\left[\frac{\partial u_1}{\partial X_3} + \frac{\partial u_3}{\partial X_1} + \frac{\partial u_k}{\partial X_1}\cdot\frac{\partial u_k}{\partial X_3}\right]$$

$$= \frac{1}{2}\left[\frac{\partial u_1}{\partial X_3} + \frac{\partial u_3}{\partial X_1} + \frac{\partial u_1}{\partial X_1}\cdot\frac{\partial u_1}{\partial X_3} + \frac{\partial u_2}{\partial X_1}\cdot\frac{\partial u_2}{\partial X_3} + \frac{\partial u_3}{\partial X_1}\cdot\frac{\partial u_3}{\partial X_3}\right]$$

$$= \frac{1}{2}\left[0 + 0 + (a_1 - 1).0 + 0.0 + 0.0\right] = 0 = r_{31}$$

1.8 Deformation Gradients, Finite Strain Tensors

$$r_{23} = \frac{1}{2}\left[\frac{\partial u_2}{\partial X_3} + \frac{\partial u_3}{\partial X_2} + \frac{\partial u_k}{\partial X_2}\cdot\frac{\partial u_k}{\partial X_3}\right]$$

$$= \frac{1}{2}\left[\frac{\partial u_2}{\partial X_3} + \frac{\partial u_3}{\partial X_2} + \frac{\partial u_1}{\partial X_2}\cdot\frac{\partial u_1}{\partial X_3} + \frac{\partial u_2}{\partial X_2}\cdot\frac{\partial u_2}{\partial X_3} + \frac{\partial u_3}{\partial X_2}\cdot\frac{\partial u_3}{\partial X_3}\right]$$

$$= \frac{1}{2}\left[0 + 0 + a_1\alpha.0 + (a_2 - 1).0 + 0.0\right] = 0 = r_{32}$$

In matrix notation, the second order Lagrangian finite, strain tensors are given by

$$(r_{ij}) = \frac{1}{2}\begin{pmatrix} a_1^2 - 1 & a_1^2\alpha & 0 \\ a_1^2\alpha & a_1^2\alpha^2 + a_2^2 - 1 & 0 \\ 0 & 0 & a_3^2 - 1 \end{pmatrix} = \frac{1}{2}\begin{pmatrix} a_1^2 & a_1^2\alpha & 0 \\ a_1^2\alpha & a_1^2\alpha^2 + a_2^2 & 0 \\ 0 & 0 & a_3^2 \end{pmatrix} - \frac{1}{2}(\delta_{ij}).$$

Inverting the given relation $x_1 = a_1(X_1 + \alpha X_2), x_2 = a_2 X_2, x_3 = a_3 X_3$, we get

$$X_2 = a_2^{-1}x_2, X_3 = a_3^{-1}x_3, X_1 = a_1^{-1}x_1 - \alpha a_2^{-1}x_2$$

The displacement components $u_i = x_i - X_i$ of a spatial point are given by

$$u_1 = x_1 - X_1 = x_1 - a_1^{-1}x_1 + \alpha a_2^{-1}x_2 = (1 - a_1^{-1})x_1 + \alpha a_2^{-1}x_2$$

$$u_2 = x_2 - X_2 = x_2 - a_2^{-1}x_2 = (1 - a_2^{-1})x_2$$

$$u_3 = x_3 - X_3 = x_3 - a_3^{-1}x_3 = (1 - a_3^{-1})x_3$$

Now the second order Eulearian finite strain tensors η_{ij} are given by Eq.(1.29). Therefore,

$$\eta_{11} = \frac{1}{2}\left[\frac{\partial u_1}{\partial x_1} + \frac{\partial u_1}{\partial x_1} - \frac{\partial u_k}{\partial x_1}\cdot\frac{\partial u_k}{\partial x_1}\right] = \frac{\partial u_1}{\partial x_1} - \frac{1}{2}\left[\left(\frac{\partial u_1}{\partial x_1}\right)^2 + \left(\frac{\partial u_2}{\partial x_1}\right)^2 + \left(\frac{\partial u_3}{\partial x_1}\right)^2\right]$$

$$= (1 - a_1^{-1}) - \frac{1}{2}\left[(1 - a_1^{-1})^2 + (0)^2 + (0)^2\right] = \frac{1}{2} - \frac{1}{2}a_1^{-2}$$

$$\eta_{22} = \frac{1}{2}\left[\frac{\partial u_2}{\partial x_2} + \frac{\partial u_2}{\partial x_2} - \frac{\partial u_k}{\partial x_2}\cdot\frac{\partial u_k}{\partial x_2}\right] = \frac{\partial u_2}{\partial x_2} - \frac{1}{2}\left[\left(\frac{\partial u_1}{\partial x_2}\right)^2 + \left(\frac{\partial u_2}{\partial x_2}\right)^2 + \left(\frac{\partial u_3}{\partial x_2}\right)^2\right]$$

$$= (1 - a_2^{-1}) - \frac{1}{2}\left[(\alpha a_2^{-1})^2 + (1 - a_2^{-1})^2 + (0)^2\right] = \frac{1}{2} - \frac{1}{2}a_2^{-2}(\alpha^2 + 1)$$

$$\eta_{33} = \frac{1}{2}\left[\frac{\partial u_3}{\partial x_3} + \frac{\partial u_3}{\partial x_3} - \frac{\partial u_k}{\partial x_3}\cdot\frac{\partial u_k}{\partial x_3}\right] = \frac{\partial u_3}{\partial x_3} - \frac{1}{2}\left[\left(\frac{\partial u_1}{\partial x_3}\right)^2 + \left(\frac{\partial u_2}{\partial x_3}\right)^2 + \left(\frac{\partial u_3}{\partial x_3}\right)^2\right]$$

$$= \left(1 - a_3^{-1}\right) - \frac{1}{2}\left[(0)^2 + (0)^2 + \left(1 - a_3^{-1}\right)^2\right] = \frac{1}{2} - \frac{1}{2}a_3^{-2}$$

$$\eta_{12} = \frac{1}{2}\left[\frac{\partial u_1}{\partial x_2} + \frac{\partial u_2}{\partial x_1} - \frac{\partial u_k}{\partial x_1}\cdot\frac{\partial u_k}{\partial x_2}\right] = \frac{1}{2}\left[\frac{\partial u_1}{\partial x_2} + \frac{\partial u_2}{\partial x_1} - \frac{\partial u_1}{\partial x_1}\cdot\frac{\partial u_1}{\partial x_2} - \frac{\partial u_2}{\partial x_1}\cdot\frac{\partial u_2}{\partial x_2} - \frac{\partial u_3}{\partial x_1}\cdot\frac{\partial u_3}{\partial x_2}\right]$$

$$= \frac{1}{2}\left[\alpha a_2^{-1} + 0 - \left(1 - a_1^{-1}\right)\cdot\alpha a_2^{-1} - 0\cdot\left(1 - a_2^{-1}\right) - 0\cdot 0\right] = -\alpha a_1^{-1} a_2^{-1} = \eta_{21}$$

$$\eta_{13} = \frac{1}{2}\left[\frac{\partial u_1}{\partial x_3} + \frac{\partial u_3}{\partial x_1} - \frac{\partial u_k}{\partial x_1}\cdot\frac{\partial u_k}{\partial x_3}\right] = \frac{1}{2}\left[\frac{\partial u_1}{\partial x_3} + \frac{\partial u_3}{\partial x_1} + \frac{\partial u_1}{\partial x_3}\cdot\frac{\partial u_1}{\partial x_1} + \frac{\partial u_2}{\partial x_1}\cdot\frac{\partial u_2}{\partial x_3} + \frac{\partial u_3}{\partial x_1}\cdot\frac{\partial u_3}{\partial x_3}\right]$$

$$= \frac{1}{2}\left[0 + 0 + \left(1 - a_1^{-1}\right)\cdot 0 + 0\cdot 0 + 0\cdot 0\right] = 0 = \eta_{31}$$

$$\eta_{23} = \frac{1}{2}\left[\frac{\partial u_2}{\partial x_3} + \frac{\partial u_3}{\partial x_2} + \frac{\partial u_k}{\partial x_2}\cdot\frac{\partial u_k}{\partial x_3}\right] = \frac{1}{2}\left[\frac{\partial u_2}{\partial x_3} + \frac{\partial u_3}{\partial x_2} + \frac{\partial u_1}{\partial x_2}\cdot\frac{\partial u_1}{\partial x_3} + \frac{\partial u_2}{\partial x_2}\cdot\frac{\partial u_2}{\partial x_3} + \frac{\partial u_3}{\partial x_2}\cdot\frac{\partial u_3}{\partial x_3}\right]$$

$$= \frac{1}{2}\left[0 + 0 + \alpha a_2^{-1}\cdot 0 + \left(1 - a_2^{-1}\right)\cdot 0 + 0\cdot 0\right] = 0 = \eta_{32}$$

In matrix notation, the second order Eulerian finite strain tensors are given by

$$\left(\eta_{ij}\right) = \frac{1}{2}\begin{pmatrix} 1 - a_1^{-2} & -\alpha a_1^{-1} a_2^{-1} & 0 \\ -\alpha a_1^{-1} a_2^{-1} & 1 - a_2^{-2}\left(\alpha^2 + 1\right) & 0 \\ 0 & 0 & 1 - a_3^{-2} \end{pmatrix}$$

$$= \frac{1}{2}\left(\delta_{ij}\right) - \frac{1}{2}\begin{pmatrix} a_1^{-2} & -\alpha a_1^{-1} a_2^{-1} & 0 \\ -\alpha a_1^{-1} a_2^{-1} & a_2^{-2}\left(\alpha^2 + 1\right) & 0 \\ 0 & 0 & a_3^{-2} \end{pmatrix}.$$

1.8.3 Deformation Gradient in Polar Co-ordinates

Consider polar co-ordinates (r, θ) such that $r = \sqrt{x^2 + y^2}$ and $\theta = \tan^{-1}(\frac{y}{x})$ (Fig. 1.5).

1.8 Deformation Gradients, Finite Strain Tensors

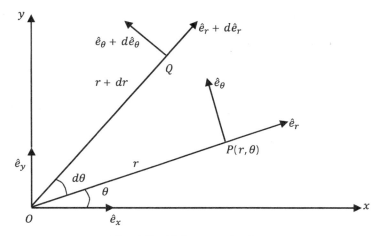

Fig. 1.5: Polar co-ordinates

The unit base vectors $\hat{e}_r$ and $\hat{e}_\theta$ can be expressed in terms of the Cartesian base vectors $\hat{e}_x$ and $\hat{e}_y$ as

$$\hat{e}_r = \cos\theta\, \hat{e}_x + \sin\theta\, \hat{e}_y;\ \hat{e}_\theta = -\sin\theta\, \hat{e}_x + \cos\theta\, \hat{e}_y.$$

Now, $d\hat{e}_x = 0$ and $d\hat{e}_y = 0$. The unit vectors $\hat{e}_r$ and $\hat{e}_\theta$ vary in direction θ as changes.

$$d\hat{e}_r = \left(-\sin\theta\, \hat{e}_x + \cos\theta\, \hat{e}_y\right) d\theta = d\theta\, \hat{e}_\theta;\ d\hat{e}_\theta = \left(-\cos\theta\, \hat{e}_x - \sin\theta\, \hat{e}_y\right) d\theta = -d\theta\, \hat{e}_r.$$

The position vector for the point P can be written as $\vec{r} = r\hat{e}_r$, where r is the magnitude of the vector $\vec{r}$. Thus

$$d\vec{r} = dr\hat{e}_r + r\, d\hat{e}_r = dr\hat{e}_r + r\, d\theta\, \hat{e}_\theta.$$

In polar co-ordinates (r, θ), the displacement vector u can be written as

$$u(r,\theta) = u_r(r,\theta)\hat{e}_r + u_\theta(r,\theta)\hat{e}_\theta a$$

where, $u_r(r, \theta)$ and $u_\theta(r, \theta)$ are the components of $u(r, \theta)$ in the directions $\hat{e}_r$ and $\hat{e}_\theta$ respectively. Since $u_r = u_r(r,\theta)$ and $u_\theta = u_\theta(r,\theta)$, so by chain rule of differentiation, we get

$$du_r = \frac{\partial u_r}{\partial r} dr + \frac{\partial u_r}{\partial \theta} d\theta;\ du_\theta = \frac{\partial u_\theta}{\partial r} dr + \frac{\partial u_\theta}{\partial \theta} d\theta.$$

Therefore,

$$du = du_r \hat{e}_r + u_r d\hat{e}_r + du_\theta \hat{e}_\theta + u_\theta d\hat{e}_\theta$$

$$= \left(\frac{\partial u_r}{\partial r}dr + \frac{\partial u_r}{\partial \theta}d\theta\right)\hat{e}_r + u_r d\theta \hat{e}_\theta + \left(\frac{\partial u_\theta}{\partial r}dr + \frac{\partial u_\theta}{\partial \theta}d\theta\right)\hat{e}_\theta - u_\theta d\theta \hat{e}_r$$

$$= \left[\frac{\partial u_r}{\partial r}dr + \left(\frac{\partial u_r}{\partial \theta} - u_\theta\right)d\theta\right]\hat{e}_r + \left[\frac{\partial u_\theta}{\partial r}dr + \left(\frac{\partial u_\theta}{\partial \theta} + u_r\right)d\theta\right]\hat{e}_\theta.$$

By definition of gradient of $u(r,\theta)$, we have

$$u(r,\theta) = \nabla u \, dr = \nabla u(dr\hat{e}_r + r \, d\theta \hat{e}_\theta) = dr \nabla u \hat{e}_r + r \, d\theta \nabla u \hat{e}_\theta$$

$$= dr\left[u_{rr}\hat{e}_r + u_{\theta r}\hat{e}_\theta\right] + r \, d\theta\left[u_{r\theta}\hat{e}_r + u_{\theta\theta}\hat{e}_\theta\right]$$

$$= \left[u_{rr}dr + u_{r\theta}r \, d\theta\right]\hat{e}_r + [u_{\theta r}dr + u_{\theta\theta}r \, d\theta]\hat{e}_\theta$$

Comparing, we get

$$u_{rr}dr + u_{r\theta}r \, d\theta = \frac{\partial u_r}{\partial r}dr + \left(\frac{\partial u_r}{\partial \theta} - u_\theta\right)d\theta$$

$$u_{\theta r}dr + u_{\theta\theta}r \, d\theta = \frac{\partial u_\theta}{\partial r}dr + \left(\frac{\partial u_\theta}{\partial \theta} + u_r\right)d\theta$$

These relations must hold for any values of dr and $d\theta$, so

$$u_{rr} = \frac{\partial u_r}{\partial r}; \quad ru_{r\theta} = \frac{\partial u_r}{\partial \theta} - u_\theta; \quad u_{\theta r} = \frac{\partial u_\theta}{\partial r}; \quad ru_{\theta\theta} = \frac{\partial u_\theta}{\partial \theta} + u_r$$

Therefore, in polar co-ordinates, the matrix form the components of the deformation gradient ∇u can be written as

$$F = [\nabla u] = \begin{pmatrix} \dfrac{\partial u_r}{\partial r} & \dfrac{1}{r}\dfrac{\partial u_r}{\partial \theta} - \dfrac{u_\theta}{r} \\ \dfrac{\partial u_\theta}{\partial r} & \dfrac{1}{r}\dfrac{\partial u_\theta}{\partial \theta} + \dfrac{u_r}{r} \end{pmatrix}. \quad (1.34)$$

The Green's deformation tensor $C = F^T.F$ is a symmetric tensor. Therefore

$$C = \begin{pmatrix} \dfrac{\partial u_r}{\partial r} & \dfrac{\partial u_\theta}{\partial r} \\ \dfrac{1}{r}\dfrac{\partial u_r}{\partial \theta} - \dfrac{u_\theta}{r} & \dfrac{1}{r}\dfrac{\partial u_\theta}{\partial \theta} + \dfrac{u_r}{r} \end{pmatrix} \begin{pmatrix} \dfrac{\partial u_r}{\partial r} & \dfrac{1}{r}\dfrac{\partial u_r}{\partial \theta} - \dfrac{u_\theta}{r} \\ \dfrac{\partial u_\theta}{\partial r} & \dfrac{1}{r}\dfrac{\partial u_\theta}{\partial \theta} + \dfrac{u_r}{r} \end{pmatrix}$$

$$C_{rr} = \left(\frac{\partial u_r}{\partial r}\right)^2 + \left(\frac{\partial u_\theta}{\partial r}\right)^2; \quad C_{r\theta} = \left(\frac{1}{r}\frac{\partial u_r}{\partial \theta} - \frac{u_\theta}{r}\right)\frac{\partial u_r}{\partial r} + \left(\frac{1}{r}\frac{\partial u_\theta}{\partial \theta} + \frac{u_r}{r}\right)\frac{\partial u_\theta}{\partial r}$$

1.8 Deformation Gradients, Finite Strain Tensors

$$C_{\theta r} = C_{r\theta}; \quad C_{\theta\theta} = \left(\frac{1}{r}\frac{\partial u_r}{\partial \theta} - \frac{u_\theta}{r}\right)^2 + \left(\frac{1}{r}\frac{\partial u_\theta}{\partial \theta} + \frac{u_r}{r}\right)^2$$

The Lagrangian finite strain tensor is $2R = C - I$, and therefore

$$R = \frac{1}{2}(C - I) = \frac{1}{2}(F^T \cdot F - I)$$

$$= \frac{1}{2}\begin{bmatrix} \left(\frac{\partial u_r}{\partial r}\right)^2 + \left(\frac{\partial u_\theta}{\partial r}\right)^2 - 1 & \left(\frac{1}{r}\frac{\partial u_r}{\partial \theta} - \frac{u_\theta}{r}\right)\frac{\partial u_r}{\partial r} + \left(\frac{1}{r}\frac{\partial u_\theta}{\partial \theta} + \frac{u_r}{r}\right)\frac{\partial u_\theta}{\partial r} \\ \left(\frac{1}{r}\frac{\partial u_r}{\partial \theta} - \frac{u_\theta}{r}\right)\frac{\partial u_r}{\partial r} + \left(\frac{1}{r}\frac{\partial u_\theta}{\partial \theta} + \frac{u_r}{r}\right)\frac{\partial u_\theta}{\partial r} & \left(\frac{1}{r}\frac{\partial u_r}{\partial \theta} - \frac{u_\theta}{r}\right)^2 + \left(\frac{1}{r}\frac{\partial u_\theta}{\partial \theta} + \frac{u_r}{r}\right)^2 - 1 \end{bmatrix}.$$

Through the chain rule of partial differentiation it is clear that $\frac{\partial X_j}{\partial x_k}\frac{\partial x_k}{\partial X_m} = \delta_{jm}$. A unique solution exists, since the jacobian of the transformation is assumed not to vanish. Using Cramer's rule of determinants, the solution for $\frac{\partial X_j}{\partial x_k}$ may be obtained in terms of $\frac{\partial x_k}{\partial X_j}$. Thus,

$$\frac{\partial X_i}{\partial x_p} = \frac{1}{J}\text{cofactor}\frac{\partial x_p}{\partial X_i}; \quad J = \text{Jacobian of transformation}$$

Here the Jacobian of transformation is

$$J = \begin{vmatrix} \frac{\partial u_r}{\partial r} & \frac{1}{r}\frac{\partial u_r}{\partial \theta} - \frac{u_\theta}{r} \\ \frac{\partial u_\theta}{\partial r} & \frac{1}{r}\frac{\partial u_\theta}{\partial \theta} + \frac{u_r}{r} \end{vmatrix} = \left(\frac{1}{r}\frac{\partial u_\theta}{\partial \theta} + \frac{u_r}{r}\right)\frac{\partial u_r}{\partial r} - \left(\frac{1}{r}\frac{\partial u_r}{\partial \theta} - \frac{u_\theta}{r}\right)\frac{\partial u_\theta}{\partial r} \neq 0.$$

Therefore, F^{-1} is given by

$$F^{-1} = \frac{1}{J}\begin{pmatrix} \frac{1}{r}\frac{\partial u_\theta}{\partial \theta} + \frac{u_r}{r} & -\frac{\partial u_\theta}{\partial r} \\ -\frac{1}{r}\frac{\partial u_r}{\partial \theta} + \frac{u_\theta}{r} & \frac{\partial u_r}{\partial r} \end{pmatrix}. \tag{1.35}$$

The Cauchy's deformation tensor $c = (F^{-1})^T \cdot F^{-1}$ is a symmetric tensor. Therefore,

$$c = (F^{-1})^T \cdot F^{-1} = \frac{1}{J^2}\begin{pmatrix} \frac{1}{r}\frac{\partial u_\theta}{\partial \theta} + \frac{u_r}{r} & -\frac{1}{r}\frac{\partial u_r}{\partial \theta} + \frac{u_\theta}{r} \\ -\frac{\partial u_\theta}{\partial r} & \frac{\partial u_r}{\partial r} \end{pmatrix}\begin{pmatrix} \frac{1}{r}\frac{\partial u_\theta}{\partial \theta} + \frac{u_r}{r} & -\frac{\partial u_\theta}{\partial r} \\ -\frac{1}{r}\frac{\partial u_r}{\partial \theta} + \frac{u_\theta}{r} & \frac{\partial u_r}{\partial r} \end{pmatrix}$$

$$= \frac{1}{J^2} \begin{pmatrix} \left(\frac{1}{r}\frac{\partial u_\theta}{\partial \theta} + \frac{u_r}{r}\right)^2 & \left(\frac{\partial u_\theta}{\partial r}\right)^2 \\ \left(-\frac{1}{r}\frac{\partial u_r}{\partial \theta} + \frac{u_\theta}{r}\right)^2 & \left(\frac{\partial u_\theta}{\partial r}\right)^2 \end{pmatrix}.$$

The Eulerian finite strain tensor is $2\eta = I - c$, and therefore

$$\eta = \frac{1}{2}(I - c) = \frac{1}{2}\begin{bmatrix} \begin{pmatrix} 1 & 0 \\ 0 & 1 \end{pmatrix} - \frac{1}{J^2} \begin{pmatrix} \left(\frac{1}{r}\frac{\partial u_\theta}{\partial \theta} + \frac{u_r}{r}\right)^2 & \left(\frac{\partial u_\theta}{\partial r}\right)^2 \\ \left(-\frac{1}{r}\frac{\partial u_r}{\partial \theta} + \frac{u_\theta}{r}\right)^2 & \left(\frac{\partial u_\theta}{\partial r}\right)^2 \end{pmatrix} \end{bmatrix}.$$

1.8.4 Deformation Gradient in Cylindrical Co-ordinates

Let the co-ordinates of the point P, in cylindrical polar co-ordinates be (r, θ, z) and $\hat{e}_r$, $\hat{e}_\theta$ and $\hat{e}_z$ be the unit vectors in the direction of increasing r, θ and z respectively (Fig. 1.6).

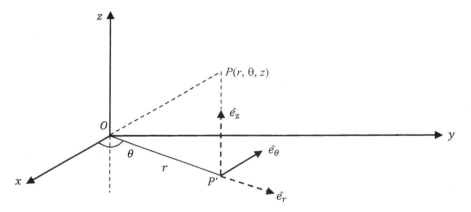

Fig. 1.6: Cylindrical co-ordinates

Let $\hat{e}_x$, $\hat{e}_y$ and $\hat{e}_z$ be the unit base vectors in the direction of increasing x, y and z respectively. Thus

$$\hat{e}_r = \cos\theta \hat{e}_x + \sin\theta \hat{e}_y; \hat{e}_\theta = -\sin\theta \hat{e}_x + \cos\theta \hat{e}_y$$

$$d\hat{e}_r = \left[-\sin\theta \hat{e}_x + \cos\theta \hat{e}_y\right] d\theta = d\theta \hat{e}_\theta; d\hat{e}_\theta = -\cos\theta \hat{e}_x - \sin\theta \hat{e}_y = -d\theta \hat{e}_r.$$

1.8 Deformation Gradients, Finite Strain Tensors

The position vector for the point P can be written as $\vec{r} = r\hat{e}_r$ where r is the magnitude of the vector $\vec{r}$. Thus

$$d\vec{r} = dr\hat{e}_r + r\,d\hat{e}_r = dr\hat{e}_r + rd\theta\hat{e}_\theta$$

In spherical polar co-ordinates (r, θ, ϕ), the displacement vector can be written as

$$u(r, \theta, z) = u_r(r, \theta, z)\hat{e}_r + u_\theta(r, \theta, z)\hat{e}_\theta + u_z(r, \theta, z)\hat{e}_z$$

Therefore

$$du = du_r\hat{e}_r + u_r d\hat{e}_r + du_\theta\hat{e}_\theta + u_\theta d\hat{e}_\theta + du_z\hat{e}_z$$

$$= \left(\frac{\partial u_r}{\partial r}dr + \frac{\partial u_r}{\partial \theta}d\theta + \frac{\partial u_r}{\partial z}dz\right)\hat{e}_r + u_r d\theta\hat{e}_\theta$$

$$+ \left(\frac{\partial u_\theta}{\partial r}dr + \frac{\partial u_\theta}{\partial \theta}d\theta + \frac{\partial u_\theta}{\partial z}dz\right)\hat{e}_\theta - u_\theta d\theta\hat{e}_r$$

$$+ \left(\frac{\partial u_z}{\partial r}dr + \frac{\partial u_z}{\partial \theta}d\theta + \frac{\partial u_z}{\partial z}dz\right)\hat{e}_z$$

$$= \left[\frac{\partial u_r}{\partial r}dr + \left(\frac{\partial u_\theta}{\partial \theta} - u_\theta\right)d\theta + \frac{\partial u_r}{\partial z}dz\right]\hat{e}_r$$

$$+ \left[\frac{\partial u_\theta}{\partial r}dr + \left(\frac{\partial u_\theta}{\partial \theta} + u_r\right)d\theta + \frac{\partial u_\theta}{\partial z}dz\right] + \left[\frac{\partial u_z}{\partial r}dr + \frac{\partial u_z}{\partial \theta}d\theta + \frac{\partial u_z}{\partial z}dz\right]\hat{e}_z.$$

Since is a scalar field, by definition of gradient

$$du = \vec{\nabla}u\,d\vec{r} = \vec{\nabla}u\left[dr\hat{e}_r + r\,d\theta\hat{e}_\theta + dz\hat{e}_z\right]$$

$$= dr\vec{\nabla}u\hat{e}_r + r\,d\theta\vec{\nabla}u\hat{e}_\theta + dz\vec{\nabla}u\hat{e}_z$$

$$= dr(u_{rr}\hat{e}_r + u_{0r}\hat{e}_\theta + u_{zr}\hat{e}_z) + r\,d\theta(u_{r\theta}\hat{e}_r + u_{\theta\theta}\hat{e}_\theta + u_{z\theta}\hat{e}_z)$$

$$+ dz(u_{rz}\hat{e}_r + u_{\theta z}\hat{e}_\theta + u_{zz}\hat{e}_z)$$

$$= \left(u_{rr}dr + ru_{r\theta}d\theta + u_{rz}dz\right)\hat{e}_r$$

$$+ \left(u_{\theta r}dr + ru_{\theta\theta}d\theta + u_{\theta z}dz\right)\hat{e}_\theta$$

$$+ \left(u_{zr}dr + ru_{z\theta}d\theta + u_{zz}dz\right)\hat{e}_z$$

Comparing, we get

$$u_{rr}dr + ru_{r\theta}d\theta + u_{rz}dz = \frac{\partial u_r}{\partial r}dr + \left(\frac{\partial u_\theta}{\partial \theta} - u_\theta\right)d\theta + \frac{\partial u_r}{\partial z}dz$$

$$u_{\theta r}dr + ru_{\theta\theta}d\theta + u_{\theta z}dz = \frac{\partial u_\theta}{\partial r}dr + \left(\frac{\partial u_\theta}{\partial \theta} + u_r\right)d\theta + \frac{\partial u_\theta}{\partial z}dz$$

$$u_{zr}dr + ru_{z\theta}d\theta + u_{zz}d = \frac{\partial u_z}{\partial r}dr + \frac{\partial u_z}{\partial \theta}d\theta + \frac{\partial u_z}{\partial z}dz$$

These equations must be valid for arbitrary values of dr, $d\theta$ and dz, therefore

$$u_{rr} = \frac{\partial u_r}{\partial r}; \quad ru_{r\theta} = \frac{\partial u_\theta}{\partial \theta} - u_\theta; \quad u_{rz} = \frac{\partial u_r}{\partial z}$$

$$u_{\theta r} = \frac{\partial u_\theta}{\partial r}; \quad ru_{\theta\theta} = \frac{\partial u_\theta}{\partial \theta} + u_r; \quad u_{\theta z} = \frac{\partial u_\theta}{\partial z}$$

$$u_{zr} = \frac{\partial u_z}{\partial r}; \quad ru_{z\theta} = \frac{\partial u_z}{\partial \theta}; \quad u_{zz} = \frac{\partial u_z}{\partial z}$$

In cylindrical co-ordinates (r, θ, z) the matrix form of the components of $\nabla \mathbf{u}$ can be written as

$$F = [\nabla u] = \begin{pmatrix} \dfrac{\partial u_r}{\partial r} & \dfrac{1}{r}\dfrac{\partial u_r}{\partial \theta} - \dfrac{u_\theta}{r} & \dfrac{\partial u_r}{\partial z} \\ \dfrac{\partial u_\theta}{\partial r} & \dfrac{1}{r}\dfrac{\partial u_\theta}{\partial \theta} + \dfrac{u_r}{r} & \dfrac{\partial u_\theta}{\partial z} \\ \dfrac{\partial u_z}{\partial r} & \dfrac{1}{r}\dfrac{\partial u_z}{\partial \theta} & \dfrac{\partial u_z}{\partial z} \end{pmatrix}. \tag{1.36}$$

The Green's deformation tensor $C = F^T.F$ is a symmetric tensor. Therefore,

$$C = \begin{pmatrix} \dfrac{\partial u_r}{\partial r} & \dfrac{\partial u_\theta}{\partial r} & \dfrac{\partial u_z}{\partial r} \\ \dfrac{1}{r}\dfrac{\partial u_r}{\partial \theta} - \dfrac{u_\theta}{r} & \dfrac{1}{r}\dfrac{\partial u_\theta}{\partial \theta} + \dfrac{u_r}{r} & \dfrac{1}{r}\dfrac{\partial u_z}{\partial \theta} \\ \dfrac{\partial u_r}{\partial z} & \dfrac{\partial u_\theta}{\partial z} & \dfrac{\partial u_z}{\partial z} \end{pmatrix} \begin{pmatrix} \dfrac{\partial u_r}{\partial r} & \dfrac{1}{r}\dfrac{\partial u_r}{\partial \theta} - \dfrac{u_\theta}{r} & \dfrac{\partial u_r}{\partial z} \\ \dfrac{\partial u_\theta}{\partial r} & \dfrac{1}{r}\dfrac{\partial u_\theta}{\partial \theta} + \dfrac{u_r}{r} & \dfrac{\partial u_\theta}{\partial z} \\ \dfrac{\partial u_z}{\partial r} & \dfrac{1}{r}\dfrac{\partial u_z}{\partial \theta} & \dfrac{\partial u_z}{\partial z} \end{pmatrix}$$

1.8 Deformation Gradients, Finite Strain Tensors

$$C_{rr} = \left(\frac{\partial u_r}{\partial r}\right)^2 + \left(\frac{\partial u_\theta}{\partial r}\right)^2 + \left(\frac{\partial u_z}{\partial r}\right)^2;$$

$$C_{r\theta} = \left(\frac{1}{r}\frac{\partial u_r}{\partial \theta} - \frac{u_\theta}{r}\right)\frac{\partial u_r}{\partial r} + \left(\frac{1}{r}\frac{\partial u_\theta}{\partial \theta} + \frac{u_r}{r}\right)\frac{\partial u_\theta}{\partial r} + \frac{1}{r}\frac{\partial u_z}{\partial \theta}\frac{\partial u_z}{\partial r}$$

$$C_{rz} = \frac{\partial u_r}{\partial r}\frac{\partial u_r}{\partial z} + \frac{\partial u_\theta}{\partial r}\frac{\partial u_\theta}{\partial z} + \frac{\partial u_z}{\partial r}\frac{\partial u_z}{\partial z}$$

$$C_{\theta r} = C_{r\theta}; C_{\theta\theta} = \left(\frac{1}{r}\frac{\partial u_r}{\partial \theta} - \frac{u_\theta}{r}\right)^2 + \left(\frac{1}{r}\frac{\partial u_\theta}{\partial \theta} + \frac{u_r}{r}\right)^2 + \frac{1}{r^2}\left(\frac{\partial u_z}{\partial \theta}\right)^2$$

$$C_{\theta z} = \left(\frac{1}{r}\frac{\partial u_r}{\partial \theta} - \frac{u_\theta}{r}\right)\frac{\partial u_r}{\partial z} + \left(\frac{1}{r}\frac{\partial u_\theta}{\partial \theta} + \frac{u_r}{r}\right)\frac{\partial u_\theta}{\partial z} + \frac{1}{r}\frac{\partial u_z}{\partial \theta}\frac{\partial u_z}{\partial z}$$

$$C_{zr} = C_{rz}; C_{z\theta} = C_{\theta z}; C_{zz} = \left(\frac{\partial u_r}{\partial z}\right)^2 + \left(\frac{\partial u_\theta}{\partial z}\right)^2 + \left(\frac{\partial u_z}{\partial z}\right)^2.$$

Here the Jacobean of transformation is

$$J = \begin{vmatrix} \frac{\partial u_r}{\partial r} & \frac{1}{r}\frac{\partial u_r}{\partial \theta} - \frac{u_\theta}{r} & \frac{\partial u_r}{\partial z} \\ \frac{\partial u_\theta}{\partial r} & \frac{1}{r}\frac{\partial u_\theta}{\partial \theta} + \frac{u_r}{r} & \frac{\partial u_\theta}{\partial z} \\ \frac{\partial u_z}{\partial r} & \frac{1}{r}\frac{\partial u_z}{\partial \theta} & \frac{\partial u_z}{\partial z} \end{vmatrix} = \left(\frac{1}{r}\frac{\partial u_\theta}{\partial \theta} + \frac{u_r}{r} - \frac{\partial u_z}{\partial z}\right)\frac{\partial u_r}{\partial r} - \left(\frac{1}{r}\frac{\partial u_r}{\partial \theta} - \frac{u_\theta}{r}\right)$$

$$\left(\frac{\partial u_\theta}{\partial r}\frac{\partial u_z}{\partial z} - \frac{\partial u_\theta}{\partial z}\frac{\partial u_z}{\partial r}\right) + \left(\frac{\partial u_\theta}{\partial r}\frac{1}{r}\frac{\partial u_z}{\partial \theta} - \left[\frac{1}{r}\frac{\partial u_\theta}{\partial \theta} + \frac{u_r}{r}\right]\frac{\partial u_z}{\partial r}\right)\frac{\partial u_r}{\partial z} \neq 0.$$

Therefore, F^{-1} is given by

$$F^{-1} = \frac{1}{J}\begin{pmatrix} \frac{\partial u_r}{\partial r} & \frac{1}{r}\frac{\partial u_r}{\partial \theta} - \frac{u_\theta}{r} & \frac{\partial u_r}{\partial z} \\ \frac{\partial u_\theta}{\partial r} & \frac{1}{r}\frac{\partial u_\theta}{\partial \theta} + \frac{u_r}{r} & \frac{\partial u_\theta}{\partial z} \\ \frac{\partial u_z}{\partial r} & \frac{1}{r}\frac{\partial u_z}{\partial \theta} & \frac{\partial u_z}{\partial z} \end{pmatrix}. \qquad (1.37)$$

1.8.5 Deformation Gradiants in Spherical Polar Coordinates

Let the co-ordinates of the point P, in spherical polar co-ordinates be (r, θ, ϕ) and $\hat{e}_r$, $\hat{e}_\theta$ and $\hat{e}_\phi$ be the unit vectors in the direction of increasing r, θ and ϕ respectively (Fig. 1.7).

Let $\hat{e}_x$, $\hat{e}_y$ and $\hat{e}_z$ be the unit vectors in the direction of increasing x, y and z respectively. Thus

$$\hat{e}_{r'} = \cos\phi\hat{e}_x + \sin\phi\hat{e}_y ; \hat{e}_\phi = -\sin\phi\hat{e}_x + \cos\phi\hat{e}_y$$

where, $\hat{e}_{r'}$ is the unit vector in the direction r' in the xy-plane. Thus

$$d\hat{e}_{r'} = \left[-\sin\phi\hat{e}_x + \cos\phi\hat{e}_y\right]d\phi = d\phi\hat{e}_\phi; \quad d\hat{e}_\phi = -\cos\phi\hat{e}_x - \sin\phi\hat{e}_y = d\phi\hat{e}_{r'}.$$

Also, the unit vectors $\hat{e}_r$, $\hat{e}_\theta$ can be written as

$$\hat{e}_r = \cos\theta\hat{e}_z + \sin\theta\hat{e}_{r'} ; \hat{e}_\theta = \cos\theta\hat{e}_{r'} - \sin\theta\hat{e}_z ; \hat{e}_{r'} = \sin\theta\hat{e}_r + \cos\theta\hat{e}_\theta$$

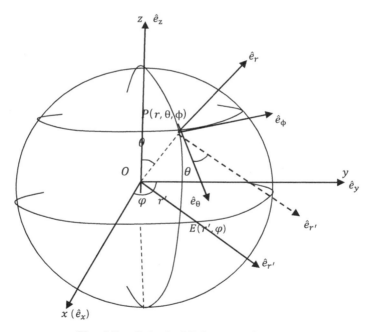

Fig. 1.7: Spherical Polar co-ordinates

Thus

$$d\hat{e}_r = \left[-\sin\theta\hat{e}_z + \cos\theta\hat{e}_{r'}\right]d\theta + \sin\theta d\hat{e}_{r'} = d\theta\hat{e}_\theta + \sin\theta d\phi\hat{e}_\phi$$

$$d\hat{e}_\theta = \left(-\sin\theta\hat{e}_{r'} - \cos\theta\hat{e}_z\right)d\theta + \cos\theta d\hat{e}_{r'} = -d\theta\hat{e}_r + \cos\theta d\phi\hat{e}_\phi$$

1.8 Deformation Gradients, Finite Strain Tensors

$$d\hat{e}_\phi = \left(-\cos\phi\,\hat{e}_{r'} - \sin\phi\,\hat{e}_y\right)d\phi = -d\phi\,\hat{e}_{r'} = -\left(\sin\theta\,\hat{e}_r + \cos\theta\,\hat{e}_\theta\right)d\phi.$$

The position vector for the point P can be written as $\vec{r} = r\hat{e}_r$ where r is the magnitude of the vector $\vec{r}$. Thus

$$d\vec{r} = dr\,\hat{e}_r + r\,d\hat{e}_r = dr\,\hat{e}_r + r\left[d\theta\,\hat{e}_\theta + \sin\theta\,d\phi\,\hat{e}_\phi\right]$$

$$= dr\,\hat{e}_r + r\,d\theta\,\hat{e}_\theta + r\sin\theta\,d\phi\,\hat{e}_\phi.$$

In spherical polar co-ordinates (r, θ, ϕ), the displacement vector can be written as

$$u(r,\theta,\phi) = u_r(r,\theta,\phi)\hat{e}_r + u_\theta(r,\theta,\phi)\hat{e}_\theta + u_\phi(r,\theta,\phi)\hat{e}_\phi$$

Therefore,

$$du = du_r\,\hat{e}_r + u_r\,d\hat{e}_r + du_\theta\,\hat{e}_\theta + u_\theta\,d\hat{e}_\theta + du_\phi\,\hat{e}_\phi + u_\phi\,d\hat{e}_\phi$$

$$= \left(\frac{\partial u_r}{\partial r}dr + \frac{\partial u_r}{\partial \theta}d\theta + \frac{\partial u_r}{\partial \phi}d\phi\right)\hat{e}_r + u_r(d\theta\,\hat{e}_\theta + \sin\theta\,d\phi\,\hat{e}_\phi)$$

$$+ \left(\frac{\partial u_\theta}{\partial r}dr + \frac{\partial u_\theta}{\partial \theta}d\theta + \frac{\partial u_\theta}{\partial \phi}d\phi\right)\hat{e}_\theta + u_\theta(-d\theta\,\hat{e}_r + \cos\theta\,d\phi\,\hat{e}_\phi)$$

$$+ \left(\frac{\partial u_\phi}{\partial r}dr + \frac{\partial u_\phi}{\partial \theta}d\theta + \frac{\partial u_\phi}{\partial \phi}d\phi\right)\hat{e}_\phi + u_\phi(-\sin\theta\,d\phi\,\hat{e}_r - \cos\theta\,d\phi\,\hat{e}_\theta)$$

$$= \left[\frac{\partial u_r}{\partial r}dr + \left(\frac{\partial u_r}{\partial \theta} - u_\theta\right)d\theta + \left(\frac{\partial u_r}{\partial \phi} - u_\phi\sin\theta\right)d\phi\right]\hat{e}_r$$

$$+ \left[\frac{\partial u_\theta}{\partial r}dr + \left(\frac{\partial u_\theta}{\partial \theta} + u_r\right)d\theta + \left(\frac{\partial u_\theta}{\partial \phi} - u_\phi\cos\theta\right)d\phi\right]\hat{e}_\theta$$

$$+ \left[\frac{\partial u_\phi}{\partial r}dr + \frac{\partial u_\phi}{\partial \theta}d\theta + \left(\frac{\partial u_\phi}{\partial \phi} + u_r\sin\theta + u_\theta\cos\theta\right)d\phi\right]\hat{e}_\phi.$$

Since u is a scalar field, by definition of gradient

$$du = \vec{\nabla}u \cdot d\vec{r} = \vec{\nabla}u\left[dr\,\hat{e}_r + r\,d\theta\,\hat{e}_\theta + r\sin\theta\,d\phi\,\hat{e}_\phi\right]$$
$$= dr\,\vec{\nabla}u\,\hat{e}_r + r\,d\theta\,\vec{\nabla}u\,\hat{e}_\theta + r\sin\theta\,d\phi\,\vec{\nabla}u\,\hat{e}_\phi$$

$$= dr(u_{rr}\hat{e}_r + u_{\theta r}\hat{e}_\theta + u_{\phi r}\hat{e}_\phi) + r\,d\theta(u_{r\theta}\hat{e}_r + u_{\theta\theta}\hat{e}_\theta + u_{\phi\theta}\hat{e}_\phi)$$

$$+ r\sin\theta, d\phi(u_{r\phi}\hat{e}_r + u_{\theta\phi}\hat{e}_\theta + u_{\phi\phi}\hat{e}_\phi)$$

$$= \left(u_{rr}\mathrm{d}r + ru_{r\theta}d\theta + r\sin\theta u_{r\phi}d\phi\right)\hat{e}_r + \left(u_{\theta r}\mathrm{d}r + ru_{\theta\theta}d\theta + r\sin\theta u_{\theta\phi}d\phi\right)\hat{e}_\theta$$

$$+ \left(u_{\phi r}\mathrm{d}r + r u_{\phi\theta}d\theta + r\sin\theta u_{\phi\phi}d\phi\right)\hat{e}_\phi.$$

Comparing, we get

$$u_{rr}\mathrm{d}r + ru_{r\theta}d\theta + r\sin\theta u_{r\phi}d\phi = \frac{\partial u_r}{\partial r}dr + \left(\frac{\partial u_\theta}{\partial \theta} - u_\theta\right)d\theta + \left(\frac{\partial u_r}{\partial \phi} - u_\phi \sin\theta\right)d\phi$$

$$u_{\theta r}\mathrm{d}r + ru_{\theta\,\theta}d\theta + r\sin\theta u_{\theta\phi}d\phi = \frac{\partial u_\theta}{\partial r}dr + \left(\frac{\partial u_\theta}{\partial \theta} + u_r\right)d\theta + \left(\frac{\partial u_\theta}{\partial \phi} - u_\phi \cos\theta\right)d\phi$$

$$u_{\phi r}\mathrm{d}r + ru_{\phi\theta}d\theta + r\sin\theta u_{\phi\phi}d\phi = \frac{\partial u_\phi}{\partial r}dr + \frac{\partial u_\phi}{\partial \theta}d\theta + \left(\frac{\partial u_\phi}{\partial \phi} + u_r \sin\theta + u_\theta \cos\theta\right)d\phi$$

These equations must be valid for arbitrary values of dr, dθ and dϕ, therefore

$$u_{rr} = \frac{\partial u_r}{\partial r}; ru_{r\theta} = \frac{\partial u_\theta}{\partial \theta} - u_\theta; r\sin\theta u_{r\phi} = \frac{\partial u_r}{\partial \phi} - u_\phi \sin\theta$$

$$u_{\theta r} = \frac{\partial u_\theta}{\partial r}; ru_{\theta\theta} = \frac{\partial u_\theta}{\partial \theta} + u_r; r\sin, u_{\theta\phi} = \frac{\partial u_\theta}{\partial \phi} - u_\phi \cos\theta$$

$$u_{\phi r} = \frac{\partial u_\phi}{\partial r}; ru_{\phi\theta} = \frac{\partial u_\phi}{\partial \theta}; r\sin, u_{\phi\phi} = \frac{\partial u_\phi}{\partial \phi} + u_r \sin\theta + u_\theta \cos\theta$$

In matrix form, we have

$$[\vec{\nabla}u] = \begin{pmatrix} \dfrac{\partial u_r}{\partial r} & \dfrac{1}{r}\left[\dfrac{\partial u_\theta}{\partial \theta} - u_\theta\right] & \dfrac{1}{r\sin\theta}\left[\dfrac{\partial u_r}{\partial \phi} - u_\phi \sin\theta\right] \\ \dfrac{\partial u_\theta}{\partial r} & \dfrac{1}{r}\left[\dfrac{\partial u_\theta}{\partial \theta} + u_r\right] & \dfrac{1}{r\sin\theta}\left[\dfrac{\partial u_\theta}{\partial \phi} - u_\phi \cos\theta\right] \\ \dfrac{\partial u_\phi}{\partial r} & \dfrac{1}{r}\dfrac{\partial u_\phi}{\partial \theta} & \dfrac{1}{r\sin\theta}\left[\dfrac{\partial u_\phi}{\partial \phi} + u_r \sin\theta + u_\theta \cos\theta\right] \end{pmatrix} \quad (1.38)$$

1.8 Deformation Gradients, Finite Strain Tensors

The Green's deformation tensor $C = F^T \cdot F$ is a symmetric tensor. Therefore,

$$C = \begin{pmatrix} \frac{\partial u_r}{\partial r} & \frac{\partial u_\theta}{\partial r} & \frac{\partial u_\phi}{\partial r} \\ \frac{1}{r}\left[\frac{\partial u_\theta}{\partial \theta} - u_\theta\right] & \frac{1}{r}\left[\frac{\partial u_\theta}{\partial \theta} + u_r\right] & \frac{1}{r}\frac{\partial u_\phi}{\partial \theta} \\ \frac{1}{r\sin\theta}\left[\frac{\partial u_r}{\partial \phi} - u_\phi \sin\theta\right] & \frac{1}{r\sin\theta}\left[\frac{\partial u_\theta}{\partial \phi} - u_\phi \cos\theta\right] & \frac{1}{r\sin\theta}\left[\frac{\partial u_\phi}{\partial \phi} + u_r \sin\theta + u_\theta \cos\theta\right] \end{pmatrix}$$

$$= \begin{pmatrix} \frac{\partial u_r}{\partial r} & \frac{1}{r}\left[\frac{\partial u_\theta}{\partial \theta} - u_\theta\right] & \frac{1}{r\sin\theta}\left[\frac{\partial u_r}{\partial \phi} - u_\phi \sin\theta\right] \\ \frac{\partial u_\theta}{\partial r} & \frac{1}{r}\left[\frac{\partial u_\theta}{\partial \theta} + u_r\right] & \frac{1}{r\sin\theta}\left[\frac{\partial u_\theta}{\partial \phi} - u_\phi \cos\theta\right] \\ \frac{\partial u_\phi}{\partial r} & \frac{1}{r}\frac{\partial u_\phi}{\partial \theta} & \frac{1}{r\sin\theta}\left[\frac{\partial u_\phi}{\partial \phi} + u_r \sin\theta + u_\theta \cos\theta\right] \end{pmatrix}$$

Therefore,

$$C_{rr} = \left(\frac{\partial u_r}{\partial r}\right)^2 + \left(\frac{\partial u_\theta}{\partial r}\right)^2 + \left(\frac{\partial u_\phi}{\partial r}\right)^2;$$

$$C_{r\theta} = \frac{1}{r}\left[\frac{\partial u_\theta}{\partial \theta} - u_\theta\right]\frac{\partial u_r}{\partial r} + \frac{1}{r}\left[\frac{\partial u_\theta}{\partial \theta} + u_r\right]\frac{\partial u_\theta}{\partial r} + \frac{1}{r}\frac{\partial u_\phi}{\partial \theta}\frac{\partial u_\phi}{\partial r}$$

$$C_{r\phi} = \frac{\partial u_r}{\partial r}\frac{1}{r\sin\theta}\left[\frac{\partial u_r}{\partial \phi} - u_\phi \sin\theta\right] + \frac{\partial u_\theta}{\partial r}\frac{1}{r\sin\theta}\left[\frac{\partial u_\theta}{\partial \phi} - u_\phi \cos\theta\right]$$

$$+ \frac{\partial u_\phi}{\partial r}\frac{1}{r\sin\theta}\left[\frac{\partial u_\phi}{\partial \phi} + u_r \sin\theta + u_\theta \cos\theta\right];$$

$$C_{\theta r} = C_{r\theta};\ C_{\theta\theta} = \left(\frac{1}{r}\frac{\partial u_\theta}{\partial \theta} - \frac{u_\theta}{r}\right)^2 + \left(\frac{1}{r}\frac{\partial u_\theta}{\partial \theta} + \frac{u_r}{r}\right)^2 + \frac{1}{r^2}\left(\frac{\partial u_\phi}{\partial \theta}\right)^2;$$

$$C_{\theta\phi} = \left(\frac{1}{r}\frac{\partial u_\theta}{\partial \theta} - \frac{u_\theta}{r}\right)\frac{1}{r\sin\theta}\left[\frac{\partial u_r}{\partial \phi} - u_\phi \sin\theta\right] + \left(\frac{1}{r}\frac{\partial u_\theta}{\partial \theta} + \frac{u_r}{r}\right)\frac{1}{r\sin\theta}\left[\frac{\partial u_\theta}{\partial \phi} - u_\phi \cos\theta\right]$$

$$+ \frac{1}{r}\frac{\partial u_\phi}{\partial \theta}\frac{1}{r\sin\theta}\left[\frac{\partial u_\phi}{\partial \phi} + u_r \sin\theta + u_\theta \cos\theta\right];\ C_{\phi r} = C_{r\phi};\ C_{\phi\theta} = C_{\theta\phi};$$

$$C_{\phi\phi} = \left(\frac{1}{r\sin\theta}\left[\frac{\partial u_r}{\partial \phi} - u_\phi \sin\theta\right]\right)^2 + \left(\frac{1}{r\sin\theta}\left[\frac{\partial u_\theta}{\partial \phi} - u_\phi \cos\theta\right]\right)^2$$

$$+ \left(\frac{1}{r\sin\theta}\left[\frac{\partial u_\phi}{\partial \phi} + u_r \sin\theta + u_\theta \cos\theta\right]\right)^2.$$

The Lagrangian finite strain tensor is $2R = C - I$, and therefore

$$R = \frac{1}{2}(C - I) = \frac{1}{2}(F^T.F - I)$$

$$= \begin{pmatrix} R_{rr} = C_{rr} - 1 & R_{r\theta} = C_{r\theta} & R_{r\phi} = C_{r\phi} \\ R_{\theta r} = R_{r\theta} & R_{\theta\theta} = C_{\theta\theta} - 1 & R_{\theta\phi} = C_{\theta\phi} \\ R_{\phi r} = R_{r\phi} & R_{\phi\theta} = R_{\theta\phi} & R_{\phi\phi} = C_{\phi\phi} - 1 \end{pmatrix}.$$

1.9 Infinitesimal Strain Tensor

There are many important engineering problems that involve structural members or machine parts for which the deformation is very small (mathematically treated as infinitesimal). In this section, we derive the tensor that characterizes the deformation of such bodies. Some common materials like metals, concrete, wood etc. undergo small changes of shape when forces of reasonable magnitude are applied to them. In this case, assume that all displacement gradients are numerically very small compared to 1, *i.e.*,
$\left|\frac{\partial u_i}{\partial X_j}\right| \leq 1; i, j = 1, 2, 3.$

We introduce the infinitesimal Lagrangian linear strain tensor E_{ij} and the Eulerian infinitesimal srtain tensor e_{ij} defined by

$$E_{ij} = \frac{\partial u_i}{\partial X_j}(X_1, X_2, X_3) + \frac{\partial u_j}{\partial X_i}(X_1, X_2, X_3) \quad (1.39)$$

$$e_{ij} = \frac{\partial u_i}{\partial x_j}(x_1, x_2, x_3) + \frac{\partial u_j}{\partial x_i}(x_1, x_2, x_3).$$

These expressions are known as the *linearized* Lagrangian and Eulerian strain tensors, respectively. If the numerical values of all the components of the displacement and the displacement gradient tensors are very small we may neglect the squares and products of these quantities in comparison to the gradients themselves. Such situations occur in large deflections of bars, plates, and shells. Furthermore, to the same order of approximation, using the relation $\frac{\partial x_k}{\partial X_i} = x_{k,i} = u_{k,i} + \delta_{ik}$ we get,

1.9 Infinitesimal Strain Tensor

$$\frac{\partial u_i}{\partial X_j} = \frac{\partial u_i}{\partial x_k} \cdot \frac{\partial x_k}{\partial X_j} = \frac{\partial u_i}{\partial x_k}\left(\frac{\partial u_k}{\partial X_j} + \delta_{kj}\right) \approx \frac{\partial u_i}{\partial x_k}\delta_{kj}.$$

Therefore, to the first order of approximation for the case of small displacement gradients, it is unimportant whether we differentiate the displacement components with respect to the material or spatial coordinates. In view of this, we may display the equivalent relative displacement gradients for small deformation theory as either $\dfrac{\partial u_i}{\partial X_j}$ or $\dfrac{\partial u_i}{\partial x_j}$. Now to the first order in the displacement gradients, the Lagrangian finite strain tensor becomes

$$r_{ij}(X_1, X_2, X_3) = \frac{1}{2}\left[\frac{\partial u_i}{\partial X_j} + \frac{\partial u_j}{\partial X_i} + \frac{\partial u_k}{\partial X_i} \cdot \frac{\partial u_k}{\partial X_j}\right](X_1, X_2, X_3)$$

$$\cong \frac{1}{2}\left[\frac{\partial u_i}{\partial X_j} + \frac{\partial u_j}{\partial X_i}\right](X_1, X_2, X_3) = E_{ij}(X_1, X_2, X_3) \quad (1.40)$$

Therefore, under the linear strain theory we get in the Lagrangian method

$$\frac{dl^2 - \delta L^2}{\delta L^2} = 2E_{ij}N_i N_j \quad (1.41)$$

and

$$\frac{dl}{dL} \cdot \frac{\delta l}{\delta L}\cos\theta - \cos\Theta = 2E_{ij}N_i M_j \quad (1.42)$$

Similarly, to the first order in the displacement gradients,

$$\eta_{ij}(x_1, x_2, x_3) = \frac{1}{2}\left[\frac{\partial u_i}{\partial x_j} + \frac{\partial u_j}{\partial x_i} - \frac{\partial u_k}{\partial x_i} \cdot \frac{\partial u_k}{\partial x_j}\right](x_1, x_2, x_3)$$

$$\cong \frac{1}{2}\left[\frac{\partial u_i}{\partial x_j} + \frac{\partial u_j}{\partial x_i}\right](x_1, x_2, x_3) = e_{ij}(x_1, x_2, x_3) \quad (1.43)$$

Therefore, under the linear strain theory we get in the Eulerian method

$$\frac{dl^2 - \delta L^2}{\delta L^2} = 2e_{ij}n_i n_j \quad \text{and} \quad \cos\theta - \cos\Theta\frac{dL}{dl} \cdot \frac{\delta L}{\delta l} = 2e_{ij}n_i m_j \quad (1.44)$$

We assume that displacement themselves are small in addition to displacement gradients so that we can neglect the product terms like $u_j \dfrac{\partial u_i}{\partial X_j}$. Since $x_i = X_i + u_i$, we have by using Taylors theorem

$$u_i(x_1, x_2, x_3) = u_i(X_1 + u_1, X_2 + u_2, X_3 + u_3)$$

$$= u_i(X_1, X_2, X_3) + u_i \frac{\partial u_i}{\partial X_j} + \ldots \cong u_i(X_1, X_2, X_3)$$

neglecting small quantities of higher order. Therefore

$$\frac{\partial u_i}{\partial X_j}(X_1, X_2, X_3) \cong \frac{\partial u_i}{\partial X_j}(x_1, x_2, x_3) = \frac{\partial u_i}{\partial x_k}(x_1, x_2, x_3) \cdot \frac{\partial x_k}{\partial X_j}$$

$$= \frac{\partial u_i}{\partial x_k}(x_1, x_2, x_3) \cdot \left[\delta_{kj} + \frac{\partial u_k}{\partial X_j}\right]$$

$$\cong \frac{\partial u_i}{\partial x_k}(x_1, x_2, x_3) \cdot \delta_{kj}, = \frac{\partial u_i}{\partial x_j}(x_1, x_2, x_3)$$

neglecting the product terms. Similarly, $\frac{\partial u_j}{\partial X_i}(X_1, X_2, X_3) \cong \frac{\partial u_j}{\partial x_i}(x_1, x_2, x_3)$. Therefore in Cartesian co-ordinates

$$E_{ij} = \frac{1}{2}\left[\frac{\partial u_i}{\partial X_j} + \frac{\partial u_j}{\partial X_i}\right] = \frac{1}{2}\left[\frac{\partial u_i}{\partial x_j} + \frac{\partial u_j}{\partial x_i}\right] = e_{ij} \qquad (1.45)$$

We observe that in the infinitesimal deformation case, Lagrangian linear strain tensor E_{ij} are identical, component by component to their counterparts in Eulerian linear strain tensor e_{ij}. This is because of the fact that it is quite immaterial whether the derivatives are to be evaluated at the position of a point before or after deformation. Material and spatial view points coalesce and we need make no distinction between Lagrangian and Eulerian small strain tensor. We simply call it small strain tensor.

Also, in general $\eta_{ij} = 0$ does not imply $e_{ij} = 0$; *i.e.*, vanishing infinitesimal strains are not sufficient for rigid deformation. We emphasize the fact that e_{ij} is not a strain measure. It is only approximately so in the infinitesimal deformation theory.

Small strain tensor is not exact measure of strain deformation. If often provides an excellent approximation to such a measure in case of material like, metal, wood. But small strain tensor is inadequate as a measure of strain deformation in case of material like rubber which is capable of undergoing large deformation.

In vector notation, Eq.(1.45) can be written as

$$E = \frac{1}{2}\left[\nabla u + (\nabla u)^T\right] = \text{symmetric part of}(\nabla u). \qquad (1.46)$$

The tensor is known as the infinitesimal strain tensor. It is a finite deformation tensor. Using the expressions derived as in above, we can obtain

1.9 Infinitesimal Strain Tensor

the matrices of infinitesimal strain tensor in terms of the components of the displacement gradients in rectangular coordinates, cylindrical coordinates, and spherical coordinates.

(a) Cartesian coordinates: The matrix of infinitesimal strain tensor E in terms of the components of the displacement gradients in rectangular Cartesian coordinates is given by

$$[E] = \begin{pmatrix} \dfrac{\partial u_1}{\partial X_1} & \dfrac{1}{2}\left[\dfrac{\partial u_1}{\partial X_2} + \dfrac{\partial u_2}{\partial X_1}\right] & \dfrac{1}{2}\left[\dfrac{\partial u_1}{\partial X_3} + \dfrac{\partial u_3}{\partial X_1}\right] \\ \dfrac{1}{2}\left[\dfrac{\partial u_1}{\partial X_2} + \dfrac{\partial u_2}{\partial X_1}\right] & \dfrac{\partial u_2}{\partial X_2} & \dfrac{1}{2}\left[\dfrac{\partial u_2}{\partial X_3} + \dfrac{\partial u_3}{\partial X_2}\right] \\ \dfrac{1}{2}\left[\dfrac{\partial u_1}{\partial X_3} + \dfrac{\partial u_3}{\partial X_1}\right] & \dfrac{1}{2}\left[\dfrac{\partial u_2}{\partial X_3} + \dfrac{\partial u_3}{\partial X_2}\right] & \dfrac{\partial u_3}{\partial X_3} \end{pmatrix}.$$

(b) Cylindrical coordinates: In matrix form the components of ∇u can be written as

$$[\nabla u] = \begin{pmatrix} \dfrac{\partial u_r}{\partial r} & \dfrac{1}{r}\dfrac{\partial u_r}{\partial \theta} - \dfrac{u_\theta}{r} & \dfrac{\partial u_r}{\partial z} \\ \dfrac{\partial u_\theta}{\partial r} & \dfrac{1}{r}\dfrac{\partial u_\theta}{\partial \theta} + \dfrac{u_r}{r} & \dfrac{\partial u_\theta}{\partial z} \\ \dfrac{\partial u_z}{\partial r} & \dfrac{1}{r}\dfrac{\partial u_z}{\partial \theta} & \dfrac{\partial u_z}{\partial z} \end{pmatrix}.$$

Using Eq.(1.46), $E = \dfrac{1}{2}\left[\nabla u + (\nabla u)^T\right]$ = symmetric part of (∇u) as the infinitesimal strain tensor E in terms of the components of the displacement gradients in cylindrical coordinates is given by

$$[E] = \begin{pmatrix} \dfrac{\partial u_r}{\partial r} & \dfrac{1}{2}\left[\dfrac{1}{r}\dfrac{\partial u_r}{\partial \theta} - \dfrac{u_\theta}{r} + \dfrac{\partial u_\theta}{\partial r}\right] & \dfrac{1}{2}\left[\dfrac{\partial u_r}{\partial z} + \dfrac{\partial z}{\partial r}\right] \\ \dfrac{1}{2}\left[\dfrac{1}{r}\dfrac{\partial u_r}{\partial \theta} - \dfrac{u_\theta}{r} + \dfrac{\partial u_\theta}{\partial r}\right] & \dfrac{1}{r}\dfrac{\partial u_\theta}{\partial \theta} + \dfrac{u_r}{r} & \dfrac{1}{2}\left[\dfrac{\partial u_\theta}{\partial z} + \dfrac{1}{r}\dfrac{\partial u_z}{\partial \theta}\right] \\ \dfrac{1}{2}\left[\dfrac{\partial u_r}{\partial z} + \dfrac{\partial z}{\partial r}\right] & \dfrac{1}{2}\left[\dfrac{\partial u_\theta}{\partial z} + \dfrac{1}{r}\dfrac{\partial u_z}{\partial \theta}\right] & \dfrac{\partial u_z}{\partial z} \end{pmatrix}$$

(c) Spherical coordinates: In matrix form the components of ∇u can be written as

$$[\nabla u] = \begin{pmatrix} \dfrac{\partial u_r}{\partial r} & \dfrac{1}{r}\dfrac{\partial u_r}{\partial \theta} - \dfrac{u_\theta}{r} & \dfrac{1}{r\sin\theta}\dfrac{\partial u_r}{\partial \phi} - \dfrac{u_\phi}{r} \\ \dfrac{\partial u_\theta}{\partial r} & \dfrac{1}{r}\dfrac{\partial u_\theta}{\partial \theta} + \dfrac{u_r}{r} & \dfrac{1}{r\sin\theta}\dfrac{\partial u_\theta}{\partial \phi} - \dfrac{u_\phi \cot\theta}{r} \\ \dfrac{\partial u_\phi}{\partial r} & \dfrac{1}{r}\dfrac{\partial u_\phi}{\partial \theta} & \dfrac{1}{r\sin\theta}\dfrac{\partial u_\phi}{\partial \phi} + \dfrac{u_r}{r} - \dfrac{u_\theta \cot\theta}{r} \end{pmatrix}.$$

Using Eq.(1.46), $E = \dfrac{1}{2}\left[\nabla u + (\nabla u)^T\right] =$ symmetric part of (∇u) as the infinitesimal strain tensor E in terms of the components of the displacement gradients in spherical polar coordinates is given by

$$[E] = \begin{pmatrix} \dfrac{\partial u_r}{\partial r} & \dfrac{1}{2}\left[\dfrac{1}{r}\dfrac{\partial u_r}{\partial \theta} - \dfrac{u_\theta}{r} + \dfrac{\partial u_\theta}{\partial r}\right] & \dfrac{1}{2}\left[\dfrac{1}{r\sin\theta}\dfrac{\partial u_r}{\partial \phi} - \dfrac{u_\phi}{r} + \dfrac{\partial u_\phi}{\partial r}\right] \\ \dfrac{1}{2}\left[\dfrac{1}{r}\dfrac{\partial u_r}{\partial \theta} - \dfrac{u_\theta}{r} + \dfrac{\partial u_\theta}{\partial r}\right] & \dfrac{1}{r}\dfrac{\partial u_\theta}{\partial \theta} + \dfrac{u_r}{r} & \dfrac{1}{2}\left[\dfrac{1}{r\sin\theta}\dfrac{\partial u_\theta}{\partial \phi} - \dfrac{u_\phi \cot\theta}{r} + \dfrac{1}{r}\dfrac{\partial u_\phi}{\partial \theta}\right] \\ \dfrac{1}{2}\left[\dfrac{1}{r\sin\theta}\dfrac{\partial u_r}{\partial \phi} - \dfrac{u_\phi}{r} + \dfrac{\partial u_\phi}{\partial r}\right] & \dfrac{1}{2}\left[\dfrac{1}{r\sin\theta}\dfrac{\partial u_\theta}{\partial \phi} - \dfrac{u_\phi \cot\theta}{r} + \dfrac{1}{r}\dfrac{\partial u_\phi}{\partial \theta}\right] & \dfrac{1}{r\sin\theta}\dfrac{\partial u_\phi}{\partial \phi} + \dfrac{u_r}{r} - \dfrac{u_\theta \cot\theta}{r} \end{pmatrix}$$

1.9.1 Geometrical Interpretation of Small Strain Components

A geometrical meaning for the strains E_{11}, E_{22}, E_{33} is provided by considering the length and angle changes as a result of the deformation. In analyzing the state of strain in undeformed body, it is natural to use the coordinates of the initial state as independent variables and follow the material description of deformation throughout.

(a) Diagonal elements of (E_{ij}): Consider a material line element $P_0 Q_0$ of length dL at $P_0(X_1, X_2, X_3)$ oriented in the direction of (N_1, N_2, N_3) in the undeformed body. After deformation has taken place, line element $P_0 Q_0$

1.9 Infinitesimal Strain Tensor

deforms into a line element PQ of length dl at $P(x_1, x_2, x_3)$ in the deformed body. We know that

$$\frac{dl^2 - dL^2}{dL^2} = 2E_{ij}N_iN_j \tag{1.47}$$

where, $E_{ij} = \frac{1}{2}\left[\frac{\partial u_i}{\partial X_j} + \frac{\partial u_j}{\partial X_i}\right]$ = infinitesimal strain tensor at $P_0(X_1, X_2, X_3)$.

From equation (1.47), we obtain

$$\frac{dl^2}{dL^2} - 1 = 2E_{ij}N_iN_j, \text{ i.e., } \frac{dl}{dL} = (1 + 2E_{ij}N_iN_j)^{1/2} = 1 + \frac{1}{2}\cdot 2E_{ij}N_iN_j + \ldots$$

The ratio $\frac{dl}{dL}$ of the lengths is called the stretch. When strain components are so small that we can neglect squares and products of E_{ij}, we find

$$\frac{dl}{dL} = 1 + E_{ij}N_iN_j, \text{ i.e. } \frac{dl}{dL} - 1 = E_{ij}N_iN_j$$

or,

$$\frac{dl - dL}{dL} = E_{ij}N_iN_j \tag{1.48}$$

Now left hand side of equation (1.48) is the extension per unit original length of a line element oriented in the direction N_1, N_2, N_3 and is called *small extensional strain* denoted it by $E_{(N)}$. Then small extensional strain = $E_{(N)} = E_{ij}N_iN_j$.

Geometrical interpretation of E_{11}, E_{22}, E_{33}: Consider a line element initially parallel to X_1 axis. Then we have $N_1 = 1, N_2 = 0, N_3 = 0$. Therefore

$$E_{(1)} = E_{11}.$$

Thus E_{11} is the extension per *unit original length of a line element which is initially parallel to X_1 axis*. Similarly, E_{22}, E_{33} represent the extension of a line element per unit original length which are initially parallel to X_2 and X_3 axes respectively. These indicate that the infinitesimal normal strains are approximately the extensions of the fibers along the coordinate axes when the deformation is small. These components E_{11}, E_{22}, E_{33} (the diagonal elements) are called extensional *strain* or *normal strain*.

Example 1.8: The deformation of a body is defined by the displacement components $u_1 = k(3X_1^2 + X_2), u_2 = k(2X_2^2 + X_3); u_3 = k(4X_3 + X_1)$, where k is a positive constant. Compute the extension of a line element that passes through the point (1, 1, 1) in the direction $\left(\frac{1}{\sqrt{3}}, \frac{1}{\sqrt{3}}, \frac{1}{\sqrt{3}}\right)$.

Solution: The infinitesimal strain tensor E_{ij} are given by Eq. (1.45). Therefore

$$E_{11} = \frac{\partial u_1}{\partial X_1} = 6kX_1; E_{22} = \frac{\partial u_2}{\partial X_2} = 4kX_2; E_{33} = \frac{\partial u_3}{\partial X_3} = 4k$$

$$E_{12} = \frac{1}{2}\left[\frac{\partial u_1}{\partial X_2} + \frac{\partial u_2}{\partial X_1}\right] = \frac{1}{2}(k+0) = \frac{k}{2} = E_{21}$$

$$E_{13} = \frac{1}{2}\left[\frac{\partial u_1}{\partial X_3} + \frac{\partial u_3}{\partial X_1}\right] = \frac{1}{2}(0+k) = \frac{k}{2} = E_{31}$$

$$E_{23} = \frac{1}{2}\left[\frac{\partial u_2}{\partial X_3} + \frac{\partial u_3}{\partial X_2}\right] = \frac{1}{2}(k+0) = \frac{k}{2} = E_{32}$$

Thus at the point (1, 1, 1) the infinitesimal strain tensors are given by

$$(E_{ij}) = \begin{pmatrix} 6kX_1 & \frac{k}{2} & \frac{k}{2} \\ \frac{k}{2} & 4kX_2 & \frac{k}{2} \\ \frac{k}{2} & \frac{k}{2} & 4k \end{pmatrix} = \begin{pmatrix} 6k & \frac{k}{2} & \frac{k}{2} \\ \frac{k}{2} & 4k & \frac{k}{2} \\ \frac{k}{2} & \frac{k}{2} & 4k \end{pmatrix}.$$

The material line element at $P_0(X_1, X_2, X_3)$ is oriented in the direction of $(N_1, N_2, N_3) = \left(\frac{1}{\sqrt{3}}, \frac{1}{\sqrt{3}}, \frac{1}{\sqrt{3}}\right)$. Then small extensional strain

$$E_{(N)} = E_{ij} N_i N_j = E_{11}(N_1)^2 + E_{22}(N_2)^2 + E_{33}(N_3)^2$$
$$+ 2\left[E_{12} N_1 N_2 + E_{13} N_1 N_3 + E_{23} N_2 N_3\right]$$

$$= 6k\left(\frac{1}{\sqrt{3}}\right)^2 + 4k\left(\frac{1}{\sqrt{3}}\right)^2 + 4k\left(\frac{1}{\sqrt{3}}\right)^2 + 2\left[\frac{k}{2}\frac{1}{\sqrt{3}}\frac{1}{\sqrt{3}} + \frac{k}{2}\frac{1}{\sqrt{3}}\frac{1}{\sqrt{3}} + \frac{k}{2}\frac{1}{\sqrt{3}}\frac{1}{\sqrt{3}}\right]$$

$$= \frac{17k}{3}$$

(b) The off diagonal elements of (E_{ij}): Consider two orthogonal material line elements $P_0 Q_0$ and $P_0 R_0$ of lengths dL and δL at $P_0(X_1, X_2, X_3)$ in the undeformed state of the body oriented in the direction (N_1, N_2, N_3) and (M_1, M_2, M_3) respectively. After deformation has taken place two line elements $P_0 Q_0$ and $P_0 R_0$ deform into two line elements PQ and PR of lengths dL and

1.9 Infinitesimal Strain Tensor

δl respectively inclined at an angle θ in the deformed state of the body. We know that

$$\frac{dl}{dL} \cdot \frac{\delta l}{\delta L} \cos\theta - \cos\frac{\pi}{2} = 2E_{ij}N_i M_j;$$

or

$$\frac{dl}{dL} \cdot \frac{\delta l}{\delta L} \sin\left(\frac{\pi}{2} - \theta\right) = 2E_{ij}N_i M_j$$

or

$$\sin\left(\frac{\pi}{2} - \theta\right) = \frac{2E_{ij}N_i M_j}{\frac{dl}{dL} \cdot \frac{\delta l}{\delta L}} \tag{1.49}$$

If E_1 denotes the extension of P_0Q_0 and E_2 that of P_0R_0, then

$$E_1 = \frac{dl - dL}{dL} \Rightarrow dl = (1 + E_1)dL$$

$$E_2 = \frac{\delta l - \delta L}{\delta L} \Rightarrow \delta l = (1 + E_2)\delta L$$

Now $\frac{\pi}{2} - \theta$ is the decrease in right angle between two orthogonal lines P_0Q_0 and P_0R_0 in the undeformed state and is called shear along the two lines. If $\gamma_{(MN)}$ denote this shear along two orthogonal line elements initially oriented in the direction (N_1, N_2, N_3) and (M_1, M_2, M_3), then

$$\gamma_{(MN)} = \frac{\pi}{2} - \theta \text{ and } \sin\gamma_{(MN)} = \frac{2E_{ij}N_i M_j}{\frac{dl}{dL} \cdot \frac{\delta l}{\delta L}}$$

or

$$\sin\gamma_{(MN)} = \frac{2E_{ij}N_i M_j}{(1+E_1) \cdot (1+E_2)} = 2E_{ij}\left[1 + E_1 + E_2 + E_1 E_2\right]^{-1} N_i M_j$$

From the above equation, it is also clear that the necessary and sufficient condition for the angle change to be zero is the vanishing of the shearing strains. For small deformation $\sin\gamma_{(MN)} \cong \gamma_{(MN)}$ and neglecting squares and products of small quantities, we get for small deformation

$$\gamma_{(MN)} = 2E_{ij}\left[1 + E_1 + E_2 + E_1 E_2\right]^{-1} N_i M_j \cong 2E_{ij}N_i M_j \tag{1.50}$$

Geometrical interpretation of E_{23}, E_{31}, E_{12}: If we consider a pair of orthogonal line elements initially parallel to X_2, X_3 axes respectively, then we have $N_1 = 0, N_2 = 1, N_3 = 0$ and $M_1 = 0, M_2 = 0, M_3 = 1$. Therefore

$$\gamma_{(23)} = 2E_{23} \text{ i.e. } E_{23} = \frac{1}{2} \cdot \gamma_{(23)}.$$

Thus E_{23} represent one-half of the shear between two linear elements which are initially parallel to X_2 and X_3 axes. A similar interpretation can be made in regard to E_{31} and E_{12}. E_{23}, E_{31}, E_{12} are called *shearing strains*. Thus E_{ij} denote increase in length of a line element per unit original length or decrease in right angle between two line elements. Therefore, the infinitesimal shearing strains are approximately one half of the angle change between the coordinate axes for small deformations. Thus for rigid deformation $E_{ij} = 0$.

Example 1.9: The strain tensor at a point is given by $(E_{ij}) = \begin{pmatrix} 5 & 3 & 0 \\ 3 & 4 & -1 \\ 0 & -1 & 2 \end{pmatrix}$.

Determine the extension of a line element in the direction of $\left(\frac{2}{3}, \frac{2}{3}, \frac{1}{3}\right)$.

What is the change of angle between two perpendicular line elements in the directions $\left(\frac{2}{3}, \frac{2}{3}, \frac{1}{3}\right)$ **and** $\left(\frac{1}{\sqrt{5}}, 0, -\frac{2}{\sqrt{5}}\right)$.

Solution: The infinitesimal strain tensors are given by $(E_{ij}) = \begin{pmatrix} 5 & 3 & 0 \\ 3 & 4 & -1 \\ 0 & -1 & 2 \end{pmatrix}$.

The material line element at $P_0(X_1, X_2, X_3)$ is oriented in the direction of $(N_1, N_2, N_3) = \left(\frac{2}{3}, \frac{2}{3}, \frac{1}{3}\right)$. Then small extensional strain

$$E_{(N)} = E_{ij} N_i N_j = E_{11}(N_1)^2 + E_{22}(N_2)^2 + E_{33}(N_3)^2$$
$$+ 2[E_{12} N_1 N_2 + E_{13} N_1 N_3 + E_{23} N_2 N_3]$$

$$= 5\left(\frac{2}{3}\right)^2 + 4\left(\frac{2}{3}\right)^2 + 2\left(\frac{1}{3}\right)^2 + 2\left[3 \frac{2}{3} \cdot \frac{2}{3} + 0 \cdot \frac{2}{3} \cdot \frac{1}{3} + (-1) \frac{2}{3} \cdot \frac{1}{3}\right] = \frac{58}{9}$$

The change of right angle between line elements in the direction of $(N_1, N_2, N_3) = \left(\frac{2}{3}, \frac{2}{3}, \frac{1}{3}\right)$ and $(M_1, M_2, M_3) = \left(\frac{1}{\sqrt{5}}, 0, \frac{2}{\sqrt{5}}\right)$ is given by

$$\gamma_{(MN)} = 2E_{ij} N_i M_j = 2[E_{11} N_1 M_1 + E_{22} N_2 M_2 + E_{33} N_3 M_3] + 2E_{12}(N_1 M_2 + N_2 M_1)$$
$$+ 2E_{13}(N_1 M_3 + N_3 M_1) + 2E_{23}(N_2 M_3 + N_3 M_2)$$

1.9 Infinitesimal Strain Tensor

$$= 2\left[5.\frac{2}{3}.\frac{1}{\sqrt{5}} + 4.\frac{2}{3}.0 + 2.\frac{1}{3}\left(-\frac{2}{\sqrt{5}}\right)\right] + 2.3\left[\frac{2}{3}.0 + \frac{2}{3}.\frac{1}{\sqrt{5}}\right]$$

$$+ 2.0\left[\frac{2}{3}\left(-\frac{2}{\sqrt{5}}\right) + \frac{1}{3}.\frac{1}{\sqrt{5}}\right] + 2.(-1)\left[\frac{2}{3}\left(-\frac{2}{\sqrt{5}}\right) + \frac{1}{3}.0\right] = \frac{32}{3\sqrt{5}}$$

Example 1.10: For the displacement field $u_1 = (X_1 - X_2)^2, u_2 = (X_2 + X_3)^2; u_3 = -X_1 X_2$, compute the change in right angle between $\vec{N} = \frac{1}{9}(8\hat{e}_1 - \hat{e}_2 + 4\hat{e}_3)$ and $\vec{M} = \frac{1}{9}(4\hat{e}_1 - 4\hat{e}_2 + 7\hat{e}_3)$ at the point $(0, 2, -1)$.

Solution: The infinitesimal strain tensor are given by Eq. (1.45). Therefore

$$E_{11} = \frac{\partial u_1}{\partial X_1} = 2(X_1 - X_2) = -4; E_{22} = \frac{\partial u_2}{\partial X_2} = 2(X_2 + X_3) = 2; E_{33} = \frac{\partial u_3}{\partial X_3} = 0$$

$$E_{12} = \frac{1}{2}\left[\frac{\partial u_1}{\partial X_2} + \frac{\partial u_2}{\partial X_1}\right] = \frac{1}{2}\left[-2(X_1 - X_2) + 0\right] = -(X_1 - X_2) = 2 = E_{21}$$

$$E_{13} = \frac{1}{2}\left[\frac{\partial u_1}{\partial X_3} + \frac{\partial u_3}{\partial X_1}\right] = \frac{1}{2}[0 - X_2] = -\frac{1}{2}X_2 = -1 = E_{31}$$

$$E_{23} = \frac{1}{2}\left[\frac{\partial u_2}{\partial X_3} + \frac{\partial u_3}{\partial X_2}\right] = \frac{1}{2}[2(X_2 + X_3) - X_1] = X_2 + X_3 - \frac{1}{2}X_1 = 1 = E_{32}$$

Thus at the point $(0, 2, -1)$ the infinitesimal strain tensors are given by

$$(E_{ij}) = \begin{pmatrix} 2(X_1 - X_2) & -(X_1 - X_2) & -\frac{1}{2}X_2 \\ -(X_1 - X_2) & 2(X_2 + X_3) & X_2 + X_3 - \frac{1}{2}X_1 \\ -\frac{1}{2}X_2 & X_2 + X_3 - \frac{1}{2}X_1 & 0 \end{pmatrix} = \begin{pmatrix} -4 & 2 & -1 \\ 2 & 2 & 1 \\ -1 & 1 & 0 \end{pmatrix}$$

The material line element at $P_0(X_1, X_2, X_3) = P_0(0, 2, -1)$ is oriented in the direction of $\vec{N} = (N_1, N_2, N_3) = \left(\frac{8}{9}, -\frac{1}{9}, \frac{4}{9}\right)$. Then small extensional strain

$$E_{(N)} = E_{ij} N_i N_j = E_{11}(N_1)^2 + E_{22}(N_2)^2 + E_{33}(N_3)^2$$
$$+ 2[E_{12} N_1 N_2 + E_{13} N_1 N_3 + E_{23} N_2 N_3]$$

$$=-4\left(\frac{8}{9}\right)^2+2\left(-\frac{1}{9}\right)^2+0\left(\frac{4}{9}\right)^2+2\left[2.\frac{8}{9}.\left(-\frac{1}{9}\right)+(-1).\frac{8}{9}.\frac{4}{9}+1.\left(-\frac{1}{9}\right).\frac{4}{9}\right]=\frac{40}{9}$$

The material line element at $P_0(X_1, X_2, X_3) = P_0(0,2,-1)$ is oriented in the direction of $\bar{M}=(M_1,M_2,M_3)=\left(\frac{4}{9},-\frac{4}{9},\frac{7}{9}\right)$. Then small extensional strain

$$E_{(M)} = E_{ij}M_iM_j = E_{11}(M_1)^2+E_{22}(M_2)^2+E_{33}(M_3)^2$$
$$+2\left[E_{12}M_1M_2+E_{13}M_1M_3+E_{23}M_2M_3\right]$$

$$=-4\left(\frac{4}{9}\right)^2+2\left(-\frac{4}{9}\right)^2+0\left(\frac{7}{9}\right)^2+2\left[2.\frac{4}{9}.\left(-\frac{4}{9}\right)+(-1).\frac{4}{9}.\frac{7}{9}+1.\left(-\frac{4}{9}\right).\frac{7}{9}\right]=-\frac{152}{81}$$

The change of right angle between $\bar{N}$ and $\bar{M}$ at the point $P_0(0,2,-1)$ is given by

$$\gamma_{(MN)} = 2E_{ij}N_iM_j = 2\left[E_{11}N_1M_1+E_{22}N_2M_2+E_{33}N_3M_3\right]+2E_{12}\left(N_1M_2+N_2M_1\right)$$
$$+2E_{13}\left(N_1M_3+N_3M_1\right)+2E_{23}\left(N_2M_3+N_3M_2\right)$$
$$=\frac{318}{81}$$

1.9.2 Change of Area

In this section we investigate the change of area with the deformation. We have already found that an infinitesimal rectangular parallelepiped with edge vectors I_1dX_1, I_2dX_2 and I_3dX_3, after deformation becomes a rectilinear parallelepiped with edge vectors C_1dX_1, C_2dX_2 and C_3dX_3 where $C_j = \frac{\partial x_j}{\partial X_k}i_j$ (Fig. 1.8).

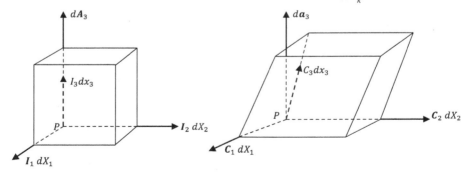

Fig. 1.8: Deformation of an infinitesimal rectangular parallelepiped

We calculate the area vector da_3 according as

1.9 Infinitesimal Strain Tensor

$$da_3 = (C_1 dX_1) \times (C_2 dX_2) = \frac{\partial x_k}{\partial X_1} \frac{\partial x_l}{\partial X_2} dX_1 dX_2 (i_k \times i_l)$$

$$= \frac{\partial x_k}{\partial X_1} \frac{\partial x_l}{\partial X_2} dA_3 \varepsilon_{klm} i_m; \quad dA_3 = dX_1 dX_2, i_k \times i_l = \varepsilon_{klm} i_m.$$

where, the alternating symbol ε_{ijk} is the permutation symbol, defined by

ε_{ijk} = 0, if any two of i, j, k are equal
= +1, if i, j, k are even permutation of 1, 2, 3
= −1, if i, j, k are odd permutation of 1, 2, 3

This may be expressed in terms of Jacobian

$$J = \left| \frac{\partial x_k}{\partial X_k} \right| = \varepsilon_{klm} \frac{\partial x_k}{\partial X_1} \frac{\partial x_l}{\partial X_2} \frac{\partial x_m}{\partial X_3}.$$

Through the chain rule of partial differentiation it is clear that

$$\frac{\partial x_k}{\partial X_j} \frac{\partial X_j}{\partial x_l} = \delta_{kl}; \quad \frac{\partial X_j}{\partial x_k} \frac{\partial x_k}{\partial X_m} = \delta_{jm}.$$

Each one of these two sets is a set of nine linear equations for the nine unknowns $\frac{\partial x_k}{\partial X_j}$ or $\frac{\partial X_j}{\partial x_k}$. A unique solution exists, since the Jacobian of the transformation is assumed not to vanish. Using Cramer's rule of determinants, the solution for $\frac{\partial X_j}{\partial x_k}$ may be obtained in terms of $\frac{\partial x_k}{\partial X_j}$. Thus

$$\frac{\partial X_i}{\partial x_p} = \frac{\text{cofactor } \frac{\partial x_p}{\partial X_i}}{J} = \frac{1}{2J} \varepsilon_{ijk} \varepsilon_{pqr} \frac{\partial x_q}{\partial X_j} \frac{\partial x_r}{\partial X_k}$$

$$\Rightarrow J \frac{\partial X_3}{\partial x_m} = \varepsilon_{klm} \frac{\partial x_k}{\partial X_1} \frac{\partial x_l}{\partial X_2}.$$

Therefore, the area vector da_3 is given by

$$da_3 = \frac{\partial x_k}{\partial X_1} \frac{\partial x_l}{\partial X_2} dA_3 \varepsilon_{klm} i_m = J \frac{\partial X_3}{\partial x_m} dA_3 i_m$$

Similar expressions are obtained for the vector area da_1 and da_2 by replacing the index 3 by 1 and 2, respectively as

$$da_1 = J \frac{\partial X_1}{\partial x_m} dA_1 i_m; \quad da_2 = J \frac{\partial X_2}{\partial x_m} dA_2 i_l$$

The area vector da is therefore given by

$$da = da_1 + da_2 + da_3 = J\left[\frac{\partial X_1}{\partial x_m}dA_1 + \frac{\partial X_2}{\partial x_m}dA_2 + \frac{\partial X_3}{\partial x_m}dA_3\right]i_m = J\frac{\partial X_k}{\partial x_m}dA_k i_m$$

Therefore, $da_m = J\dfrac{\partial X_k}{\partial x_m}dA_k$. The magnitude of the area, da, is calculated by

$$da^2 = da\,.da = J^2 \frac{\partial X_k}{\partial x_m}\frac{\partial X_l}{\partial x_m}dA_k dA_l$$

The Cauchy's deformation tensor $c_{ij} = \dfrac{\partial X_k}{\partial x_i}\cdot\dfrac{\partial X_k}{\partial x_j}$ is a symmetric tensor. The symmetric tensor $C_{ij} = \dfrac{\partial x_k}{\partial X_i}\cdot\dfrac{\partial x_k}{\partial X_j}$ is known as the Green's deformation tensor and

$$|C_{ij}| = \left|\frac{\partial x_k}{\partial X_i}\cdot\frac{\partial x_k}{\partial X_j}\right| = \left|\frac{\partial x_k}{\partial X_i}\right|^2 = J^2 = \begin{vmatrix} C_{11} & C_{12} & C_{13} \\ C_{21} & C_{22} & C_{23} \\ C_{31} & C_{32} & C_{33} \end{vmatrix}.$$

Thus, the expression of magnitude of the area, da, is given by

$$da^2 = J^2 \frac{\partial X_k}{\partial x_m}\frac{\partial X_l}{\partial x_m}dA_k dA_l = |C_{ij}|c_{kl}dA_k dA_l. \tag{1.51}$$

which is analogous to the expression of the element of length dl^2. Thus we conclude that, with respect to the measure of area, $|C_{ij}|c_{kl}$ plays a role analogous to that which the tensor C_{ij} plays with respect to the measure of length.

1.9.3 Dilation

Consider a continuum body occupying at time $t = 0$, (see Fig. 1.9) an elementary rectangular parallelepiped $P_0\,Q_0\,R_0\,S_0$ at $P_0\,(X_1,X_2,X_3)$ of the volume element dV_0 with its edges P_0Q_0, P_0R_0, P_0S_0 of lengths dX_1, dX_2, dX_3 parallel to the coordinate axes and with one of its vertices at a point $P_0\,(X_1,X_2,X_3)$ in the undeformed state. Therefore

$$dV_0 = dX_1.dX_2.dX_3 \tag{1.52}$$

Due to deformation, material points initially at P_0, Q_0, R_0, S_0 move to P, Q, R, S with position vectors $\vec{x}$, $\vec{x}+d\vec{x}^{(1)}$, $\vec{x}+d\vec{x}^{(2)}$, $\vec{x}+d\vec{x}^{(3)}$, so that rectangular parallelepiped $P_0\,Q_0\,R_0\,S_0$ deforms into skewed parallelepiped $PQRS$ of volume element dV in the deformed state at time t. Therefore,

$$dV = \left[d\vec{x}^{(1)}.\left(d\vec{x}^{(2)}\times d\vec{x}^{(3)}\right)\right]. \tag{1.53}$$

1.9 Infinitesimal Strain Tensor

If x_i are coordinates of the point P, $x_i + dx_i^{(1)}$ coordinates of Q, $x_i + dx_i^{(2)}$ coordinates of R and $x_i + dx_i^{(3)}$ coordinates of S then

$$dV = \varepsilon_{ijk} dx_i^{(1)} dx_j^{(2)} dx_k^{(3)} \tag{1.54}$$

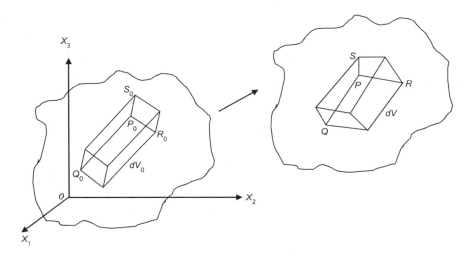

Fig. 1.9: Change of volume due to strain deformation

where, the alternating symbol ε_{ijk} is the permutation symbol.

In the material method, this deformation is characterized by the equation $x_i = x_i (X_1, X_2, X_3)$. The line of material points that form dX_1 now form the differential line element $dx_i^{(1)}$

$$dx_i^{(1)} = \frac{\partial x_i}{\partial X_1} . dX_1. \tag{1.55}$$

Similarly, dX_1, dX_2 become $dx_i^{(2)}$, $dx_i^{(3)}$, respectively, so that

$$dx_j^{(2)} = \frac{\partial x_j}{\partial X_2} . dX_2; \quad dx_k^{(3)} = \frac{\partial x_k}{\partial X_3} . dX_3.$$

From equation (1.54), we get

$$dV = \varepsilon_{ijk} \frac{\partial x_i}{\partial X_1} \frac{\partial x_j}{\partial X_2} \frac{\partial x_k}{\partial X_3} dX_1 dX_2 dX_3 = J dV_0$$

where,

$$J = \left| \frac{\partial x_k}{\partial X_x} \right| = \varepsilon_{klm} \frac{\partial x_k}{\partial X_1} \frac{\partial x_l}{\partial X_2} \frac{\partial x_m}{\partial X_3}.$$

In terms of the displacement u_i of a spatial point from its position P to P_0, we have $x_i = X_i + u_i$ and so for small strain

$$J = \left|\frac{\partial x_i}{\partial X_j}\right| = \left|\frac{\partial}{\partial X_j}(X_i + u_i)\right| = \left|\frac{\partial X_i}{\partial X_j} + \frac{\partial u_i}{\partial X_j}\right| = \left|\delta_{ij} + u_{ij}\right|$$

$$= \begin{vmatrix} 1 + \frac{\partial u_1}{\partial X_1} & \frac{\partial u_2}{\partial X_1} & \frac{\partial u_3}{\partial X_1} \\ \frac{\partial u_1}{\partial X_2} & 1 + \frac{\partial u_2}{\partial X_2} & \frac{\partial u_3}{\partial X_2} \\ \frac{\partial u_1}{\partial X_3} & \frac{\partial u_2}{\partial X_3} & 1 + \frac{\partial u_3}{\partial X_3} \end{vmatrix}$$

$$\cong \left(1 + \frac{\partial u_1}{\partial X_1}\right)\left(1 + \frac{\partial u_2}{\partial X_2}\right)\left(1 + \frac{\partial u_3}{\partial X_3}\right)$$

$$\cong 1 + \frac{\partial u_1}{\partial X_1} + \frac{\partial u_2}{\partial X_2} + \frac{\partial u_3}{\partial X_3} \cong 1 + u_{i,i} = 1 + E_{11} + E_{22} + E_{33}$$

$$\Rightarrow E_{11} + E_{22} + E_{33} = E_{kk} = J - 1$$

$$= \frac{dV}{dV_0} - 1 = \frac{dV - dV_0}{dV_0} \tag{1.56}$$

Thus, for small strain, $trace\,(E_{ij}) = E_{kk} = E_{11} + E_{22} + E_{33}$ geometrically represents the change in volume per unit original volume and is called volumetric infinitesimal strain. The volumetric strain is a measure of volume changes, and for small strains is related to the Jacobian of the deformation gradient $E_{kk} = J - 1$.

1.9.4 The Infinitesimal Rotation Tensor

Consider two neighbouring material points (see Fig. 1.10) at positions P_0 and Q_0 of the continuum in the undeformed state with coordinates X_i and $X_i + dX_i$ respectively. As a result of deformation, the material point at P_0 undergoes a displacement u_i and moves to the position P and let the material point at position Q_0 experiences a displacement $u_i + du_i$ and moves to the position Q in the deformed state. If we draw Q_0Q' equal and parallel to P_0P then the relative displacement of material point originally at Q_0 with respect to the material point originally at P_0 will be represented by $\overrightarrow{Q'Q}$. Now $\overrightarrow{P_0P} = \overrightarrow{Q_0Q'} = \vec{u}$ and $\overrightarrow{Q_0Q} = \vec{u} + d\vec{u}$, i.e.,

$$\overrightarrow{Q_0Q_0'} + \overrightarrow{Q'Q} = \vec{u} + d\vec{u}, \text{ i.e., } \vec{u} + \overrightarrow{Q'Q} = \vec{u} + d\vec{u}$$

1.9 Infinitesimal Strain Tensor

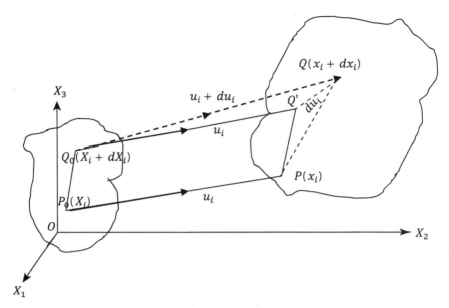

Fig. 1.10: Small rotation vector

so that $\overrightarrow{Q'Q} = \vec{du}$. Relative displacement of material point at Q_0 with respect to material point at P_0 will be represented by du_i. The material method of description is

$$u_i = F_i(X_1, X_2, X_3) \tag{1.57}$$

so that we can write

$$u_i + du_i = F_i(X_1 + dX_1, X_2 + dX_2, X_3 + dX_3)$$

Since positions P_0 and Q_0 are very close together, dX_i are small. Using Taylors's series and neglect terms containing powers of dX_i higher than first, we get

$$u_i + du_i = F_i(X_1, X_2, X_3) + \frac{\partial F_i}{\partial X_1}.dX_1 + \frac{\partial F_i}{\partial X_2}.dX_2 + \frac{\partial F_i}{\partial X_3}.dX_3$$

$$= F_i(X_1, X_2, X_3) + \frac{\partial F_i}{\partial X_j}.dX_j = u_i + \frac{\partial F_i}{\partial X_j}.dX_j$$

or,

$$du_i = \frac{\partial F_i}{\partial X_j}.dX_j = \frac{\partial u_i}{\partial X_j}.dX_j \tag{1.58}$$

which can be expressed in the form

$$du_i = \frac{\partial u_i}{\partial X_j}.dX_j = \left(\frac{1}{2}\left[\frac{\partial u_i}{\partial X_j} + \frac{\partial u_j}{\partial X_i}\right] + \frac{1}{2}\left[\frac{\partial u_i}{\partial X_j} - \frac{\partial u_j}{\partial X_i}\right]\right).dX_j = (R_{ij} + E_{ij}).dX_j$$

$$= R_{ij}.dX_j + E_{ij}.dX_j = du_i^{(1)} + du_i^{(2)} \qquad (1.59)$$

where, $\qquad du_i^{(1)} = R_{ij}.dX_j \qquad du_i^{(2)} = E_{ij}.dX_j$. Now,

$$E_{ij} = \frac{1}{2}\left[\frac{\partial u_i}{\partial X_j} + \frac{\partial u_j}{\partial X_i}\right] = \text{Symmetric tensor of order } 2 = E_{ji}$$

$$R_{ij} = \frac{1}{2}\left[\frac{\partial u_i}{\partial X_j} - \frac{\partial u_j}{\partial X_i}\right] = \text{Skew-symmetric tensor of order } 2 = -R_{ji}$$

Thus the relative displacement du_i consists of two parts; $du_i^{(1)}$ order $du_i^{(2)}$.

In order to study $du_i^{(1)}$ we form a vector R_i by setting $R_i = e_{ijk} R_{kj}$, therefore

$$e_{ijk}R_i = e_{ijk}e_{ipq}R_{qp} = (\delta_{jp}\delta_{kq} - \delta_{jq}\delta_{kp})R_{qp}$$

$$= R_{kj} - R_{jk}$$

$$= R_{kj} + R_{kj} = 2R_{kj} \quad \text{since} \quad R_{jk} = -R_{kj}$$

or, $\qquad R_{kj} = \frac{1}{2}e_{ijk}R_i. \qquad (1.60)$

Therefore, $du_i^{(1)}$ becomes

$$du_k^{(1)} = R_{kj}.dX_j = \frac{1}{2}e_{ijk}R_i.dX_j$$

or, $\qquad d\vec{u}^{(1)} = \frac{1}{2}\vec{R} \times d\vec{X} \qquad (1.61)$

where $d\vec{X}$ is the vector connecting the positions P_0 and Q_0 of the continuum. Now,

$$R_i = e_{ijk}R_{kj} = \frac{1}{2}e_{ijk}\left[\frac{\partial u_i}{\partial X_j} - \frac{\partial u_j}{\partial X_i}\right] = \frac{1}{2}\left(e_{ijk}u_{k,j} - e_{ijk}u_{j,k}\right)$$

1.9 Infinitesimal Strain Tensor

$$= \frac{1}{2}\left(e_{ijk}u_{k,j} - e_{ikj}u_{k,j}\right)$$

$$= \frac{1}{2}\left(e_{ijk}u_{k,j} + e_{ijk}u_{k,j}\right)$$

$$\Rightarrow \quad R_i = e_{ijk}u_{k,j} = (rot\,\vec{u})_i, \text{ i.e., } \vec{R} = rot\,\vec{u}. \tag{1.62}$$

Therefore, we have the following conclusions:

The part $du_i^{(1)} = R_{ij}.dX_j$ represents a relative displacement involving small rigid body rotation of the neighbourhood element of P_0 through an angle $\frac{1}{2}\vec{R} = \frac{1}{2}rot\,\vec{u}$. The $\vec{R} = rot\,\vec{u}$ is called small rotation vector and R_{ij} are called small rotation tensor. Now pure rotation does not bring about any strain deformation in the body.

We know, E_{ij} is a symmetric small strain tensor indicating the increase in the length of a line element per unit original length or decrease in right angle between two linear elements. Thus the part $du_i^{(2)} = E_{ij}.dX_j$ represents a relative displacement involving a strain deformation causing a change in shape in contrast to rigid body deformation.

The absolute displacement of a material point at $Q_0(X_i + dX_i)$ in the neighbourhood of $P_0(X_i)$ is given by

$$u_i + du_i = u_i + du_i^{(1)} + du_i^{(2)} = u_i + R_{ij}.dX_j + E_{ij}.dX_j$$

Therefore, the displacement of the neighbourhood of a point $P_0(X_i)$ now appears decomposed into three parts:

   (*i*) The displacement due to rigid body translation which carries the element as a whole with the displacement u_i at the point $P_0(X_i)$.
   (*ii*) The displacement due to rigid body rotation determined by R_{ij} which rotates the element as a whole through an angle $\frac{1}{2}rot\,\vec{u}$ and
   (*iii*) The displacement due to straining determined by E_{ij} which causes a change in the length and orientation of every line element and thereby causing a change in shape.

When the displacement component $du_i^{(1)}$ due to rotation vanishes, *i.e.*, $rot\,\vec{u} = \vec{0}$, displacement is called irrotational. In this case, there exists a scalar potential function Φ, called displacement potential, such that

$$\vec{u} = -\nabla\Phi.$$

When the displacement component $du_i^{(1)}$ due to rotation does not vanish, i.e., $rot\,\vec{u} \neq \vec{0}$, displacement is called rotational.

Example 1.11: The displacement field for small deformation theory is given by $u_1 = (X_1 - X_3)^2, u_2 = (X_2 + X_3)^2; u_3 = -X_1 X_2$. Determine the strain tensor, rotation tensor and rotation vector at the point $(0, 2, -1)$.

Solution: The infinitesimal strain tensor are given by $E_{ij} = \dfrac{1}{2}\left[\dfrac{\partial u_i}{\partial X_j} + \dfrac{\partial u_j}{\partial X_i}\right]$. Therefore,

$$E_{11} = \frac{\partial u_1}{\partial X_1} = 2(X_1 - X_3); E_{22} = \frac{\partial u_2}{\partial X_2} = 2(X_2 + X_3); E_{33} = \frac{\partial u_3}{\partial X_3} = 0$$

$$E_{12} = \frac{1}{2}\left[\frac{\partial u_1}{\partial X_2} + \frac{\partial u_2}{\partial X_1}\right] = \frac{1}{2}(0+0) = 0 = E_{21}$$

$$E_{13} = \frac{1}{2}\left[\frac{\partial u_1}{\partial X_3} + \frac{\partial u_3}{\partial X_1}\right] = \frac{1}{2}\left[2(X_2 + X_3) - X_1\right] = X_2 + X_3 - \frac{1}{2}X_1 = E_{31}$$

$$E_{13} = \frac{1}{2}\left[\frac{\partial u_1}{\partial X_3} + \frac{\partial u_3}{\partial X_1}\right] = \frac{1}{2}\left[2(X_2 + X_3) - X_1\right] = X_2 + X_3 - \frac{1}{2}X_1 = E_{31}$$

$$E_{23} = \frac{1}{2}\left[\frac{\partial u_2}{\partial X_3} + \frac{\partial u_3}{\partial X_2}\right] = \frac{1}{2}\left[-2(X_1 - X_3) - X_2\right] = -X_1 + X_3 - \frac{1}{2}X_2 = E_{32}$$

Thus at the point $(0, 2, -1)$ the infinitesimal strain tensors are given by

$$(E_{ij}) = \begin{pmatrix} 2(X_1 - X_3) & 0 & X_2 + X_3 - \dfrac{1}{2}X_1 \\ 0 & 2(X_2 + X_3) & -X_1 + X_3 - \dfrac{1}{2}X_2 \\ X_2 + X_3 - \dfrac{1}{2}X_1 & -X_1 + X_3 - \dfrac{1}{2}X_2 & 0 \end{pmatrix} = \begin{pmatrix} 2 & 0 & -2 \\ 0 & 2 & 1 \\ -2 & 1 & 0 \end{pmatrix}$$

The infinitesimal rotation tensor are given by $R_{ij} = \dfrac{1}{2}\left[\dfrac{\partial u_i}{\partial X_j} - \dfrac{\partial u_j}{\partial X_i}\right]$. Therefore,

$$R_{11} = \frac{1}{2}\left[\frac{\partial u_1}{\partial X_1} - \frac{\partial u_1}{\partial X_1}\right] = 0; R_{22} = \frac{1}{2}\left[\frac{\partial u_2}{\partial X_2} - \frac{\partial u_2}{\partial X_2}\right] = 0; R_{33} = \frac{1}{2}\left[\frac{\partial u_3}{\partial X_3} - \frac{\partial u_3}{\partial X_3}\right] = 0$$

$$R_{12} = \frac{1}{2}\left[\frac{\partial u_1}{\partial X_2} - \frac{\partial u_2}{\partial X_1}\right] = \frac{1}{2}(0-0) = 0 = R_{21}$$

$$R_{13} = \frac{1}{2}\left[\frac{\partial u_1}{\partial X_3} - \frac{\partial u_3}{\partial X_1}\right] = \frac{1}{2}\left[2(X_2 + X_3) + X_1\right] = X_2 + X_3 + \frac{1}{2}X_1 = R_{13}$$

1.9 Infinitesimal Strain Tensor

$$R_{23} = \frac{1}{2}\left[\frac{\partial u_2}{\partial X_3} - \frac{\partial u_3}{\partial X_2}\right] = \frac{1}{2}\left[-2(X_1 - X_3) + X_2\right] = -X_1 + X_3 + \frac{1}{2}X_2 = R_{32}$$

Thus, at the point $(0, 2, -1)$ the infinitesimal rotation tensors are given by

$$(R_{ij}) = \begin{pmatrix} 0 & 0 & X_2 + X_3 + \frac{1}{2}X_1 \\ 0 & 0 & -X_1 + X_3 + \frac{1}{2}X_2 \\ X_2 + X_3 + \frac{1}{2}X_1 & -X_1 + X_3 + \frac{1}{2}X_2 & 0 \end{pmatrix} = \begin{pmatrix} 0 & 0 & 1 \\ 0 & 0 & 0 \\ 1 & 0 & 0 \end{pmatrix}$$

at $(0, 2, -1)$

The infinitesimal rotation vector $\vec{R}$ is given by

$$\vec{R} = rot\vec{u} = \begin{vmatrix} \vec{e}_1 & \vec{e}_2 & \vec{e}_3 \\ \dfrac{\partial}{\partial X_1} & \dfrac{\partial}{\partial X_2} & \dfrac{\partial}{\partial X_3} \\ (X_1 - X_3)^2 & (X_2 + X_3)^2 & -X_1 X_2 \end{vmatrix}$$

$$= \left[-X_1 - 2(X_2 + X_3)\right]\vec{e}_1 - \left[-2(X_1 - X_3) + X_2\right]\vec{e}_2 + [0 - 0]\vec{e}_3$$

$$= \left[-X_1 - 2(X_2 + X_3)\right]\vec{e}_1 - \left[-2(X_1 - X_3) + X_2\right]\vec{e}_2 = (-2)\vec{e}_1 \text{ at } (0, 2, -1).$$

1.9.5 Stretch Ratios

An important measure of the extensional strain of a differential line element is the ratio dl/dL, known as the stretch or stretch ratio. This ratio may be expressed in terms of either N or n and may be defined at either the point $P(z_i)$ in the undeformed configuration or at the point $P_0(x_i)$ in the deformed configuration. We denote the stretch either by $\Lambda_{(N)}$ or by $\lambda_{(n)}$. In particular, for the differential element in the direction of the unit vector N at $P_0(X_i)$, we write

$$\Lambda_{(N)} = \frac{dl}{dL}$$

where dl is the deformed magnitude of $dX = dL\,N$. Thus the squared stretch at point $P_0(X_i)$ for the line element along the unit vector $N = \dfrac{dX}{dX}$, is given by

$$\left(\frac{dl}{dL}\right)^2 = \Lambda_{(N)}^2 = \frac{dX}{dX} \cdot C \cdot \frac{dX}{dX} = N \cdot C \cdot N \tag{1.63}$$

for any element originally in the direction N. In an analogous way, we define the stretch ratio, $\lambda_{(n)}$ in the direction of the unit vector $n = \dfrac{dx}{dx}$, at the point $P(x_i)$ is given by

$$\frac{1}{\lambda_{(n)}} = \frac{dL}{dl}$$

Similarly, the reciprocal of the squared stretch for the line element at $P(x_i)$ along the unit vector $n = \dfrac{dx}{dx}$ is given by

$$\left(\frac{dL}{dl}\right)^2 = \frac{1}{\lambda_{(n)}^2} = \frac{dx}{dx}.c.\frac{dx}{dx} = n.c.n \qquad (1.64)$$

In general, $\Lambda_{(N)} \neq \lambda_{(n)}$, since the element originally along the X axis will not likely lie along the x axis after deformation. However, if n is a unit vector in the direction that N assumes in the deformed configuration, the two stretches are the same. From equations (1.56) and (1.57) it is clear that the normal components of C and c in the directions of N and n are respectively, the squares and inverse squares of stretches in these directions. This point is clarified further if we select N and n along one of the coordinate axes. For an element originally along the local X_2 axis $N = I_2$ and therefore $\dfrac{dX_1}{dX} = \dfrac{dX_3}{dX} = 0$, $\dfrac{dX_2}{dX} = 1$ so that equation (1.56) yields for such an element

$$\Lambda_{(N)}^2 = \Lambda_{(I_2)}^2 = I_2.C.I_2 = C_{22} = 1 + 2E_{22} \qquad (1.65)$$

Similar results may be determined for $\Lambda_{(I_1)}^2$, and $\Lambda_{(I_3)}^2$. For an element parallel to the x_2 axis $n = i_2$ after deformation, equation (1.64) yields the result

$$\frac{1}{\lambda_{(n)}^2} = \frac{1}{\lambda_{(i_2)}^2} = i_2.c.i_2 = c_{22} = 1 - 2e_{22} \qquad (1.66)$$

with similar expressions for the quantities $\dfrac{1}{\lambda_{(i_1)}^2}$, and $\dfrac{1}{\lambda_{(i_3)}^2}$. The stretch ratio provides a basis for interpretation of the finite strain tensors. Thus the change of length per unit of original length is

$$e_{(N)} = \frac{dx - dX}{dX} = \Lambda_{(N)} - 1 = \sqrt{N.C.N} - 1 \qquad (1.67)$$

and for the element P_0Q_0 along the X_1 axis, the unit extension (longitudinal strain) is therefore

1.9 Infinitesimal Strain Tensor

$$L_{(2)} = \sqrt{i_2.c.i_2} - 1 = \sqrt{C_{22}} - 1 = \sqrt{1+2E_{22}} - 1 \qquad (1.68)$$

Also, the unit extensions $L_{(1)}$ and $L_{(2)}$ are given by analogous equations in terms of E_{11} and E_{33} respectively. From this it follows that

$$2E_{22} = (1+L_{(2)})^2 - 1 = \Lambda_{(1)}^2 - 1.$$

Thus, the normal component of the Lagrangian strain is one half of the square of the stretch, minus one. When the extension is small, $L_{(1)} \ll 1$, by expanding (1.61), neglecting the square of $L_{(1)}$, we get

$$E_{11} \cong L_{(1)} \cong \tilde{E}_{11}$$

Similar results are of course valid for E_{22} and E_{33}, which indicates that the infinitesimal normal strains are approximately the extensions of the fibers along the coordinate axes when the deformation is small. The corresponding results for the Eulerian strains are

$$e_{(1)} = \lambda_{(1)} - 1 = \frac{1}{\sqrt{1-2e_{11}}} - 1 \Rightarrow 2e_{11} = 1 - (1+e_{(1)})^{-2}$$

so that for small extensions as compared to unity, we obtaitn

$$e_{11} \cong e_{(1)} \cong \tilde{e}_{11}$$

The change in angle between any two line elements may also be given in terms of stretch. The geometrical meaning of the shear strain E_{12}, E_{23}, E_{31} is found by considering the angle changes between two directions N_1, and N_2. To simplify the presentation, we consider two vectors $I_1 dX_1$, and $I_2 dX_2$ which after deformation become $C_1 dX_1$ and $C_2 dX_2$, having an angle θ_{12} given by

$$\cos\theta_{(12)} = \frac{N_1.C.N_2}{\Lambda_{(N_1)}\Lambda_{(N_2)}} = \frac{(C_1 dX_1).(C_2 dX_2)}{|C_1 dX_1|.|C_2 dX_2|} = \frac{C_{12}}{\sqrt{C_{11}C_{22}}} \qquad (1.69)$$

The change in angle between any two line elements may also be given in terms of stretch. For the two differential line elements, the change in angle $\gamma_{(23)} = \frac{\pi}{2} - \theta$ is given in terms of $\Lambda_{(I_2)}$ and $\Lambda_{(I_3)}$ by

$$\sin\gamma_{(23)} = \frac{2L_{23}}{\Lambda_{(I_2)}\Lambda_{(I_3)}} = \frac{2L_{23}}{\sqrt{1+E_{22}}\sqrt{1+E_{33}}} \qquad (1.70)$$

In a similar fashion, if Θ be the angle between dx_1 and dx_2, we obtain

$$\cos\tilde{} = \frac{n_1.c.n_2}{\lambda_{(n_1)}\lambda_{(n_2)}} = \frac{n_1.c.n_2}{\sqrt{n_1.c.n_1}\sqrt{n_2.c.n_2}} = \lambda_{(n_1)}\lambda_{(n_2)}(n_1.c.n_2) \quad (1.71)$$

which gives the original angle between elements in the directions n_1 and n_2 of the current configuration. It is evident from Eq. (1.62) that if the coordinate axes are chosen in the principal directions of C, the deformed angle θ_{12} is a right angle ($C_{12}= 0$ in this case) and there has been no change in the angle between elements in the X_1 and X_2 directions. By the same argument, any three mutually perpendicular *principal* axes of C at $P_0(X_i)$ are deformed into three mutually perpendicular axes at $P(X_i)$. Consider, therefore, the volume element of a rectangular parallelepiped whose edges are in the principal directions of C (and thus also of E). Since there is no shear strain between any two of these edges, the new volume is still a rectangular parallelopiped, and in the edge directions N_i ($i = 1, 2, 3$) the unit strains are

$$e_{(N_i)} = \Lambda_{(N_i)} - 1; \quad i=1,2,$$

so that now

$$dx_i = dX_i + dX_i\left[\Lambda_{(N_i)} - 1\right] = dX_i\Lambda_{(N_i)}; i = 1,2,3 \quad (1.72)$$

and the ratio of the deformed volume to the original becomes

$$\frac{dV}{dV_0} = \frac{dx_1 dx_2 dx_3}{dX_1 dX_2 dX_3} = \Lambda_{(N_1)}\Lambda_{(N_2)}\Lambda_{(N_3)} = \sqrt{C_1 C_2 C_3} \quad (1.73)$$

The importance of the second form of Eq. (1.66) is that it is an invariant expression and can be calculated without reference to principal axes of C.

Example 1.12: A homogeneous deformation is given by the mapping equation $x_1 = X_1 - X_2 + X_3$, $x_2 = X_1 + X_2 - X_3$ and $x_3 = -X_1 + X_2 + X_3$. Determine

(a) the stretch ratio in the direction of $N_1 = \frac{1}{\sqrt{2}}(I_1 + I_2)$, and

(b) the angle in the deformed configuration between elements that were originally in the directions of N_1 and $N_1 = I_2$.

Solution: The deformation gradient F has the matrix form

1.9 Infinitesimal Strain Tensor

$$(F_{ij}) = \begin{pmatrix} \dfrac{\partial x_1}{\partial X_1} & \dfrac{\partial x_1}{\partial X_2} & \dfrac{\partial x_1}{\partial X_3} \\ \dfrac{\partial x_2}{\partial X_1} & \dfrac{\partial x_2}{\partial X_2} & \dfrac{\partial x_2}{\partial X_3} \\ \dfrac{\partial x_3}{\partial X_1} & \dfrac{\partial x_3}{\partial X_2} & \dfrac{\partial x_3}{\partial X_3} \end{pmatrix} = \begin{pmatrix} 1 & -1 & 1 \\ 1 & 1 & -1 \\ -1 & 1 & 1 \end{pmatrix}.$$

The Green's deformation tensor $C = F^T.F$ has the matrix form

$$(C_{ij}) = \begin{pmatrix} 1 & -1 & 1 \\ 1 & 1 & -1 \\ -1 & 1 & 1 \end{pmatrix}^T \begin{pmatrix} 1 & -1 & 1 \\ 1 & 1 & -1 \\ -1 & 1 & 1 \end{pmatrix} = \begin{pmatrix} 3 & -1 & -1 \\ -1 & 3 & -1 \\ -1 & -1 & 3 \end{pmatrix}$$

(a) Therefore, from the relation $\Lambda^2_{(N)} = N.C.N$, we get

$$\Lambda^2_{(N_1)} = \begin{bmatrix} \dfrac{1}{\sqrt{2}} & \dfrac{1}{\sqrt{2}} & 0 \end{bmatrix} \begin{pmatrix} 3 & -1 & -1 \\ -1 & 3 & -1 \\ -1 & -1 & 3 \end{pmatrix} \begin{bmatrix} \dfrac{1}{\sqrt{2}} \\ \dfrac{1}{\sqrt{2}} \\ 0 \end{bmatrix} = 2 \Rightarrow \Lambda_{(N_1)} = \sqrt{2}$$

(b) For $N_1 = I_2$, $\Lambda^2_{(I_2)} = I_2.C.I_2 = C_{22} = 3$ so that

$$\cos\theta_{(12)} = \dfrac{N_1.C.N_2}{\Lambda_{(N_1)}\Lambda_{(N_2)}} = \dfrac{C_{12}}{\sqrt{C_{11}C_{22}}} = \dfrac{2/\sqrt{2}}{\sqrt{6}} \Rightarrow \theta_{(12)} = 54.7^0$$

Thus the original 45° angle is enlarged by 9.7°.

1.9.6 Relation between Strain Vector and Strain Tensor

Consider a material line element P_0Q_0, joining a pair of neighboring points P_0Q_0, of length dL oriented in the direction $N = (N_1, N_2, N_3)$ in the initial undeformed region B_0 at time $t = 0$. If P_0 has coordinates $X = (X_1, X_2, X_3)$ and Q_0 has coordinates $(X_1 + dX_1, X_2 + dX_2, X_3 + dX_3)$ with respect to an orthogonal set of coordinate axes fixed in space, then

$$N_i = \dfrac{X_i + dX_i - X_i}{dL} = \dfrac{dX_i}{dL}.$$

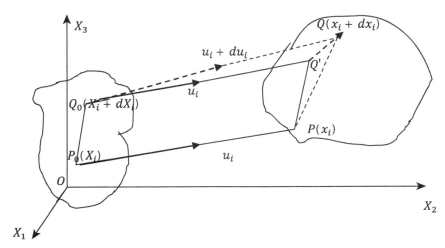

Fig. 1.11: Relation between Strain Vector and Strain Tensor

When the deformation has taken place, let the material point at P_0 undergoes a displacement u_i and move to the position P and another material point at Q_0 experiences a displacement $u_i + du_i$ and move to the position Q in the deformed state(Fig. 1.11) . Then the strain vector $\vec{E}^{(N)}$ at P_0 is defined by

$$\vec{E}^{(N)} = \frac{\overrightarrow{PQ} - |\overrightarrow{P_0Q_0}|}{|\overrightarrow{P_0Q_0}|}$$

If we draw Q_0Q' equal and parallel to P_0P, then

$$\vec{E}^{(N)} = \frac{1}{dL}\left[\overrightarrow{PQ} - \overrightarrow{PQ'}\right] = \frac{1}{dL}\overrightarrow{Q'Q} = \frac{1}{dL}\left[\overrightarrow{Q_0Q} - \overrightarrow{Q_0Q'}\right]$$

$$E_i^{(N)} = \frac{u_i + du_i - u_i}{dL} = \frac{du_i}{dL}$$

If the deformation consists of strain deformation only involving no rigid body deformation; then relative displacement is given by

$$du_i = E_{ij}dX_j$$

where E_{ij} is the small strain tensor at P_0. Therefore

$$E_i^{(N)} = E_{ij}\frac{dX_j}{dL} = E_{ij}N_j,$$

which is the relation between strain vector and strain tensor at P_0. The normal component of strain vector $\vec{E}^{(N)}$ in the direction $N = (N_1, N_2, N_3)$ will be

1.10 Strain Quadric

$$E_i^{(N)} N_i = E_{ij} N_j N_i = E_{(N)}$$

$$= E_{11}; \text{when } N_1 = 1, N_2 = 0, N_3 = 0$$

$$= E_{22}; \text{when } N_1 = 0, N_2 = 1, N_3 = 0$$

$$= E_{33}; \text{when } N_1 = 0, N_2 = 0, N_3 = 1$$

For this reason extensional strain E_{11}, E_{22}, E_{33} are also called *normal strain*.

1.10 Strain Quadric

Consider a point $P_0(X_i)$ in the undeformed state of a continuum body. We now study the deformation of an infinitesimal sphere $dL=$ const, at $P_0(X_i)$.

The sphere of radius k at $P_0(X_i)$ swept by a vector, dX, that is,

$$E_{kl} dX_k dX_l = dL^2 = k^2 \tag{1.74}$$

after deformation becomes a quadric surface swept by dx at x, that is,

$$e_{kl} dx_k dx_l = dL^2 = k^2 \tag{1.75}$$

Since (1.74) is positive definite, (1.75) is also positive definite. Consequently, the quadric surface that this sphere deforms into, (1.75), is an ellipsoid. Similarly, a sphere swept by dx_k at x, before deformation, was an ellipsoid, that is, the sphere at $P(x_i)$ of the deformed body

$$E_{kl} dx_k dx_l = dl^2 = k^2 \tag{1.76}$$

was an ellipsoid in the undeformed body at $P(x_i)$

$$e_{kl} dX_k dX_l = dl^2 = k^2 \tag{1.77}$$

These ellipsoids are called the strain ellipsoids of Cauchy. The ellipsoid (1.75) is known as the material ellipsoid of Cauchy, and (1.77) as the spatial ellipsoid of Cauchy. The study of the basic characteristics of one of these ellipsoids will clarify the character of the deformation in the neighborhood of a point.

Let E_{ij} be the small strain tensor at $P_0(X_i)$ with respect to a system of axes OX_1, OX_2, OX_3 fixed in space (see Fig. 1.12). We introduce a local system of axes $P_0\xi_1$, $P_0\xi_2$, $P_0\xi_3$

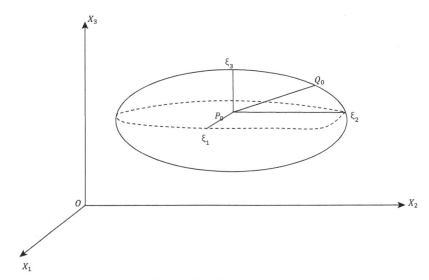

Fig. 1.12: Strain quadric

with the origin at $P_0(X_i)$ and parallel to the axes OX_1, OX_2, OX_3 respectively. For a given set of strain tensor E_{ij}, we can construct a quadric surface with its centre at $P_0(X_i)$ given by

$$E_{ij}\xi_i\xi_j = 1 \tag{1.78}$$

The quadric surface is known as strain quadric, and every straight line meets the quadric surface in two points.

Properties of Strain Quadric:

Property 1: The extensional strain of a line element through the centre of a strain quadratic in the direction of any central radius vector is equal to the inverse of the square of the radius vector.

Consider a point $P_0(X_i)$ in the undeformed state of a continuum body. Let E_{ij} be the small strain tensor at $P_0(X_i)$ with respect to a fixed system of axes OX_1, OX_2, OX_3 fixed in space. We introduce a local system of axes $P_0\xi_1$, $P_0\xi_2$, $P_0\xi_3$ with the origin at $P_0(X_i)$ and parallel to the axes OX_1, OX_2, OX_3 respectively. Let the equation of the strain quadric with its centre at $P_0(X_i)$ be

$$E_{ij}\xi_i\xi_j = 1$$

Draw any line P_0Q_0 through the centre P_0 (Fig. 1.12) to intersect the quadric surface at a point $Q_0(\xi_1,\xi_2,\xi_3)$. Let L denote the length P_0Q_0, (N_1,N_2,N_3) be direction cosines of P_0Q_0. For the point Q_0

$$\xi_i = LN_i \quad i$$

$$N_i = \frac{\xi_i}{L}$$

1.10 Strain Quadric

Let $E_{(N)}$ be the extension of line element $P_0 Q_0$ in the direction of $P_0 Q_0$, then

$$E_{(N)} = E_{ij} N_i N_j = \frac{1}{L^2} E_{ij} \xi_i \xi_j = \frac{1}{L^2}$$

as Q lies on the strain quadric whose coordinates ξ_1, ξ_2, ξ_3 satisfies $E_{ij}\xi_i\xi_j = 1$.

Property 2: The displacement of a material point at any point on the strain quadratic relative to that at the centre is directed along the normal to the surface of the quadric at that point.

Let the equation of the strain quadric with its centre at $P_0(X_i)$ be

$$E_{ij}\xi_i\xi_j = 1.$$

Draw any line $P_0 Q_0$ through the centre P_0 to intersect the quadric surface at $Q_0(\xi_1, \xi_2, \xi_3)$. Let $\bar{u}_i$ be the displacement of the material point at $Q_0(\xi_1, \xi_2, \xi_3)$ relative to that at $P_0(X_1, X_2, X_3)$ due to strain deformation only. Since (ξ_1, ξ_2, ξ_3) are the relative coordinates of Q_0 relative to that at $P_0(X_1, X_2, X_3)$, we have

$$\bar{u}_i = E_{ij}\xi_j$$

Let us consider a quadric function

$$2G(\xi_1, \xi_2, \xi_3) = E_{ij}\xi_i\xi_j$$

So the equation of the strain quadric becomes $2G(\xi_1, \xi_2, \xi_3) = 1$. It follows from the above result that

$$\frac{\partial}{\partial \xi_i}\left[2G(\xi_1,\xi_2,\xi_3)\right] = \frac{\partial}{\partial \xi_i}\left[E_{kl}\xi_k\xi_l\right] = E_{kl}\left[\frac{\partial \xi_k}{\partial \xi_i}\xi_l + \frac{\partial \xi_l}{\partial \xi_i}\xi_k\right]$$

$$= E_{kl}\left[\delta_{ki}\xi_l + \delta_{li}\xi_k\right]$$

$$= E_{kl}\delta_{ki}\xi_l + E_{kl}\delta_{li}\xi_k$$

$$= E_{il}\xi_l + E_{ki}\xi_k$$

$$= E_{ij}\xi_j + E_{ji}\xi_j = E_{ij}\xi_j + E_{ij}\xi_j$$

$$= 2E_{ij}\xi_j$$

Therefore, $\quad \dfrac{\partial}{\partial \xi_i}\left[G(\xi_1,\xi_2,\xi_3)\right] = E_{ij}\xi_j = \bar{u}_i$

But $\dfrac{\partial}{\partial \xi_i}\left[G(\xi_1,\xi_2,\xi_3)\right]$ are direction ratios of the normal to the quadric surface $2G(\xi_1, \xi_2, \xi_3) = 1$ at the point $Q_0(\xi_1, \xi_2, \xi_3)$. It follows that relative displacement is directed along the normal to the quadric surface at $Q_0(\xi_1, \xi_2, \xi_3)$.

1.11 Principal Strain

The direction of a line element generally changes due to strain deformation. In particular, if the direction of a line element at a given point of a continuum body remains unchanged by strain deformation then this direction is know as the principal direction of strain or principal axis of strain and the extension that occurs along the principal direction is called principal strain.

(a) Principal strain and its direction: Consider two neighbouring material points at the positions $P_0(X_1, X_2, X_3)$ and $Q_0(X_1+ dX_1, X_2+ dX_2, X_3+ dX_3)$ (Fig. 1.11) of the continuum in the undeformed state, with respect to an orthogonal set of coordinate axes fixed in space. The material line element $P_0 Q_0$ has the length dL oriented in the direction (N_1, N_2, N_3) in the initial undeformed state of a continuum body, then

$$N_i = \frac{dX_i}{dL}; \text{ and } N_i N_i = 1, i = 1, 2, 3 \tag{1.79}$$

Let $E_{ij} = \frac{1}{2}\left[\frac{\partial u_i}{\partial X_j} + \frac{\partial u_j}{\partial X_i}\right]$ = infinitesimal strain tensor at $P_0(X_1, X_2, X_3)$.

In the deformed state let $P_0(X_1, X_2, X_3)$ and $Q_0(X_1+ dX_1, X_2+ dX_2, X_3+ dX_3)$ move to the point $P(x_1, x_2, x_3)$ and $Q(x_1+ dx_1, x_2+ dx_2, x_3+ dx_3)$ respectively. If the line element P_0Q_0 is to be the principle direction of strain at P_0, then $\overrightarrow{P_0Q_0}$ must be parallel to $\overrightarrow{PQ}$. If u_i be the displacement of P_0 and $u_i + du_i$ be the displacement of Q_0 then du_i is lie along PQ, then du_i will be proportional to dX_i, i.e.,

$$du_i = \lambda dX_i \tag{1.80}$$

where, λ is the constant of proportionality. Thus

$$\lambda = \frac{du_i}{dX_i} = \frac{dx_i - dX_i}{dX_i} \tag{1.81}$$

= extension of each component of dX_i per unit length.

Hence, λ is the extension of the line element P_0Q_0 in the direction of $\overrightarrow{P_0Q_0}$ and λ is called the principle strain. We know that the strain vector is given by

$$E_i^{(N)} = \frac{du_i}{dL} = \lambda \frac{dX_i}{dL} = \lambda N_i \tag{1.82}$$

Also the strain vector is related to the strain tensor by the equation

$$E_i^{(N)} = E_{ij} N_j \tag{1.83}$$

From equations (1.82) and (1.83) we get

1.11 Principal Strain

$$E_{ij}N_j = \lambda N_i = \lambda \delta_{ij} N_j$$

or, $\quad (E_{ij} - \lambda \delta_{ij})N_j = 0; j = 1,2,3$

By expanding we get

$$(E_{11} - \lambda)N_1 + E_{12}N_2 + E_{13}N_3 = 0$$

$$E_{21}N_1 + (E_{22} - \lambda)N_2 + E_{23}N_3 = 0 \qquad (1.84)$$

$$E_{31}N_1 + E_{32}N_2 + (E_{33} - \lambda)N_3 = 0$$

This is a set of three homogeneous linear equations for N_1, N_2, N_3 which has to satisfy the condition $N_i N_i = 1$, i.e.,

$$N_1^2 + N_2^2 + N_3^2 = 1. \qquad (1.85)$$

The trivial solution of equation (1.84) i.e., solution $N_1 = N_2 = N_3 = 0$ does not satisfy the condition (1.85) and should be rejected.

The condition for the existence for a non-trivial solution of the equation (1.84) is

$$\begin{vmatrix} E_{11} - \lambda & E_{12} & E_{13} \\ E_{21} & E_{22} - \lambda & E_{23} \\ E_{31} & E_{32} & E_{33} - \lambda \end{vmatrix} = 0. \qquad (1.86)$$

This is a cubic equation is λ and is called the characteristic equation which has three roots E_1, E_2, E_3, called the principle strain. Corresponding to each E_i we can solve the system of equation (1.84) subject to (1.85) and find (N_1, N_2, N_3) which gives the corresponding principal direction.

Note:

(i) Since E_{ij} is symmetric second order tensor the roots of the characteristic equation (1.86) are real and hence all principal strains are real.

(ii) For a given matrix $E = (E_{ij})$, the principal strains are to be found from the characteristic equation of (E_{ij}). The unit elongations along the principal directions are the eigenvalues of E, or principal strains. They include the maximum and the minimum normal strains among all directions emanating from the particle.

(iii) Also, as the strain tensor is symmetric, there exist at least three mutually perpendicular directions N_1, N_2, N_3 with respect to which the matrix of (E_{ij}) is diagonal. Geometrically, this means that infinitesimal line elements in the directions remain mutually perpendicular after deformation. These directions are known as principal directions.

(iv) Principal directions of strain corresponding to distinct principal strains are orthogonal to each other.

(v) When two roots say E_1, E_2 are equal we calculate the principle axis corresponding to $\lambda = E_3$. The other two axes are given by any two mutually perpendicular line which are also perpendicular to the third axis.

(vi) When $E_1 = E_2 = E_3$ any three mutually perpendicular lines to the point P_0 may be taken as the principal axis.

Theorem 1.1: **All principal strains are real.**

Proof: Here, we have to show that the three roots E_1, E_2, E_3 of the equation (1.86) and corresponding $N_i = \left(N_1^{(i)}, N_2^{(i)}, N_3^{(i)}\right)$ vectors are all real. Let one root, say E_1, of the equation (1.86) be complex. Since the coefficients of the equation (1.86) are all real so the complex conjugate E_1^* of E_1 is also a root of the equation (1.86). Corresponding to these roots we obtain complex direction $N_i^{(1)}$ and its complex conjugate $N_i^{*(1)}$ satisfying the system of linears equation (1.84), i.e.,

$$E_{ij}N_j^{(1)} = E_1 N_i^{(1)} \text{ and } E_{ij}N_j^{*(1)} = E_1^* N_i^{*(1)}; i = 1, 2, 3. \quad (1.87)$$

Multiplying both sides of the first equation of (1.87) by $N_i^{*(1)}$ and second by $N_i^{(1)}$, we get

$$E_{ij}N_j^{(1)}N_i^{*(1)} = E_1 N_i^{(1)}N_i^{*(1)} \text{ and } E_{ij}N_j^{*(1)}N_i^{(1)} = E_1^* N_i^{*(1)}N_i^{(1)}; i = 1, 2, 3. \quad (1.88)$$

Now,

$$E_{ij}N_j^{*(1)}N_i^{(1)} = E_{ji}N_i^{*(1)}N_j^{(1)}; \quad \text{interchanging dummy indices}$$
$$= E_{ij}N_i^{*(1)}N_j^{(1)}; \quad \text{since } E_{ij} \text{ is symmetric}$$

It follows from equation (1.88) that

$$E_1 N_i^{(1)}N_i^{*(1)} = E_1^* N_i^{*(1)}N_i^{(1)} \Rightarrow \left(E_1 - E_1^*\right)N_i^{*(1)}N_i^{(1)} = 0.$$

Since $N_i^{*(1)}N_i^{(1)} \neq 0$, which is being a sum of squares of real numbers. Hence

$$E_1 = E_1^* \Rightarrow \text{Imaginary part of } E_1 = 0.$$

Therefore, E_1 is real. Hence, the three roots E_1, E_2, E_3 of the equation (1.86) are all real and the corresponding values of $N_i^{(1)}$ of equation (1.84) are all real.

Theorem 1.2: **Principal directions of strain corresponding to distinct principal strains are orthogonal to each other.**

1.11 Principal Strain

Proof: We consider the following three cases:

Case I: Three principal strains E_1, E_2, E_3 are distinct: First let E_1 and E_2 be any two distinct real roots of the characteristic equation (1.86); $N_i^{(1)}$ and $N_i^{(2)}$ be corresponding direction cosines obtained from equation (1.84). Therefore,

$$E_{ij}N_j^{(1)} = E_1 N_i^{(1)} \text{ and } E_{ij}N_j^{(2)} = E_2 N_i^{(2)}; \; i = 1,2,3. \qquad (1.89)$$

Multiplying both sides of the first equation of (1.89) by $N_i^{(2)}$ and second by $N_i^{(1)}$, we get

$$E_{ij}N_j^{(1)}N_i^{(2)} = E_1 N_i^{(1)} N_i^{(2)} \text{ and } E_{ij}N_j^{(2)}N_i^{(1)} = E_2 N_i^{(2)} N_i^{(1)}; \; i = 1,2,3. \qquad (1.90)$$

Now, $E_{ij}N_j^{(2)}N_i^{(1)} = E_{ji}N_i^{(2)}N_j^{(1)};$ interchanging dummy indices

$\qquad\qquad = E_{ij}N_j^{(1)}N_i^{(2)};$ since E_{ij} is symmetric

Therefore, from equation (1.90), we have

$$E_1 N_i^{(1)} N_i^{(2)} = E_2 N_i^{(2)} N_i^{(1)} \Rightarrow (E_1 - E_2) N_i^{(1)} N_i^{(2)} = 0$$

$$\Rightarrow N_i^{(1)} N_i^{(2)} = 0; \text{ as } E_1 \neq E_2. \qquad (1.91)$$

This shows that $N_i^{(1)}$ and $N_i^{(2)}$ are orthogonal. Thus two principal directions of strain corresponding to two distinct principal strains are orthogonal. Similar results are obtained for other set of pair of roots. Consequently, three principal directions of strain are mutually perpendicular provided three principal strains are distinct.

Thus, in general, at any point in the undeformed state, there exist a single system of three real mutually perpendicular principal directions of strain whose orientations are left unaltered by strain deformation provided three principal strains are distinct.

Case II: Two principal strains are equal: Let the two roots E_1 and E_2 of the characteristic equation (1.86) be equal so that we can write $E_1 = E_2$. We know that the equation (1.86) has at least one real root, say E_3. For $E = E_3$, equation (1.84) has one real solution $N_i^{(3)}$. This solution defines the third principal direction of strain. Choose the associated direction as X_3 axis and any pair of mutually perpendicular lines each of which is perpendicular to X_3 axis are taken as X_1 and X_2 axes. In the new system of coordinates

$$N_1^{(3)} = 0, \; N_2^{(3)} = 0, \; N_3^{(3)} = 1.$$

It follows from the equation $E_{ij} N_j = E N_i$ that $E_{ij} N_j^{(3)} = E_3 N_i^{(3)}$. Therefore,

$$E_{1j}N_j^{(3)} = E_3 N_1^{(3)} = 0; \ i=1 \Rightarrow E_{13} = 0.$$

$$E_{2j}N_j^{(3)} = E_3 N_2^{(3)} = 0; \ i=2 \Rightarrow E_{23} = 0.$$

$$E_{3j}N_j^{(3)} = E_3 N_3^{(3)} = E_3; \ i=2 \Rightarrow E_{33} = E_3.$$

Therefore, the equation (1.86) reduces to

$$\begin{vmatrix} E_{11}-E & E_{12} & 0 \\ E_{21} & E_{22}-E & 0 \\ 0 & 0 & E_3-E \end{vmatrix} = 0$$

$$\Rightarrow (E_3 - E)\left[(E_{11}-E)(E_{22}-E) - E_{21}^2\right] = 0.$$

This is a cubic equation in E. $E = E_3$ is one of the roots of this equation, Other two roots of this cubic equation are given by

$$(E_{11}-E)(E_{22}-E) - E_{21}^2 = 0$$

$$\Rightarrow E^2 - (E_{11}+E_{22})E + (E_{11}E_{22} - E_{21}^2) = 0.$$

Since $E_1 = E_2$, the above equation would have equal roots. The condition for equal roots is

$$(E_{11}+E_{22})^2 - 4(E_{11}E_{22} - E_{21}^2) = 0$$

$$\Rightarrow (E_{11}+E_{22})^2 - 4E_{11}E_{22} + 4E_{12}^2 = 0$$

$$\Rightarrow (E_{11}-E_{22})^2 + 4E_{12}^2 = 0 \Rightarrow E_{11} = E_{22}, \ E_{12} = 0.$$

Consequently, $E_{13} = E_{23} = E_{12} = 0$. It follows that for $E_1 = E_2$, coordinate system $N_i^{(3)}$ and in consequence any coordinate system containing the third principal axis corresponding to E_3 as X_3 defines one parameter family of systems of principal directions of strain. Thus, two roots are equal, any two mutually orthogonal directions lying in a plane perpendicular to the principal direction corresponding to simple root may be taken as corresponding principal directions of strain.

Case III: All the principal strains are equal: Consider the case when all three principal strains E_1, E_2, E_3 are equal, i.e., $E_1 = E_2 = E_3$.

When, $E_1 = E_2$ any system of coordinate axis with third principal axis $N_i^{(3)}$ corresponding to E_3 as X_3 defines a system of principal directions of strain. When, $E_2 = E_3$ any system of coordinate axis with first principal axis $N_i^{(1)}$ corresponding to E_1 as X_1 defines a system of principal directions of strain.

1.11 Principal Strain

When, $E_3 = E_1$ any system of coordinate axis with second principal axis $N_i^{(2)}$ corresponding E_2 to X_2 as defines a system of principal directions of strain.

Consequently, for $E_1 = E_2 = E_3$, every direction of space is a principal direction of strain.

1.11.1 Strain invariants

There are a number of components of strain tensor E_{ij} which remain unaltered by the rotation of the coordinate system. These components are called *strain invariants*. Now, the three principal strains E_1, E_2, E_3 are the roots of the characteristic equation

$$\begin{vmatrix} E_{11} - E & E_{12} & E_{13} \\ E_{21} & E_{22} - E & E_{23} \\ E_{31} & E_{32} & E_{33} - E \end{vmatrix} = 0. \quad (1.92)$$

Expanding, we get the cubic equation

$$E^3 - (E_{11} + E_{22} + E_{33})E^2 + (E_{11}E_{22} + E_{22}E_{33} + E_{33}E_{11} - E_{23}^2 - E_{12}^2 - E_{31}^2)E$$

$$- \left[E_{23}^2 E_{11} + E_{12}^2 E_{33} + E_{31}^2 E_{22} - E_{11}E_{22}E_{33} - 2E_{12}E_{23}E_{31} \right] = 0$$

$$\Rightarrow E^3 - \theta_1 E^2 + \theta_2 E - \theta_3 = 0. \quad (1.93)$$

where,
$$\theta_1 = E_{11} + E_{22} + E_{33}. \quad (1.94)$$

$$\theta_2 = \begin{vmatrix} E_{11} & E_{12} \\ E_{21} & E_{22} \end{vmatrix} + \begin{vmatrix} E_{22} & E_{23} \\ E_{32} & E_{33} \end{vmatrix} + \begin{vmatrix} E_{33} & E_{31} \\ E_{13} & E_{11} \end{vmatrix}. \quad (1.95)$$

$$\theta_3 = |E_{ij}| = \begin{vmatrix} E_{11} & E_{12} & E_{13} \\ E_{21} & E_{22} & E_{23} \\ E_{31} & E_{32} & E_{33} \end{vmatrix}. \quad (1.96)$$

Since E_1, E_2, E_3 are the roots of the equation (1.93), by the relation between the roots θ_1, θ_2, θ_3 and coefficients, we have

$$\theta_1 = E_1 + E_2 + E_3; \; \theta_2 = E_1 E_2 + E_2 E_3 + E_3 E_1; \; \theta_3 = E_1 E_2 E_3. \quad (1.97)$$

Since the principal strains E_1, E_2, E_3 at a point have a geometrical meaning independent of the choice of coordinate system, it is clear from equation (1.97) that θ_1, θ_2, θ_3 given by equation (1.94)-(1.96) are invariant with respect to orthogonal transformations of coordinates and are respectively called first, second and third strain invariants.

Geometrical interpretation of first strain invariant: We consider the change in volume element in initial undeformed and subsequent deformed states. Let

P_0 be a point in the initial state. Let E_1, E_2, E_3 be the three principal strains at the point P_0. Let us consider a volume element of continuum occupying initially a rectangular parallelepiped of volume dV_0 with one of its vertices at the point P_0, with edges parallel to the principal direction of strains at P_0 and lengths L_1, L_2, L_3 so that

$$dV_0 = L_1.L_2.L_3$$

On deformation, this element becomes again a rectangular parallelepiped of volume dV because orientation of principal directions of strain remain unchanged. If l_1, l_2, l_3 be the lengths of its edges then

$$dV = l_1.l_2.l_3$$

Now,

$$E_1 = \frac{l_1 - L_1}{L_1} \Rightarrow l_1 = (1+E_1)L_1;$$

$$E_2 = \frac{l_2 - L_2}{L_2} \Rightarrow l_2 = (1+E_2)L_2;$$

$$E_3 = \frac{l_3 - L_3}{L_3} \Rightarrow l_3 = (1+E_3)L_3.$$

Therefore,

$$\frac{dV - dV_0}{dV_0} = \frac{l_1.l_2.l_3 - L_1.L_2.L_3}{L_1.L_2.L_3} = \frac{(1+E_1)L_1.(1+E_2)L_2.(1+E_3)L_3 - L_1.L_2.L_3}{L_1.L_2.L_3}$$

$$= (1+E_1).(1+E_2).(1+E_3) - 1 = E_1 + E_2 + E_3$$

(neglecting the higher order terms)

$$= E_{11} + E_{22} + E_{33} = \theta_1$$

Thus the first strain invariant θ_1 represents the change in volume per unit original volume. The deformation in which volume element remains unaltered is called isochoric deformation. Thus far isochoric deformation

$$\theta_1 = E_{11} + E_{22} + E_{33} = 0$$

(i) For small deformation strain in general

$$\theta_1 = E_{ii} = \frac{\partial u_i}{\partial x_i} = \text{div}\,\mathbf{u}.$$

This unit volume change is known as *dilatation*.

(ii) In terms of displacements, we have

1.11 Principal Strain

Cartesian coordinates:

$$\theta_1 = \frac{\partial u_1}{\partial x_1} + \frac{\partial u_2}{\partial x_2} + \frac{\partial u_3}{\partial x_3}.$$

Polar co-ordinates: In polar co-ordinates, the matrix form of the components of ∇u can be written as

$$[\nabla u] = \begin{pmatrix} \dfrac{\partial u_r}{\partial r} & \dfrac{1}{r}\dfrac{\partial u_r}{\partial \theta} - \dfrac{u_\theta}{r} \\ \dfrac{\partial u_\theta}{\partial r} & \dfrac{1}{r}\dfrac{\partial u_\theta}{\partial \theta} + \dfrac{u_r}{r} \end{pmatrix}$$

Using the components of ∇u given above, we have

$$\theta_1 = \mathrm{div}\,u = \mathrm{tr}(\nabla u) = \frac{\partial u_r}{\partial r} + \frac{1}{r}\frac{\partial u_\theta}{\partial \theta} + \frac{u_r}{r}.$$

Cylindrical coordinates: In cylindrical co-ordinates (r, θ, z) the matrix form the components of ∇u can be written as

$$[\nabla u] = \begin{pmatrix} \dfrac{\partial u_r}{\partial r} & \dfrac{1}{r}\dfrac{\partial u_r}{\partial \theta} - \dfrac{u_\theta}{r} & \dfrac{\partial u_r}{\partial z} \\ \dfrac{\partial u_\theta}{\partial r} & \dfrac{1}{r}\dfrac{\partial u_\theta}{\partial \theta} + \dfrac{v_r}{r} & \dfrac{\partial u_\theta}{\partial z} \\ \dfrac{\partial u_z}{\partial r} & \dfrac{1}{r}\dfrac{\partial u_z}{\partial \theta} & \dfrac{\partial u_z}{\partial z} \end{pmatrix}$$

Using the components of ∇u given above, we have

$$\theta_1 = \mathrm{div}\,u = \mathrm{tr}(\nabla u) = \frac{\partial u_r}{\partial r} + \frac{1}{r}\left(\frac{\partial u_\theta}{\partial \theta} + u_r\right) + \frac{\partial u_z}{\partial z}.$$

Spherical coordinates: In spherical polar co-ordinates (r, θ, ϕ), the matrix form the components of ∇u can be written as

$$[\nabla u] = \begin{pmatrix} \dfrac{\partial u_r}{\partial r} & \dfrac{1}{r}\dfrac{\partial u_r}{\partial \theta} - \dfrac{u_\theta}{r} & \dfrac{1}{r\sin\theta}\dfrac{\partial u_r}{\partial \phi} - \dfrac{u_\phi}{r} \\ \dfrac{\partial u_\theta}{\partial r} & \dfrac{1}{r}\dfrac{\partial u_\theta}{\partial \theta} + \dfrac{u_r}{r} & \dfrac{1}{r\sin\theta}\dfrac{\partial u_\theta}{\partial \phi} - \dfrac{u_\phi \cot\theta}{r} \\ \dfrac{\partial u_\phi}{\partial r} & \dfrac{1}{r}\dfrac{\partial u_\phi}{\partial \theta} & \dfrac{1}{r\sin\theta}\dfrac{\partial u_\phi}{\partial \phi} + \dfrac{u_r}{r} - \dfrac{u_\theta \cot\theta}{r} \end{pmatrix}$$

Using the components of ∇u given above, we have

$$\theta_1 = \text{div } u = \text{tr}(\nabla u) = \frac{\partial u_r}{\partial r} + \frac{1}{r}\frac{\partial u_\theta}{\partial \theta} + \frac{2}{r}u_r + \frac{1}{r\sin\theta}\frac{\partial u_\phi}{\partial \phi} + \frac{u_\theta \cot\theta}{r}.$$

Example 1.13: Given the strain field $(E_{ij}) = \begin{pmatrix} K_1 X_2 & 0 & 0 \\ 0 & -K_2 X_2 & 0 \\ 0 & 0 & -K_2 X_2 \end{pmatrix}$.

What should be the relation between K_1 and K_2 such that there will be no volume change or isochoric deformation.

Solution: In the given strain field $E_{11} = K_1 X_2$, $E_{22} = -K_2 X_2$, $E_{33} = -K_2 K_2$. Isochoric deformation is one in which volume element remains unaltered. Thus for isochoric deformation $\theta_1 = E_{11} + E_{22} + E_{33} = 0$. Therefore

$$E_{11} + E_{22} + E_{33} = 0 \Rightarrow K_1 X_2 - K_2 X_2 - K_2 X_2 = 0 \Rightarrow K_1 = 2K_2.$$

Example 1.14: Determine principal strains and corresponding direction from the strain tensors

$$(E_{ij}) = \begin{pmatrix} 1 & 1 & 3 \\ 1 & 5 & 1 \\ 3 & 1 & 1 \end{pmatrix}.$$

Solution: The principal strains E_1, E_2, E_3 at the point P are the roots of the characteristic equation

$$\begin{vmatrix} 1-E & 1 & 3 \\ 1 & 5-E & 1 \\ 3 & 1 & 1-E \end{vmatrix} = 0$$

or,

$$E^3 + 11E^2 + 36 = 0 \Rightarrow E = -2, 3, 6.$$

The principal directions of stress at P can be obtained by solving the following system of equations

$$(1-E)N_1 + N_2 + 3N_3 = 0$$

$$N_1 + (5-E)N_2 + N_3 = 0$$

$$3N_1 + N_2 + (1-E)N_3 = 0$$

For $E = E_1 = -2$, the above system of equations become

1.11 Principal Strain

$$\begin{pmatrix} 1+2 & 1 & 3 \\ 1 & 5+2 & 1 \\ 3 & 1 & 1+2 \end{pmatrix} \begin{pmatrix} N_1^{(1)} \\ N_2^{(1)} \\ N_3^{(1)} \end{pmatrix} = \begin{pmatrix} 0 \\ 0 \\ 0 \end{pmatrix} \Rightarrow \begin{pmatrix} N_1^{(1)} \\ N_2^{(1)} \\ N_3^{(1)} \end{pmatrix} = \begin{pmatrix} 1 \\ 0 \\ -1 \end{pmatrix}$$

For $E = E_2 = 3$, the above system of equations become

$$\begin{pmatrix} 1-3 & 1 & 3 \\ 1 & 5-3 & 1 \\ 3 & 1 & 1-3 \end{pmatrix} \begin{pmatrix} N_1^{(2)} \\ N_2^{(2)} \\ N_3^{(2)} \end{pmatrix} = \begin{pmatrix} 0 \\ 0 \\ 0 \end{pmatrix} \Rightarrow \begin{pmatrix} N_1^{(2)} \\ N_2^{(2)} \\ N_3^{(2)} \end{pmatrix} = \begin{pmatrix} 1 \\ 1 \\ 1 \end{pmatrix}$$

For $E = E_3 = 6$, the above system of equations become

$$\begin{pmatrix} 1-6 & 1 & 3 \\ 1 & 5-6 & 1 \\ 3 & 1 & 1-6 \end{pmatrix} \begin{pmatrix} N_1^{(3)} \\ N_2^{(3)} \\ N_3^{(3)} \end{pmatrix} = \begin{pmatrix} 0 \\ 0 \\ 0 \end{pmatrix} \Rightarrow \begin{pmatrix} N_1^{(3)} \\ N_2^{(3)} \\ N_3^{(3)} \end{pmatrix} = \begin{pmatrix} 1 \\ 2 \\ 1 \end{pmatrix}$$

Therefore, the principal directions are given by $\frac{1}{\sqrt{2}}(1, 0, -1); \frac{1}{\sqrt{3}}(1, 1, 1); \frac{1}{\sqrt{6}}(1, 2, 1)$.

Example 1.15: The strain tensor at a point is given by $(E_{ij}) = \begin{pmatrix} a & b & 0 \\ b & -a & 0 \\ 0 & 0 & 0 \end{pmatrix}$.

Determine principal axes of strain and corresponding direction ratios of principal strains.

Solution: The principal strains E_1, E_2, E_3 at the point P are the roots of the characteristic equation

$$\begin{vmatrix} a-E & b & 0 \\ b & -a-E & 0 \\ 0 & 0 & 0-E \end{vmatrix} = 0$$

$$\Rightarrow E(a^2 + b^2 - E^2) = 0 \Rightarrow E = 0, \pm\sqrt{a^2 + b^2}.$$

The principal directions of stress at P are given by the system of equations

$$(a-E) N_1 + bN_2 = 0$$
$$bN_1 + (-a-E)N_2 = 0$$
$$-EN_3 = 0$$

For $E = E_1 = 0$, the above system of equations become

$$\begin{pmatrix} a-0 & b & 0 \\ b & -a-0 & 0 \\ 0 & 0 & 0-0 \end{pmatrix} \begin{pmatrix} N_1^{(1)} \\ N_2^{(1)} \\ N_3^{(1)} \end{pmatrix} = \begin{pmatrix} 0 \\ 0 \\ 0 \end{pmatrix} \Rightarrow \begin{pmatrix} N_1^{(1)} \\ N_2^{(1)} \\ N_3^{(1)} \end{pmatrix} = \begin{pmatrix} 0 \\ 0 \\ 1 \end{pmatrix}$$

For $E = E_2 = \sqrt{a^2 + b^2}$, the above system of equations become

$$\begin{pmatrix} a - \sqrt{a^2+b^2} & b & 0 \\ b & -a-\sqrt{a^2+b^2} & 0 \\ 0 & 0 & 0-\sqrt{a^2+b^2} \end{pmatrix} \begin{pmatrix} N_1^{(2)} \\ N_2^{(2)} \\ N_3^{(2)} \end{pmatrix} = \begin{pmatrix} 0 \\ 0 \\ 0 \end{pmatrix}$$

$$\Rightarrow \begin{pmatrix} N_1^{(2)} \\ N_2^{(2)} \\ N_3^{(2)} \end{pmatrix} = \frac{1}{b} \begin{pmatrix} a + \sqrt{a^2+b^2} \\ b \\ 0 \end{pmatrix}$$

For $E = E_3 = -\sqrt{a^2 + b^2}$, the above system of equations become

$$\begin{pmatrix} a + \sqrt{a^2+b^2} & b & 0 \\ b & -a+\sqrt{a^2+b^2} & 0 \\ 0 & 0 & 0+\sqrt{a^2+b^2} \end{pmatrix} \begin{pmatrix} N_1^{(3)} \\ N_2^{(3)} \\ N_3^{(3)} \end{pmatrix} = \begin{pmatrix} 0 \\ 0 \\ 0 \end{pmatrix}$$

$$\Rightarrow \begin{pmatrix} N_1^{(3)} \\ N_2^{(3)} \\ N_3^{(3)} \end{pmatrix} = \frac{1}{b} \begin{pmatrix} a - \sqrt{a^2+b^2} \\ b \\ 0 \end{pmatrix}$$

The direction ratios of principal strains are $\begin{pmatrix} 0 \\ 0 \\ 1 \end{pmatrix}$, $\frac{1}{b}\begin{pmatrix} a+\sqrt{a^2+b^2} \\ b \\ 0 \end{pmatrix}$, $\frac{1}{b}\begin{pmatrix} a+\sqrt{a^2+b^2} \\ b \\ 0 \end{pmatrix}$

Example 1.16: Determine the principal directions and principal strains for $(E_{ij}) = \begin{pmatrix} e & e & e \\ e & e & e \\ e & e & e \end{pmatrix}$.

1.11 Principal Strain

Solution: The principal strains E_1, E_2, E_3 at the point P are the roots of the characteristic equation

$$\begin{vmatrix} e-E & e & e \\ e & e-E & e \\ e & e & e-E \end{vmatrix} = 0$$

$$\Rightarrow E^2(3e-E) = 0 \Rightarrow E = 0, 0, 3e.$$

The principal directions of strain at P are given by the system of equations

$$(e-E)N_1 + eN_2 + eN_3 = 0$$
$$eN_1 + (e-E)N_2 + eN_3 = 0$$
$$eN_1 + eN_2 + (e-E)N_3 = 0$$

For $E = E_1 = 3e$, the system of equations become

$$\begin{pmatrix} e-3e & e & e \\ e & e-3e & e \\ e & e & e-3e \end{pmatrix} \begin{pmatrix} N_1^{(1)} \\ N_2^{(1)} \\ N_3^{(1)} \end{pmatrix} = \begin{pmatrix} 0 \\ 0 \\ 0 \end{pmatrix} \Rightarrow \begin{pmatrix} N_1^{(1)} \\ N_2^{(1)} \\ N_3^{(1)} \end{pmatrix} = \frac{1}{\sqrt{3}} \begin{pmatrix} 1 \\ 1 \\ 1 \end{pmatrix}$$

For $E = E_2 = 0$, the above system of equations become

$$\begin{pmatrix} e-0 & e & e \\ e & e-0 & e \\ e & e & e-0 \end{pmatrix} \begin{pmatrix} N_1^{(2)} \\ N_2^{(2)} \\ N_3^{(2)} \end{pmatrix} = \begin{pmatrix} 0 \\ 0 \\ 0 \end{pmatrix} \Rightarrow N_1^{(2)} + N_2^{(2)} + N_3^{(2)} = 0.$$

These equations together with $N_1^{(2)2} + N_2^{(2)2} + N_3^{(2)2} = 1$ are insufficient to determine the principal direction corresponding to $E_2 = 0$. Since two principal strains are equal, any pair of lines perpendicular to each other and each perpendicular to $\left(\frac{1}{\sqrt{3}}, \frac{1}{\sqrt{3}}, \frac{1}{\sqrt{3}}\right)$ may serve as principal direction of strain.

1.11.2 Extremum of Strain Values

Theorem 1.3: (**Extermun normal stress**): The extremum values of normal strain or extensional strain at a point of a continuum are principal strains.

Proof: Consider a point $P_0(X_i)$ in the undeformed (initial) state of a continuum body. Let us take the rectangular coordinate axes along the principal directions of strain at the point $P_0(X_i)$. Let E_1, E_2, E_3 be the principal strains at the point

$P_0(X_i)$ and ordered so that $E_1 > E_2 > E_3$. If E_{ij} be the strain tensors at a point $P_0(X_i)$, then

$$E_{11} = E_1, E_{22} = E_2, E_{33} = E_3, E_{12} = E_{31} = E_{23} = 0. \qquad (1.98)$$

If $E_{(N)}$ be the extensional strain at the point P acting in the direction $\vec{N} = (N_1, N_2, N_3)$, then

$$E_{(N)} = E_{ij} N_i N_j = E_{11} N_1^2 + E_{22} N_2^2 + E_{33} N_3^2 = E_1 N_1^2 + E_2 N_2^2 + E_3 N_3^2. \qquad (1.99)$$

The direction cosines $\vec{N} = (N_1, N_2, N_3)$ satisfy the equation

$$N_i N_j = N_1^2 + N_2^2 + N_3^2 = 1. \qquad (1.100)$$

We require the extremum values of the extensional strain $E_{(N)}$ for variations of N_1, N_2, N_3 subject to the constraint (1.100). For this, let us construct a Lagrangian function

$$F(N_1, N_2, N_3) = E_{(N)} - \lambda(N_1^2 + N_2^2 + N_3^2 - 1)$$
$$= E_1 N_1^2 + E_2 N_2^2 + E_3 N_3^2 - \lambda\left(N_1^2 + N_2^2 + N_3^2 - 1\right), \qquad (1.101)$$

where, λ is the Lagrangian parameter.
For extreme value of F, we have

$$\frac{\partial F}{\partial N_1} = 0 \Rightarrow E_1 N_1 - \lambda N_1 = 0;$$

$$\frac{\partial F}{\partial N_2} = 0 \Rightarrow E_2 N_2 - \lambda N_2 = 0;$$

$$\frac{\partial F}{\partial N_3} = 0 \Rightarrow E_3 N_3 - \lambda N_3 = 0.$$

Therefore, we have

$$\Rightarrow N_1(E_1 N_1 - \lambda N_1) + N_2\left(E_2 N_2 - \lambda N_2\right) + N_3\left(E_3 N_3 - \lambda N_3\right) = 0$$

$$\Rightarrow \lambda\left(N_1^2 + N_2^2 + N_3^2\right) = E_1 N_1^2 + E_2 N_2^2 + E_3 N_3^2 \Rightarrow \lambda = E_{(N)}. \qquad (1.102)$$

Therefore,

$$E_1 N_1 - \lambda N_1 = 0 \Rightarrow \left(E_1 - E_{(N)}\right) N_1 = 0 \qquad (1.103)$$

$$E_2 N_2 - \lambda N_2 = 0 \Rightarrow \left(E_2 - E_{(N)}\right) N_2 = 0 \qquad (1.104)$$

$$E_3 N_3 - \lambda N_3 = 0 \Rightarrow \left(E_3 - E_{(N)}\right) N_3 = 0 \qquad (1.105)$$

1.11 Principal Strain

Equations (1.103)-(1.105) determine three unknowns N_1, N_2, N_3 for which $E_{(N)}$ is extremum. The trivial zero solution of the above system of equations is $N_1 = 0, N_2 = 0, N_3 = 0$ which is not compatible with the condition (1.100), and should be rejected. One type of non-trivial solution is

$$N_1 = 0, N_2 = 0, N_3 \neq 0,$$

so that, from equation (1.100), $N_3^2 = 1 \Rightarrow N_3 = \pm 1$. Equations (1.103) and (1.104) are automatically satisfied. Equation (1.105) gives $E_{(N)} = E_3$. Similarly, the solution $N_1 = 0, N_2 \neq 0, N_3 = 0$, of equations (1.103)-(1.105) gives $E_{(N)} = E_2$ and the solution $N_1 \neq 0, N_2 = 0, N_3 = 0$, of equations (1.103)-(1.105) gives $E_{(N)} = E_1$. Since $E_1 > E_2 > E_3$, maximum value of the external strain $E_{(N)} = E_1$ and minimum value of $E_{(N)} = E_3$. Therefore, extremum values of the external strain at a point are always principal stresses acting across planes for which shearing strain components vanish identically.

Theorem 1.4: (Extermum shearing strain): **The maximum shearing strain at any point of the continuum is equal to one-half the difference between algebraically, the largest and smallest principal strains and acts on the plane that bisects the angle between the directions corresponding to largest and smallest principal directions.**

Proof: Consider a point $P_0(X_i)$ in the undeformed (initial) state of a continuum body. Let us take the rectangular coordinate axes along the principal directions of strain at the point $P_0(X_i)$. Let E_1, E_2, E_3 be the principal strains at the point $P_0(X_i)$ and ordered so that $E_1 > E_2 > E_3$. If E_{ij} be the strain tensors at a point $P_0(X_i)$ then

$$E_{11} = E_1, E_{22} = E_2, E_{33} = E_3, E_{12} = E_{31} = E_{23} = 0. \quad (1.106)$$

If $E_{(N)}$ be the extensional strain at the point P acting in the direction $\vec{N} = (N_1, N_2, N_3)$, then

$$E_{(N)} = E_{ij} N_i N_j = E_{11} N_1^2 + E_{22} N_2^2 + E_{33} N_3^2 = E_1 N_1^2 + E_2 N_2^2 + E_3 N_3^2. \quad (1.107)$$

The direction cosines $\vec{N} = (N_1, N_2, N_3)$ satisfy the equation

$$N_i N_j = N_1^2 + N_2^2 + N_3^2 = 1. \quad (1.108)$$

Let the Cartesian coordinate frame of axes $P_0 X_1 X_2 X_3$ be rotated about $P_0(X_i)$ to obtain a new Cartesian coordinate frame $P_0 X_1' X_2' X_3'$. Let L_1, L_2, L_3 be the direction cosines of $P_0 X_1'$ axis; M_1, M_2, M_3 be the direction cosines of $P_0 X_2'$ axis; N_1, N_2, N_3 be the direction cosines of $P_0 X_3'$ axis respectively with respect to original coordinate frame $P_0 X_1 X_2 X_3$. Let E_{ij}' be components of strain at P_0 with respect to new axes $P_0 X_i'$. Now normal strain in the direction $P_0 X_i'$ is given by

$$E'_{11} = E_{ij}L_iL_j = E_1L_1^2 + E_2L_2^2 + E_3L_3^2. \quad (1.109)$$

Normal strain in the direction P_0X_2' is given by

$$E'_{22} = E_{ij}M_iM_j = E_1M_1^2 + E_2M_2^2 + E_3M_3^2. \quad (1.110)$$

Shearing strain in the plane $X_1'X_2'$ is given by

$$E'_{12} = E_{ij}L_iM_j = E_1L_1M_1 + E_2L_2M_2 + E_3L_3M_3. \quad (1.111)$$

The direction cosines of new axes satisfy the equations

$$L_1M_1 + L_2M_2 + L_3M_3 = 0; L_1^2 + L_2^2 + L_3^2 = 1; M_1^2 + M_2^2 + M_3^2 = 1. \quad (1.112)$$

It is obvious from equation (1.111) that if new axes are directed along the principal directions of strain so that

$$L_1 = 1, L_2 = 0, L_3 = 0; M_1 = 0, M_2 = 1, M_3 = 0; N_1 = 0, N_2 = 0, N_3 = 1,$$

then $E'_{12} = 0$. Thus the minimum value of the shearing strain is zero and is associated with principal directions of strain. To determine the directions associated with the maximum value of E'_{12} we have to maximize E'_{12} subject to constraint equations (1.112). For this, let us construct a Lagrangian function

$$F = E'_{12} - \lambda(L_1M_1 + L_2M_2 + L_3M_3) - \lambda_1(L_1^2 + L_2^2 + L_3^2 - 1) - \lambda_2(M_1^2 + M_2^2 + M_3^2 - 1)$$

$$= E_1L_1M_1 + E_2L_2M_2 + E_3L_3M_3 - \lambda(L_1M_1 + L_2M_2 + L_3M_3)$$

$$- \lambda_1(L_1^2 + L_2^2 + L_3^2 - 1) - \lambda_2(M_1^2 + M_2^2 + M_3^2 - 1)$$

$$= (E_1 - \lambda)L_1M_1 + (E_2 - \lambda)L_2M_2 + (E_3 - \lambda)L_3M_3$$

$$- \lambda_1(L_1^2 + L_2^2 + L_3^2 - 1) - \lambda_2(M_1^2 + M_2^2 + M_3^2 - 1)$$

where, $\lambda, \lambda_1, \lambda_2$ are the unknown Lagrangian parameters. For extreme value of E'_{12}, we have

$$\frac{\partial F}{\partial L_1} = 0 \Rightarrow (E_1 - \lambda)M_1 - 2\lambda_1 L_1 = 0;$$

$$\frac{\partial F}{\partial L_2} = 0 \Rightarrow (E_2 - \lambda)M_2 - 2\lambda_1 L_2 = 0$$

$$\frac{\partial F}{\partial L_3} = 0 \Rightarrow (E_3 - \lambda)M_3 - 2\lambda_1 L_3 = 0$$

$$\frac{\partial F}{\partial M_1} = 0 \Rightarrow (E_1 - \lambda)L_1 - 2\lambda_2 M_1 = 0;$$

1.11 Principal Strain

and
$$\frac{\partial F}{\partial M_2} = 0 \Rightarrow (E_2 - \lambda)L_2 - 2\lambda_2 M_2 = 0$$

$$\frac{\partial F}{\partial M_3} = 0 \Rightarrow (E_3 - \lambda)L_3 - 2\lambda_2 M_3 = 0$$

Therefore, from the first set of equations, we get

$$L_1\left[(E_1-\lambda)M_1 - 2\lambda_1 L_1\right] + L_2\left[(E_2-\lambda)M_2 - 2\lambda_1 L_2\right] + L_3\left[(E_3-\lambda)M_3 - 2\lambda_1 L_3\right] = 0$$

$$\Rightarrow (E_1-\lambda)L_1 M_1 + (E_2-\lambda)L_2 M_2 + (E_3-\lambda)L_3 M_3 = 2\lambda_1 (L_1^2 + L_2^2 + L_3^2)$$

$$\Rightarrow E_1 L_1 M_1 + E_2 L_2 M_2 + E_3 L_3 M_3 - \lambda(L_1 M_1 + L_2 M_2 + L_3 M_3) = 2\lambda_1$$

$$\Rightarrow \lambda_1 = \frac{1}{2}[E_1 L_1 M_1 + E_2 L_2 M_2 + E_3 L_3 M_3] = \frac{1}{2}E'_{12} \qquad (1.113)$$

From the second set of equations, we get

$$M_1\left[(E_1-\lambda)L_1 - 2\lambda_2 M_1\right] + M_2\left[(E_2-\lambda)L_2 - 2\lambda_2 M_2\right] + M_3\left[(E_3-\lambda)L_3 - 2\lambda_2 M_3\right] = 0$$

$$\Rightarrow (E_1-\lambda)L_1 M_1 + (E_2-\lambda)L_2 M_2 + (E_3-\lambda)L_3 M_3 = 2\lambda_2 (M_1^2 + M_2^2 + M_3^2)$$

$$\Rightarrow E_1 L_1 M_1 + E_2 L_2 M_2 + E_3 L_3 M_3 - \lambda(L_1 M_1 + L_2 M_2 + L_3 M_3) = 2\lambda_2$$

$$\Rightarrow \lambda_2 = \frac{1}{2}[E_1 L_1 M_1 + E_2 L_2 M_2 + E_3 L_3 M_3] = \frac{1}{2}E'_{22} \qquad (1.114)$$

Again from the first set of equations

$$M_1\left[(E_1-\lambda)M_1 - 2\lambda_1 L_1\right] + M_2\left[(E_2-\lambda)M_2 - 2\lambda_1 L_2\right] + M_3\left[(E_3-\lambda)M_3 - 2\lambda_1 L_3\right] = 0$$

$$\Rightarrow (E_1-\lambda)M_1^2 + (E_2-\lambda)M_2^2 + (E_3-\lambda)M_3^2 = 2\lambda_1(L_1 M_1 + L_2 M_2 + L_3 M_3) = 0$$

$$\Rightarrow \lambda(M_1^2 + M_2^2 + M_3^2) = E_1 M_1^2 + E_2 M_2^2 + E_3 M_3^2$$

$$\Rightarrow \lambda = E_1 M_1^2 + E_2 M_2^2 + E_3 M_3^2 = E'_{22} \qquad (1.115)$$

Also from the second set of equations, we get

$$L_1\left[(E_1-\lambda)L_1 - 2\lambda_2 M_1\right] + L_2\left[(E_2-\lambda)L_2 - 2\lambda_2 M_2\right] + L_3\left[(E_3-\lambda)L_3 - 2\lambda_2 M_3\right] = 0$$

$$\Rightarrow (E_1-\lambda)L_1^2 + (E_2-\lambda)L_2^2 + (E_3-\lambda)L_3^2 = 2\lambda_2(L_1 M_1 + L_2 M_2 + L_3 M_3) = 0$$

$$\Rightarrow \lambda(L_1^2 + L_2^2 + L_3^2) = E_1 L_1^2 + E_2 L_2^2 + E_3 L_3^2 \Rightarrow \lambda = E'_{11} \qquad (1.116)$$

Therefore, $\lambda = E'_{11} = E'_{22}; \lambda_1 = \lambda_2 = \frac{1}{2} E'_{12}$. Substituting the values of $\lambda, \lambda_1, \lambda_2$, in the following sets of equation,

$$(E_1 - \lambda)M_1 - 2\lambda_1 L_1 = 0; \ (E_2 - \lambda)M_2 - 2\lambda_1 L_2 = 0; \ (E_3 - \lambda)M_3 - 2\lambda_1 L_3 = 0$$

and

$$(E_1 - \lambda)L_1 - 2\lambda_2 M_1 = 0; \ (E_2 - \lambda)L_2 - 2\lambda_2 M_2 = 0; \ (E_3 - \lambda)L_3 - 2\lambda_2 M_3 = 0$$

We get

$$(E_1 - E'_{11})M_1 - E'_{12}L_1 = 0; \ (E_2 - E'_{11})M_2 - E'_{12}L_2 = 0; \ (E_3 - E'_{11})M_3 - E'_{12}L_3 = 0 \tag{1.117}$$

and

$$(E_1 - E'_{11})L_1 - E'_{12}M_1 = 0; \ (E_2 - E'_{11})L_2 - E'_{12}M_2 = 0; \ (E_3 - E'_{11})L_3 - E'_{12}M_3 = 0 \tag{1.118}$$

The above six equation given by (1.117) and 1.118) determine six unknowns $L_1, L_2, L_3; M_1, M_2, M_3$ for which shearing strain E'_{12} is maximum. The trivial zero solution of the above system of equations is $L_1 = 0, L_2 = 0, L_3 = 0; M_1 = 0, M_2 = 0, M_3 = 0$ which is not compatible with the condition (1.112), and should be rejected. One type of non-trivial solution is

$$L_1 = 0, L_2 = 0, L_3 \neq 0; \ M_1 = 0, M_2 = 0, M_3 \neq 0$$

so that,

$$(E_3 - E'_{11})M_3 - E'_{12}L_3 = 0; \ (E_3 - E'_{11})L_3 - E'_{12}M_3 = 0$$

$$\Rightarrow \frac{L_3}{M_3} = \frac{M_3}{L_3} \Rightarrow L_3^2 = M_3^2 \Rightarrow L_3 = \pm M_3 \tag{1.119}$$

which is also not compatible with the condition (1.112), and should be rejected. We may consider another type of solution by assuming

$$L_1 \neq 0, L_2 \neq 0, L_3 = 0; \ M_1 \neq 0, M_2 \neq 0, M_3 = 0. \tag{1.120}$$

Obviously, the equations $(E_3 - E'_{11})M_3 - E'_{12}L_3 = 0; \ (E_3 - E'_{11})L_3 - E'_{12}M_3 = 0$ are satisfied. Also

$$(E_1 - E'_{11})M_1 - E'_{12}L_1 = 0; (E_1 - E'_{11})L_1 - E'_{12}M_1 = 0$$

$$\Rightarrow \frac{L_1}{M_1} = \frac{M_1}{L_1} \Rightarrow L_1^2 = M_1^2 \Rightarrow L_1 = \pm M_1 \tag{1.121}$$

and

$$(E_2 - E'_{11})L_2 - E'_{12}M_2 = 0; \ (E_2 - E'_{11})L_2 - E'_{12}M_2 = 0$$

1.11 Principal Strain

$$\Rightarrow \frac{L_2}{M_2} = \frac{M_2}{L_2} \Rightarrow L_2^2 = M_2^2 \Rightarrow L_2 = \pm M_2 \quad (1.122)$$

Now L_1 and L_2 have both positive and negative values. If we take the signs of all L_i's and M_i's be the same, say positive, then

$$L_1 = M_1, L_2 = M_2 \text{ and equation (1.112) gives } M_1^2 + M_2^2 = 1.$$

This results in $M_1 = 0$ and $M_2 = 0$ which is not possible as they violate the equation (1.120). Hence we require that signs of all L_i's and M_i's be the same except one. We may take L_1, M_1, M_2 be positive and L_2 be negative. Thus we take $L_1 = M_1$, $L_2 = -M_2$ and so the equation $L_1 M_1 + L_2 M_2 + L_3 M_3 = 0$ gives $M_1^2 = M_2^2$. Then we have from equations (1.121) and (1.122),

$$L_1^2 = L_2^2 = M_1^2 = M_2^2.$$

Equation (1.112) gives

$$L_1^2 + L_2^2 + 0^2 = 1;\ M_1^2 + M_2^2 + 0^2 = 1$$

$$\Rightarrow 2L_1^2 = 1;\ 2M_1^2 = 1;\ L_3 = M_3 = 0$$

$$\Rightarrow L_1 = \frac{1}{\sqrt{2}} = M_1 = M_2;\ L_2 = -\frac{1}{\sqrt{2}};\ L_3 = M_3 = 0$$

Maximum value of the shearing strain E'_{12} in $X_1'X_2'$ plane

$$E_1 L_1 M_1 + E_2 L_2 M_2 + E_3 L_3 M_3 = \frac{1}{2}(E_1 - E_2)$$

and corresponds to the direction $\left(\frac{1}{\sqrt{2}}, \pm\frac{1}{\sqrt{2}}, 0\right)$ which bisects the angle between X_1 and X_2 axes. Similarly, the maximum value of the shearing strain E'_{23} in plane $X_2'X_3'$ and the maximum value of the shearing strain E'_{31} in $X_3'X_1'$ plane will be $\frac{1}{2}(E_1 - E_3)$ and $\frac{1}{2}(E_1 - E_3)$ respectively. Since $E_1 > E_2 > E_3$, maximum value of the shearing strain E'_{12} is given by $\frac{1}{2}(E_1 - E_3)$ and corresponding direction bisects the angle between X_1 and X_3 axes. Therefore, the maximum shearing strain at any point of the continuum is equal to one-half the difference between algebraically, the largest and smallest principal stresses and acts on the plane that bisects the angle between the directions corresponding to largest and smallest principal stresses.

Example 1.17: The strain tensor at a point is given by $(E_{ij}) = \begin{pmatrix} 1 & 3 & -2 \\ 3 & 1 & -2 \\ -2 & -2 & 6 \end{pmatrix}$.

Determine principal strains and principal directions of strain. Find the maximum value of normal strain and shearing strain.

Solution: The principal strains E_1, E_2, E_3 at the point P are the roots of the characteristic equation

$$\begin{vmatrix} 1-E & 3 & -2 \\ 3 & 1-E & -2 \\ -2 & -2 & 6-E \end{vmatrix} = 0$$

$\Rightarrow E^3 - 8E^2 - 4E + 32 = 0 \Rightarrow (E-8)(E+2)(E-2) = 0 \Rightarrow E = 8, \pm 2$.

Therefore, principal stains are $E_1 = 8$, $E_2 = 2$, $E_3 = -2$. The principal direction of strain $(N_1^{(1)}, N_2^{(1)}, N_3^{(1)})$ corresponding to $E_1 = 8$ is given by

$$\begin{pmatrix} -7 & 3 & -2 \\ 3 & -7 & -2 \\ -2 & -2 & -2 \end{pmatrix} \begin{pmatrix} N_1^{(1)} \\ N_2^{(1)} \\ N_3^{(1)} \end{pmatrix} = \begin{pmatrix} 0 \\ 0 \\ 0 \end{pmatrix} \quad \begin{pmatrix} N_1^{(1)} \\ N_2^{(1)} \\ N_3^{(1)} \end{pmatrix} = \begin{pmatrix} -\dfrac{1}{\sqrt{6}} \\ -\dfrac{1}{\sqrt{6}} \\ -\dfrac{2}{\sqrt{6}} \end{pmatrix}$$

Corresponding to $E_2 = 2$ and $E_3 = -2$ other two principal directions are given by

$$\begin{pmatrix} N_1^{(2)} \\ N_2^{(2)} \\ N_3^{(3)} \end{pmatrix} = \begin{pmatrix} \dfrac{1}{\sqrt{3}} \\ \dfrac{1}{\sqrt{3}} \\ -\dfrac{1}{\sqrt{3}} \end{pmatrix} \text{ and } \begin{pmatrix} N_1^{(3)} \\ N_2^{(3)} \\ N_3^{(3)} \end{pmatrix} = \begin{pmatrix} \dfrac{1}{\sqrt{2}} \\ -\dfrac{1}{\sqrt{2}} \\ 0 \end{pmatrix}.$$

Maximum value of normal strain = maximum principal strain = 8

Maximum value of shearing strain = $\dfrac{1}{2}(E_1 - E_3)$ since $E_1 > E_2 > E_3$

$$= \dfrac{1}{2}(8-(-2)) = 5.$$

1.12 Strain Equations of Compatibility for Infinitesimal Strains

It is sometimes necessary to *invert* the relations between strain and displacement—that is to say, given the strain field, to compute the displacements. In this section, we outline how this is done, for the special case of *infinitesimal deformations*. For infinitesimal motions the relation between strain and displacement is

$$E_{ij} = \frac{1}{2}\left[\frac{\partial u_i}{\partial X_j} + \frac{\partial u_j}{\partial X_i}\right] = \frac{1}{2}(u_{ijj} + u_{jii}) = \frac{1}{2}(u_{i,j} + u_{j,i}) \quad (1.123)$$

for determining three unknown displacement components u_i. Given the six strain components E_{ij}, (since $E_{ij} = E_{ji}$) we wish to determine the three displacement components u_i.

One way of finding the compatibility equations is the elimination of displacements u_i from six equations (1.123) by partial differentiation. From equation (1.123), we get

$$E_{kl} = \frac{1}{2}(u_{k,l} + u_{l,k})$$

Therefore,
$$E_{kl,ij} = \frac{1}{2}(u_{k,lji} + u_{l,kij}). \quad (1.124)$$

Again,
$$E_{ij,kl} = \frac{1}{2}(u_{i,jkl} + u_{j,ikl}). \quad (1.125)$$

Adding equations (1.124) and (1.125), we get

$$E_{ij,kl} + E_{kl,ij} = \frac{1}{2}(u_{i,jkl} + u_{j,ikl} + u_{k,lji} + u_{l,kij}). \quad (1.126)$$

Interchanging j and k, we get

$$E_{ik,jl} + E_{jl,ik} = \frac{1}{2}(u_{i,kjl} + u_{k,ijl} + u_{j,lik} + u_{l,jik}). \quad (1.127)$$

Substituting (1.127) in (1.126), we get

$$E_{ij,kl} + E_{kl,ij} - E_{ik,jl} - E_{jl,ik} = 0. \quad (1.128)$$

This is Saint-Venant's compatibility equation for strain components and is necessary condition for the existence of single-valued displacement. This is a system of $3^4 = 81$ number of equations of which only 6 are linearly independent others being dependent or identities because of symmetry in ij and kl and the order of differentiation can be changed. To be integrable, the strains must satisfy the compatibility conditions, which may be expressed as

$$\varepsilon_{ipm}\varepsilon_{jpn}\frac{\partial^2 E_{mn}}{\partial X_p \partial X_q} = 0$$

or, equivalently

$$\frac{\partial^2 E_{ij}}{\partial X_k \partial X_l} + \frac{\partial^2 E_{kl}}{\partial X_i \partial X_j} - \frac{\partial^2 E_{il}}{\partial X_j \partial X_k} - \frac{\partial^2 E_{jk}}{\partial X_i \partial X_l} = 0$$

or, once more equivalently, the 6 equations of compatibility written in details are given by

$$\frac{\partial^2 E_{11}}{\partial X_2 \partial X_3} = \frac{\partial}{\partial X_1}\left[-\frac{\partial E_{23}}{\partial X_1} + \frac{\partial E_{31}}{\partial X_2} + \frac{\partial E_{12}}{\partial X_3}\right]$$

$$\frac{\partial^2 E_{22}}{\partial X_3 \partial X_1} = \frac{\partial}{\partial X_2}\left[\frac{\partial E_{23}}{\partial X_1} - \frac{\partial E_{31}}{\partial X_2} + \frac{\partial E_{12}}{\partial X_3}\right]$$

$$\frac{\partial^2 E_{33}}{\partial X_1 \partial X_2} = \frac{\partial}{\partial X_3}\left[\frac{\partial E_{23}}{\partial X_1} + \frac{\partial E_{31}}{\partial X_2} - \frac{\partial E_{12}}{\partial X_3}\right]$$

$$2\frac{\partial^2 E_{23}}{\partial X_2 \partial X_3} = \frac{\partial^2 E_{22}}{\partial X_3^2} + \frac{\partial^2 E_{33}}{\partial X_2^2};$$

$$2\frac{\partial^2 E_{31}}{\partial X_3 \partial X_1} = \frac{\partial^2 E_{33}}{\partial X_1^2} + \frac{\partial^2 E_{11}}{\partial X_3^2};$$

$$2\frac{\partial^2 E_{12}}{\partial X_1 \partial X_2} = \frac{\partial^2 E_{11}}{\partial X_2^2} + \frac{\partial^2 E_{22}}{\partial X_1^2}$$

It is easy to show that all strain fields must satisfy these conditions - one simply need to substitute for the strains in terms of displacements and show that the appropriate equation is satisfied. For example,

$$\frac{\partial^2 E_{11}}{\partial X_2^2} + \frac{\partial^2 E_{22}}{\partial X_1^2} - 2\frac{\partial^2 E_{12}}{\partial X_1 \partial X_2} = \frac{\partial^4 u_1}{\partial X_1 \partial X_2^2} + \frac{\partial^4 u_2}{\partial X_1 \partial X_2^2} - 2\frac{\partial^2}{\partial X_2 \partial X_1^2}\frac{1}{2}\left[\frac{\partial u_1}{\partial X_2} + \frac{\partial u_2}{\partial X_1}\right] = 0$$

and similarly for the other expressions. Not that for planar problems for which $E_{13} = E_{13} = 0$ and $\frac{dE_{ij}}{dX_3} = 0$ all of these compatibility equations are satisfied trivially, with the exception of the first:

$$\frac{\partial^2 E_{11}}{\partial X_2^2} + \frac{\partial^2 E_{22}}{\partial X_1^2} - 2\frac{\partial^2 E_{12}}{\partial X_1 \partial X_2} = 0.$$

1.12 Strain Equations of Compatibility for Infinitesimal Strains

It can be shown that
  (i) If the strains do not satisfy the equations of compatibility, then a displacement vector can not be integrated from the strains.
  (ii) If the strains satisfy the compatibility equations, and the continuum simply connected (i.e. it contains no holes that go all the way through its thickness), then a displacement vector can be integrated from the strains.
  (iii) If the continuum is not simply connected, a displacement vector can be calculated, but it may not be single valued i.e. one may get different solutions depending on how the path of integration encircles the holes.

Example 1.18: Given $E_{11} = k(X_1^2 - X_2^2)$, $E_{12} = -kX_1X_2$, $E_{22} = kX_1X_2$; $E_{13} = E_{33} = E_{23} = 0$, **a possible state of strain. Find the displacement components.**

Solution: The given strain is compatible. For this two dimensional problem, we only need to determine the displacement components $u_1(X_1, X_2)$, and $u_2(X_1, X_2)$ such that

$$\frac{\partial u_1}{\partial X_1} = E_{11} = k(X_1^2 - X_2^2), \frac{\partial u_2}{\partial X_2} = E_{22} = kX_1X_2, \frac{\partial u_1}{\partial X_2} + \frac{\partial u_2}{\partial X_1} = 2E_{12} = -2kX_1X_2$$

We can integrate the first equation with respect to X_1 and the second equation with respect to X_2 to get

$$u_1 = k\left(\frac{1}{3}X_1^3 - X_1X_2^2\right) + f(X_2); \quad u_2 = \frac{k}{2}X_1X_2^2 + g(X_1)$$

where $f(X_2)$ and $g(X_1)$ are two functions of X_2 and X_1 respectively, which are yet to be determined. We can find these functions by substituting the formulas for $u_1(X_1, X_2)$, and $u_2(X_1, X_2)$ into the expression for shear strain

$$\frac{\partial u_1}{\partial X_2} + \frac{\partial u_2}{\partial X_1} = -2kX_1X_2 \Rightarrow g'(X_1) + f'(X_2) = -\frac{k}{2}X_2^2$$

$$\Rightarrow g'(X_1) = -\frac{k}{2}X_2^2 - f'(X_2).$$

Since left hand side is a function of X_1 only and right hand side is a function X_2 of alone, each side must be constant equal to say ω, this means that

$$g'(X_1) = \omega; \quad -\frac{k}{2}X_2^2 - f'(X_2) = \omega$$

where, ω is an arbitrary constant. We can now integrate these expressions to see that

$$g(X_1) = \omega X_1 + d; \quad f(X_2) = -\frac{k}{6}X_2^3 - \omega X_2 + c$$

where, c and d are two more arbitrary constants. Finally, the displacement field follows as

$$u_1(X_1, X_2) = k\left(\frac{1}{3}X_1^3 - X_1 X_2^2\right) - \frac{k}{6}X_2^3 - \omega X_2 + c; \quad u_2(X_1 X_2) = \frac{k}{2}X_1 X_2^2 + \omega X_1 + d$$

The three arbitrary constants ω, c and d can be seen to represent a small rigid rotation through angle ω about the X_3 axis, together with a displacement (c, d) parallel to X_1, X_2 axes, respectively. Omitting the linear part of the displacements (which corresponds to rigid motion), we have for pure deformation

$$u_1(X_1, X_2) = k\left(\frac{1}{3}X_1^3 - X_1 X_2^2\right) - \frac{k}{6}X_2^3; \quad u_2(X_1, X_2) = \frac{k}{2}X_1 X_2^2.$$

Example 1.19: A displacement field is given by $u_1 = 3x_1 x_2^2$, $u_2 = 2x_1 x_3$, $u_3 = x_3^2 - x_1 x_2$. Determine the strain components and check whether they satisfy the compatibility equations. Also find the strain components.

Solution: The displacement field is given by : $u_1 = 3x_1 x_2^2, u_2 = 2x_1 x_3, u_3 = x_3^2 - x_1 x_2$. Therefore, the strain components are given by

$$e_{11} = \frac{\partial u_1}{\partial x_1} = 3x_2^2, \quad e_{22} = \frac{\partial u_2}{\partial x_2} = 0, \quad e_{33} = \frac{\partial u_3}{\partial x_3} = 2x_3$$

$$e_{23} = \frac{1}{2}\left(\frac{\partial u_2}{\partial x_3} + \frac{\partial u_3}{\partial x_2}\right) = x_1; \quad e_{31} = \frac{1}{2}\left(\frac{\partial u_1}{\partial x_3} + \frac{\partial u_3}{\partial x_1}\right) = -x_2;$$

$$e_{12} = \frac{1}{2}\left(\frac{\partial u_1}{\partial x_2} + \frac{\partial u_2}{\partial x_1}\right) = 6x_1 x_2 + 2x_3.$$

Also,

$$\frac{\partial e_{11}}{\partial x_2} = 6x_2; \quad \frac{\partial^2 e_{11}}{\partial x_2^2} = 6; \quad \frac{\partial e_{22}}{\partial x_1} = 0; \quad \frac{\partial^2 e_{22}}{\partial x_1^2} = 0; \quad \frac{\partial^2 e_{12}}{\partial x_1 \partial x_2} = \frac{\partial}{\partial x_1}\left(\frac{\partial e_{12}}{\partial x_2}\right) = \frac{\partial}{\partial x_1}(6x_1) = 6$$

$$\frac{\partial^2 e_{11}}{\partial x_2^2} + \frac{\partial^2 e_{22}}{\partial x_1^2} = 6 + 0 = 6 = \frac{\partial^2 e_{12}}{\partial x_1 \partial x_2}$$

$$\frac{\partial e_{22}}{\partial x_3} = 0; \quad \frac{\partial^2 e_{22}}{\partial x_3^2} = 0; \quad \frac{\partial e_{33}}{\partial x_2} = 0; \quad \frac{\partial^2 e_{33}}{\partial x_2^2} = 0; \quad \frac{\partial^2 e_{23}}{\partial x_2 \partial x_3} = \frac{\partial}{\partial x_2}\left(\frac{\partial e_{23}}{\partial x_3}\right) = \frac{\partial}{\partial x_2}(0) = 0$$

1.12 Strain Equations of Compatibility for Infinitesimal Strains

$$\frac{\partial^2 e_{11}}{\partial x_2^2} + \frac{\partial^2 e_{22}}{\partial x_1^2} = 0 = \frac{\partial^2 e_{12}}{\partial x_1 \partial x_2}$$

$$\frac{\partial e_{33}}{\partial x_1} = 0; \quad \frac{\partial^2 e_{33}}{\partial x_1^2} = 0; \quad \frac{\partial e_{11}}{\partial x_3} = 0; \quad \frac{\partial^2 e_{11}}{\partial x_3^2} = 0; \quad \frac{\partial^2 e_{31}}{\partial x_3 \partial x_1} = \frac{\partial}{\partial x_3}\left(\frac{\partial e_{31}}{\partial x_1}\right) = \frac{\partial}{\partial x_3}(0) = 0$$

$$\frac{\partial^2 e_{33}}{\partial x_1^2} + \frac{\partial^2 e_{11}}{\partial x_3^2} = 0 = \frac{\partial^2 e_{31}}{\partial x_3 \partial x_1}$$

So, the first group of compatibility conditions are satisfied. Now

$$\frac{\partial}{\partial x_1}\left[-\frac{\partial e_{23}}{\partial x_1} + \frac{\partial e_{31}}{\partial x_2} + \frac{\partial e_{12}}{\partial x_3}\right] = \frac{\partial}{\partial x_1}[-1 - 1(-1) + 2] = 0$$

$$\frac{\partial}{\partial x_2}\left[-\frac{\partial e_{31}}{\partial x_2} + \frac{\partial e_{12}}{\partial x_3} + \frac{\partial e_{23}}{\partial x_1}\right] = \frac{\partial}{\partial x_2}[-1(-1) + 2 + 1] = 0$$

$$\frac{\partial}{\partial x_3}\left[-\frac{\partial e_{12}}{\partial x_3} + \frac{\partial e_{23}}{\partial x_1} + \frac{\partial e_{31}}{\partial x_2}\right] = \frac{\partial}{\partial x_3}[-2 + 1 + (-1)] = 0$$

Also,

$$\frac{\partial^2 e_{11}}{\partial x_2 \partial x_3} = \frac{\partial}{\partial x_2}\left(\frac{\partial e_{11}}{\partial x_3}\right) = \frac{\partial}{\partial x_3}(0) = 0 = \frac{\partial}{\partial x_1}\left[-\frac{\partial e_{23}}{\partial x_1} + \frac{\partial e_{31}}{\partial x_2} + \frac{\partial e_{12}}{\partial x_3}\right]$$

$$\frac{\partial^2 e_{22}}{\partial x_3 \partial x_1} = \frac{\partial}{\partial x_3}\left(\frac{\partial e_{22}}{\partial x_1}\right) = \frac{\partial}{\partial x_3}(0) = 0 = \frac{\partial}{\partial x_2}\left[-\frac{\partial e_{31}}{\partial x_2} + \frac{\partial e_{12}}{\partial x_3} + \frac{\partial e_{23}}{\partial x_1}\right]$$

$$\frac{\partial^2 e_{33}}{\partial x_1 \partial x_2} = \frac{\partial}{\partial x_1}\left(\frac{\partial e_{33}}{\partial x_2}\right) = \frac{\partial}{\partial x_1}(0) = 0 = \frac{\partial}{\partial x_3}\left[-\frac{\partial e_{12}}{\partial x_3} + \frac{\partial e_{23}}{\partial x_1} + \frac{\partial e_{31}}{\partial x_2}\right]$$

So, the second group of compatibility conditions are satisfied.

Example 1.20: The beam has a rectangular cross-section with height $2a$ and out-of-plane width b. The strain field in the beam is $E_{11} = 2kX_1X_2$, $E_{12} = (1+\gamma)k(a^2 - X_2^2)$, $E_{22} = -2\gamma k X_1 X_2$; a possible state of strain. Find the displacement components.

Solution: We first check that the strain is compatible. Now

$$\frac{\partial^2 E_{11}}{\partial X_2^2} + \frac{\partial^2 E_{22}}{\partial X_1^2} - 2\frac{\partial^2 E_{12}}{\partial X_1 \partial X_2} = 0 + 0 - 2.0 = 0$$

which is clearly satisfied. For this two dimensional problem, we only need to determine the displacement components $u_1(X_1, X_2)$, and $u_2(X_1, X_2)$ such that

$$\frac{\partial u_1}{\partial X_1} = E_{11} = 2kX_1X_2, \frac{\partial u_2}{\partial X_2} = E_{22} = -2\gamma kX_1X_2,$$

$$\frac{\partial u_1}{\partial X_2} + \frac{\partial u_2}{\partial X_1} = 2E_{12} = 2(1+\gamma)k(a^2 - X_2^2).$$

We can integrate the first equation with respect to X_1 and the second equation with respect to X_2 to get

$$u_1 = kX_1^2 X_2 + f(X_2); u_2 = -k\gamma X_1 X_2^2 + g(X_1)$$

where $f(X_2)$ and $g(X_1)$ are two functions of X_2 and X_1 respectively, which are yet to be determined. We can find these functions by substituting the formulas for $u_1(X_1, X_2)$, and $u_2(X_1, X_2)$ into the expression for shear strain

$$\frac{\partial u_1}{\partial X_2} + \frac{\partial u_2}{\partial X_1} = 2E_{12} = 2(1+\gamma)k(a^2 - X_2^2)$$

$$\Rightarrow \frac{1}{2}\left[kX_1^2 - k\gamma X_2^2 + f'(X_2) + g'(X_1)\right] = (1+\gamma)k(a^2 - X_2^2)$$

$$\Rightarrow g'(X_1) + kX_1^2 = -\left[f'(X_2) - k\gamma X_2^2 - 2(1+\gamma)k(a^2 - X_2^2)\right]$$

Since left hand side is a function of X_1 only and right hand side is a function of X_2 alone, each side must be constant and equal to, say, ω, this means that

$$g'(X_1) + kX_1^2 = \omega; f'(X_2) - k\gamma X_2^2 - 2(1+\gamma)k(a^2 - X_2^2) = -\omega$$

where, ω is an arbitrary constant. We can now integrate these expressions to see that

$$g(X_1) = \omega X_1 - \frac{k}{3}X_1^3 + d; \ f(X_2) = \left[2(1+\gamma)ka^2 - \omega\right]X_2 - \frac{k}{3}(2+\gamma)X_2^3 + c$$

where, c and d are two more arbitrary constants. Finally, the displacement field follows as

$$u_1 = kX_1^2 X_2 + \left[2(1+\gamma)ka^2 - \omega\right]X_2 - \frac{k}{3}(2+\gamma)X_2^3 + c;$$

$$u_2 = -k\gamma X_1 X_2^2 + \omega X_1 - \frac{k}{3}X_1^3 + d$$

The three arbitrary constants ω, c and d can be seen to represent a small rigid rotation through angle ω about the X_3 axis, together with a displacement (c,d) parallel to (X_1, X_2) axes, respectively.

1.13 Stretch and Rotation Tensors

In previous section, we noted that an arbitrary second-order tensor may be resolved by an additive decomposition into its symmetric and skew-symmetric parts. Here, we introduce a multiplicative decomposition known as the *polar decomposition* by which any non-singular tensor can be decomposed into a product of two component tensors. Recall that the deformation gradient is a non-singular (invertible) tensor.

Tensors **F** and **G** are the building blocks of various deformation measures used in nonlinear continuum mechanics. The whole subject is dominated by the *polar decomposition theorem*: which states that any particle deformation can be expressed as a pure deformation followed by a rotation, or by a rotation followed by a pure deformation. Because of this nonsingularity, mathematically the deformation gradient can have multiplicative decompositions:

$$F = R.U = V.R \qquad (1.129)$$

where R is the orthogonal *rotation tensor,* and U and V are symmetric, positive-definite tensors called the *right stretch tensor* and the *left stretch tensor,* respectively. Moreover, U and V have the same eigenvalues. If the deformation is a pure rotation, $U = V = I$.

The deformation gradient can be thought of as a mapping of the infinitesimal vector dX of the reference configuration into the infinitesimal vector dx of the current configuration. Note that the first decomposition in (1.129) replaces the linear transformation $dx = F.\, dX$ by two sequential transformations,

$$dx' = U.dX$$

followed by

$$dx = R.dx'$$

The tensor U has three positive eigenvalues, $U_{(1)}$, $U_{(1)}$, and $U_{(3)}$ called the principal stretches, and associated with each is a principal stretch direction, N_1, N_2, and N_3, respectively. These unit vectors form an orthogonal triad known as the *right principal directions of stretch.* By the transformation $dx' = U.dX$, line elements along the directions $N_{(i)}$, ($i = 1,2,3$) are stretched by an amount $U_{(i)}$, ($i = 1,2,3$), respectively, with no change in direction. This is followed by a rigid body rotation given by equation $dx=R.dx'$. The second decomposition of Eq. (1.129) reverses the sequence: first a rotation by R, then the stretching by V. In a general deformation, a rigid body translation may also be involved as well as the rotation and stretching described here.

As a preliminary to determining the rotation and stretch tensors, we note that an arbitrary tensor T is positive definite if $v.T.v > 0$ for all vectors $v \neq 0$. A necessary and sufficient condition for T to be positive definite is for all its eigenvalues to be positive. In this regard, consider the tensor $C = F^T.F$. Inasmuch as F is non-singular (det $F \neq 0$) and $F.v \neq 0$ if $v \neq 0$, so that $(F.v) \cdot (F.v)$ is a sum of squares and hence greater than zero. Thus

$$(F.v).(F.v) = v.F^T.F.v = v.C; \quad v > 0$$

and C is positive definite. Furthermore,

$$\left(F^T.F\right)^T = F^T.(F^T)^T = F^T.F$$

which proves that C is also symmetric. By the same arguments we may show that $C = \left(F^{-1}\right)^T \cdot (F^{-1})$ is also symmetric and positive definite. The combinations $C = F^T.F$ and $C = F.F^T$ are symmetric positive definite matrices that are called the right and left Cauchy-Green stretch tensors, respectively. Now let C be given in principal axes form by the matrix

$$\left[C^*_{AB}\right] = \begin{bmatrix} C_{(1)} & 0 & 0 \\ 0 & C_{(2)} & 0 \\ 0 & 0 & C_{(3)} \end{bmatrix}$$

and let $[a_{MN}]$ be the orthogonal transformation that relates the components of C^* to the components of C in any other set of axes through the equation expressed here in both indicial and matrix form

$$C^*_{AB} = a_{AQ}a_{BP}C_{QP} \text{ or } C^* = ACA^T$$

We define U as the square root of C— that is, $U = \sqrt{C}$, or $U.U = C$— and since the principal values $C_{(i)}$, (i = 1,2,3) are all positive we may write

$$\left[\sqrt{C^*_{AB}}\right] = \begin{bmatrix} \sqrt{C_{(1)}} & 0 & 0 \\ 0 & \sqrt{C_{(2)}} & 0 \\ 0 & 0 & \sqrt{C_{(3)}} \end{bmatrix} = \left[U^*_{AB}\right]$$

1.13 Stretch and Rotation Tensors

$$[(U^*_{AB})^{-1}] = \begin{bmatrix} 1/\sqrt{C_{(1)}} & 0 & 0 \\ 0 & 1/\sqrt{C_{(2)}} & 0 \\ 0 & 0 & 1/\sqrt{C_{(3)}} \end{bmatrix}$$

Note that both and are symmetric positive definite tensors given by

$$U_{AB} = a_{QA} a_{PB} U^*_{QP} \text{ or } U = A^T U^* A \text{ and}$$

$$U_{AB}^{-1} = a_{QA} a_{PB} (U^*_{QP})^{-1} \text{ or } U^{-1} = A^T (U^*)^{-1} A$$

respectively. Therefore, now, from the first decomposition in equation (1.129), $R = F \cdot U^{-1}$ so that

$$R^T \cdot R = (F \cdot U^{-1})^T \cdot (F \cdot U^{-1}) = (U^{-1})^T \cdot F^T \cdot F \cdot U^{-1}$$

$$= U^{-1} \cdot C \cdot U^{-1} = U^{-1} \cdot U \cdot U \cdot U^{-1} = I$$

which shows that R is proper orthogonal. The second decomposition may be confirmed by the similar development using $C^{-1} = F.F^T = V^2$.

Example 1.21: A homogeneous deformation is given by the equations $x_1 = 2X_1 - 2X_2$, $x_2 = X_1 + X_2$, and $x_3 = X_3$. Determine the polar decomposition $F = R \cdot U$ for this deformation.

Solution: The deformation gradient F has the matrix form

$$(F_{ki}) = \begin{pmatrix} \frac{\partial x_1}{\partial X_1} & \frac{\partial x_1}{\partial X_2} & \frac{\partial x_1}{\partial X_3} \\ \frac{\partial x_2}{\partial X_1} & \frac{\partial x_2}{\partial X_2} & \frac{\partial x_2}{\partial X_3} \\ \frac{\partial x_3}{\partial X_1} & \frac{\partial x_3}{\partial X_2} & \frac{\partial x_3}{\partial X_3} \end{pmatrix} = \begin{pmatrix} 2 & -2 & 0 \\ 1 & 1 & 0 \\ 0 & 0 & 1 \end{pmatrix}.$$

The Green's deformation tensor $C = F^T.F$ has the matrix form

$$(C_{ij}) = \begin{pmatrix} 2 & -2 & 0 \\ 1 & 1 & 0 \\ 0 & 0 & 1 \end{pmatrix}^T \begin{pmatrix} 2 & -2 & 0 \\ 1 & 1 & 0 \\ 0 & 0 & 1 \end{pmatrix} = \begin{pmatrix} 5 & -3 & 0 \\ -3 & 5 & 0 \\ 0 & 0 & 1 \end{pmatrix}$$

The characteristic equation of the Green's deformation matrix $C = (C_{ij})$ is given by

$$\begin{vmatrix} C_{11}-\lambda & C_{12} & C_{13} \\ C_{21} & C_{22}-\lambda & C_{23} \\ C_{31} & C_{32} & C_{33}-\lambda \end{vmatrix} = \begin{vmatrix} 5-\lambda & -3 & 0 \\ -3 & 5-\lambda & 0 \\ 0 & 0 & 1-\lambda \end{vmatrix} = 0$$

$$\Rightarrow (5-\lambda)(5-\lambda)(1-\lambda)-9(1-\lambda)=0 \Rightarrow (1-\lambda)(8-\lambda)(2-\lambda)=0$$
$$\Rightarrow \lambda = 8,2,1.$$

For $\lambda = \lambda_1 = 8$, the system of equations become

$$\begin{pmatrix} 5-8 & -3 & 0 \\ -3 & 5-8 & 0 \\ 0 & 0 & 1-8 \end{pmatrix} \begin{pmatrix} N_1^{(1)} \\ N_2^{(1)} \\ N_3^{(1)} \end{pmatrix} = \begin{pmatrix} 0 \\ 0 \\ 0 \end{pmatrix} \Rightarrow \begin{pmatrix} N_1^{(1)} \\ N_2^{(1)} \\ N_3^{(1)} \end{pmatrix} = \frac{1}{\sqrt{2}}\begin{pmatrix} 1 \\ 1 \\ 0 \end{pmatrix}$$

For $\lambda = \lambda_2 = 2$, the system of equations become

$$\begin{pmatrix} 5-2 & -3 & 0 \\ -3 & 5-2 & 0 \\ 0 & 0 & 1-2 \end{pmatrix} \begin{pmatrix} N_1^{(2)} \\ N_2^{(2)} \\ N_3^{(2)} \end{pmatrix} = \begin{pmatrix} 0 \\ 0 \\ 0 \end{pmatrix} \Rightarrow \begin{pmatrix} N_1^{(2)} \\ N_2^{(2)} \\ N_3^{(2)} \end{pmatrix} = \frac{1}{\sqrt{2}}\begin{pmatrix} -1 \\ 1 \\ 0 \end{pmatrix}$$

For $\lambda = \lambda_3 = 1$, the above system of equations become

$$\begin{pmatrix} 5-1 & -3 & 0 \\ -3 & 5-1 & 0 \\ 0 & 0 & 1-1 \end{pmatrix} \begin{pmatrix} N_1^{(3)} \\ N_2^{(3)} \\ N_3^{(3)} \end{pmatrix} = \begin{pmatrix} 0 \\ 0 \\ 0 \end{pmatrix} \Rightarrow \begin{pmatrix} N_1^{(3)} \\ N_2^{(3)} \\ N_3^{(3)} \end{pmatrix} = \begin{pmatrix} 0 \\ 0 \\ 1 \end{pmatrix}$$

In the principal axes form, the matrix becomes

$$[C_{AB}^*] = \begin{pmatrix} \frac{1}{\sqrt{2}} & -\frac{1}{\sqrt{2}} & 0 \\ \frac{1}{\sqrt{2}} & \frac{1}{\sqrt{2}} & 0 \\ 0 & 0 & 1 \end{pmatrix}^T \begin{pmatrix} 5 & -3 & 0 \\ -3 & 5 & 0 \\ 0 & 0 & 1 \end{pmatrix} \begin{pmatrix} \frac{1}{\sqrt{2}} & -\frac{1}{\sqrt{2}} & 0 \\ \frac{1}{\sqrt{2}} & \frac{1}{\sqrt{2}} & 0 \\ 0 & 0 & 1 \end{pmatrix}$$

$$= \begin{pmatrix} 8 & 0 & 0 \\ 0 & 2 & 0 \\ 0 & 0 & 1 \end{pmatrix} = \begin{pmatrix} C_{(1)} & 0 & 0 \\ 0 & C_{(2)} & 0 \\ 0 & 0 & C_{(3)} \end{pmatrix}$$

1.13 Stretch and Rotation Tensors

with an orthogonal transformation matrix found to be

$$[a_{MN}] = \begin{pmatrix} \dfrac{1}{\sqrt{2}} & -\dfrac{1}{\sqrt{2}} & 0 \\ \dfrac{1}{\sqrt{2}} & \dfrac{1}{\sqrt{2}} & 0 \\ 0 & 0 & 1 \end{pmatrix}$$

We define U as the square root of C — that is, $U = \sqrt{C}$, or $U.U = C$ — and since the principal values $C_{(i)}$, (i = 1,2,3) are all positive we may write

$$\left[\sqrt{C^*_{AB}}\right] = \begin{bmatrix} \sqrt{C_{(1)}} & 0 & 0 \\ 0 & \sqrt{C_{(2)}} & 0 \\ 0 & 0 & \sqrt{C_{(3)}} \end{bmatrix} = \begin{bmatrix} \sqrt{8} & 0 & 0 \\ 0 & \sqrt{2} & 0 \\ 0 & 0 & \sqrt{1} \end{bmatrix} = \begin{bmatrix} 2\sqrt{2} & 0 & 0 \\ 0 & \sqrt{2} & 0 \\ 0 & 0 & 1 \end{bmatrix} = [U^*_{AB}]$$

and

$$\left[(U^*_{AB})^{-1}\right] = \begin{bmatrix} 1/\sqrt{C_{(1)}} & 0 & 0 \\ 0 & 1/\sqrt{C_{(2)}} & 0 \\ 0 & 0 & 1/\sqrt{C_{(3)}} \end{bmatrix} = \begin{bmatrix} 1/2\sqrt{2} & 0 & 0 \\ 0 & 1/\sqrt{2} & 0 \\ 0 & 0 & 1 \end{bmatrix}$$

$$= \dfrac{1}{2\sqrt{2}} \begin{bmatrix} 1 & 0 & 0 \\ 0 & 2 & 0 \\ 0 & 0 & 2\sqrt{2} \end{bmatrix}$$

By use of the transformation equations $U = A^T U^* A$ and $U^{-1} = A^T (U^*)^{-1} A$ we get

$$(U_{ij}) = \begin{pmatrix} \dfrac{1}{\sqrt{2}} & -\dfrac{1}{\sqrt{2}} & 0 \\ \dfrac{1}{\sqrt{2}} & \dfrac{1}{\sqrt{2}} & 0 \\ 0 & 0 & 1 \end{pmatrix}^T \begin{pmatrix} 2\sqrt{2} & 0 & 0 \\ 0 & \sqrt{2} & 0 \\ 0 & 0 & 1 \end{pmatrix} \begin{pmatrix} \dfrac{1}{\sqrt{2}} & -\dfrac{1}{\sqrt{2}} & 0 \\ \dfrac{1}{\sqrt{2}} & \dfrac{1}{\sqrt{2}} & 0 \\ 0 & 0 & 1 \end{pmatrix} = \begin{pmatrix} \dfrac{3}{\sqrt{2}} & -\dfrac{1}{\sqrt{2}} & 0 \\ -\dfrac{1}{\sqrt{2}} & \dfrac{3}{\sqrt{2}} & 0 \\ 0 & 0 & 1 \end{pmatrix}$$

$$(U_{ij}^{-1}) = \begin{pmatrix} \frac{1}{\sqrt{2}} & -\frac{1}{\sqrt{2}} & 0 \\ \frac{1}{\sqrt{2}} & \frac{1}{\sqrt{2}} & 0 \\ 0 & 0 & 1 \end{pmatrix}^T \begin{pmatrix} 1/2\sqrt{2} & 0 & 0 \\ 0 & 1/\sqrt{2} & 0 \\ 0 & 0 & 1 \end{pmatrix} \begin{pmatrix} \frac{1}{\sqrt{2}} & -\frac{1}{\sqrt{2}} & 0 \\ \frac{1}{\sqrt{2}} & \frac{1}{\sqrt{2}} & 0 \\ 0 & 0 & 1 \end{pmatrix}$$

$$= \frac{1}{4\sqrt{2}} \begin{pmatrix} 3 & 1 & 0 \\ 1 & 3 & 0 \\ 0 & 0 & 4\sqrt{2} \end{pmatrix}$$

Finally, the proper orthogonal rotation tensor is given by $R = F \cdot U^{-1}$, so that

$$R = F \cdot U^{-1} = \begin{pmatrix} 2 & -2 & 0 \\ 1 & 1 & 0 \\ 0 & 0 & 1 \end{pmatrix} \frac{1}{4\sqrt{2}} \begin{pmatrix} 3 & 1 & 0 \\ 1 & 3 & 0 \\ 0 & 0 & 4\sqrt{2} \end{pmatrix} = \begin{pmatrix} \frac{1}{\sqrt{2}} & -\frac{1}{\sqrt{2}} & 0 \\ \frac{1}{\sqrt{2}} & \frac{1}{\sqrt{2}} & 0 \\ 0 & 0 & 1 \end{pmatrix}$$

It is easily confirmed using these results that $F = R \cdot U$ and that $R \cdot R^T = I$.

1.14 Plane Strain and Mohr's Circles for Strain

When one and only one of the principal strains at a point in a continuum is zero, a state of plane strain is said to exist at that point. In the Eulerian description (the Lagrangian description follows exactly the same pattern), it is taken as the direction of the zero principal strain, a state of plane strain parallel to the $x_1 x_2$ plane exists and the linear strain tensor is given by

$$(e_{ij}) = \begin{pmatrix} e_{11} & e_{12} & 0 \\ e_{21} & e_{22} & 0 \\ 0 & 0 & 0 \end{pmatrix}$$

When x_1 and x_2 are also principal directions, the strain tensor has the form

$$e_{ij}^* = \begin{pmatrix} e_1 & 0 & 0 \\ 0 & e_2 & 0 \\ 0 & 0 & 0 \end{pmatrix}$$

Sometimes plane strain is referred to as plane deformation since the deformation field is identical in all planes perpendicular to the direction of the zero principal strain. For plane strain perpendicular to the x_3 axis, the displacement vector

1.14 Plane Strain and Mohr's Circles for Strain

may be taken as a function of x_1 and x_2 only. The appropriate displacement components for this case of plane strain are designated by

$$u_1 = u_1(x_1, x_2); \quad u_2 = u_2(x_1, x_2); \quad u_3 = k,$$

where, k is a constant, taken as zero. Inserting these expressions into the definition of e_{ij} given by $e_{ij} = \frac{1}{2}\left[\frac{\partial u_i}{\partial x_j} + \frac{\partial u_j}{\partial x_i}\right]$ produces the plane strain tensor in the same form shown in (e_{ij}^*). A graphical description of the state of strain at a point is provided by the Mohr's circles for strain. For this purpose the strain tensor is often displayed in the form

$$(e_{ij}) = \begin{pmatrix} e_{11} & \frac{1}{2}\gamma_{12} & \frac{1}{2}\gamma_{13} \\ \frac{1}{2}\gamma_{21} & e_{22} & \frac{1}{2}\gamma_{23} \\ \frac{1}{2}\gamma_{31} & \frac{1}{2}\gamma_{32} & e_{33} \end{pmatrix}$$

Here the γ_{ij} (with $i \neq j$) are the so-called "engineering" shear strain components, which are twice the tensorial shear strain components. The state of strain at an unloaded point on the bounding surface of a continuum body is locally plane strain. Frequently, in experimental studies involving strain measurements at such a surface point, Mohr's strain circles are useful for reporting the observed data. Usually three normal strains are measured at the given point by means of a strain rosette, and the Mohr's circles diagram constructed from these. Corresponding to the plane strain Mohr's circles, a typical case of plane strain diagram is shown in Fig. (1.13).

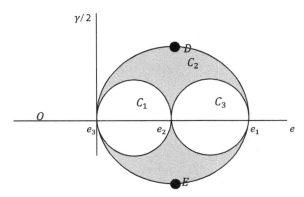

Fig. 1.13: Mohr's circle for three-dimensional state of strain.

The principal normal strains are labeled as such in the diagram, and the maximum shear strain values are represented by points D and E. For a state of plane strain parallel to the $x_2 x_3$ axes, the expressions for the normal strain e'_{22} and the shear strain e'_{23} when the primed and unprimed axes are oriented is shown in Fig. 1.14.

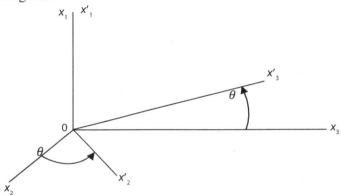

Fig. 1.14: Representative rotation of axes for plane strain.

Example 1.22: Construct the Mohr's circles for the case of plane strain
$$(e_{ij}) = \begin{pmatrix} 0 & 0 & 0 \\ 0 & 5 & \sqrt{3} \\ 0 & \sqrt{3} & 3 \end{pmatrix}$$
and determine the maximum shear strain. Verify the result analytically.

Solution: With the given state of strain referred to the x_i axes, the points $B(e_{22} = 5, e_{23} = \sqrt{3})$ and D are established as the diameter of the larger inner circle in Fig. 1.15.

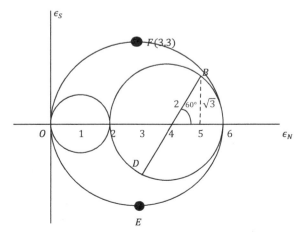

Fig 1.15: Mohr's circle

1.14 Plane Strain and Mohr's Circles for Strain

Since $e_{(1)} = 0$ is a principal value for plane strain, the other circles are drawn as shown. A rotation of 30° about the x_i axis (equivalent to 60° in the Mohr's diagram) results in the principal strain axes with the principal strain tensor e_{ij}^* given by

$$\begin{pmatrix} 1 & 0 & 0 \\ 0 & \sqrt{3}/2 & 1/2 \\ 0 & -1/2 & \sqrt{3}/2 \end{pmatrix} \begin{pmatrix} 0 & 0 & 0 \\ 0 & 5 & \sqrt{3} \\ 0 & \sqrt{3} & 3 \end{pmatrix} \begin{pmatrix} 1 & 0 & 0 \\ 0 & \sqrt{3}/2 & -1/2 \\ 0 & 1/2 & \sqrt{3}/2 \end{pmatrix} = \begin{pmatrix} 0 & 0 & 0 \\ 0 & 6 & 0 \\ 0 & 0 & 2 \end{pmatrix}$$

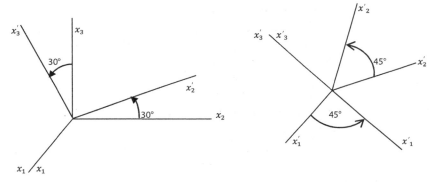

Next a rotation of 45° about the x_3^* axis (90° in the Mohr diagram) results in the x_i' axes and the associated strain tensor e_{ij}' given by

$$\begin{pmatrix} 1/\sqrt{2} & 1/\sqrt{2} & 0 \\ -1/\sqrt{2} & 1/\sqrt{2} & 0 \\ 0 & 0 & 1 \end{pmatrix} \begin{pmatrix} 0 & 0 & 0 \\ 0 & 6 & 0 \\ 0 & 0 & 2 \end{pmatrix} \begin{pmatrix} 1/\sqrt{2} & -1/\sqrt{2} & 0 \\ 1/\sqrt{2} & 1/\sqrt{2} & 0 \\ 0 & 0 & 1 \end{pmatrix} = \begin{pmatrix} 3 & 3 & 0 \\ 3 & 3 & 0 \\ 0 & 0 & 2 \end{pmatrix}$$

the first two rows of which represent the state of strain specified by point in Fig. 1.15. Note that a rotation of 45° about x_3^* would correspond to point E in Fig. 1.15.

Exercise

Theoretical

1. Explain the difference between Eulerian and Lagrangian description of deformation.
2. Deduce the expression for Lagrangian strain components in the form

$$r_{ij} = \frac{1}{2}\left[\frac{\partial u_j}{\partial x_i} + \frac{\partial u_i}{\partial x_j} + \frac{\partial u_k}{\partial x_j}\cdot\frac{\partial u_k}{\partial x_i}\right]$$

where, u_i's are the components of displacement vector at a point of the medium.

3. Deduce the expression for Eulerian strain components in the form

$$\eta_{ij} = \frac{1}{2}\left[\frac{\partial u_j}{\partial x_i} + \frac{\partial u_i}{\partial x_j} - \frac{\partial u_k}{\partial x_j}\cdot\frac{\partial u_k}{\partial x_i}\right]$$

where, u_i's are the components of displacement vector at a point of the medium.

4. Define strain quadratic of Cauchy at a point of a continuous medium and prove the following as its properties:
   (*i*) The extensional strain of a line element through the centre of a strain quadratic in the direction of any central radius vector is equal to the inverse of the square of the radius vector.
   (*ii*) Prove that, the displacement of a material point at any point on the strain quadratic relative to that at the centre is directed along the normal to the surface of the quadratic at that point.

5. Give geometrical interpretation of the components of infinitesimal strains at a point of a continuous medium. Write the expressions for the three independent strain invariants.

6. Show that the expressions in Eulerian and Lagrangian description of deformation of a continuous medium are identical in infinitesimal theory.

7. Give the geometrical interpretation of the normal and shearing strain components for infinitesimal deformation.

8. Give the geometrical interpretation of the first strain invariant for infinitesimal deformation.

9. Obtain the relation between strain vector and strain tensor.

10. Define principal directions of strain and principal strains at a point of a continuous medium.

11. Prove that the set of quantities representing the infinitesimal strain components in a continuous medium form a Cartesian tensor of rank two.

12. Prove the following regarding extremum values of normal strain:
    (*i*) The extremum values of normal strain or extensional strain at a point of a continuum are principal strains.
    (*ii*) The maximum shearing strain at any point of the continuum is equal to one-half the difference between algebraically, the largest and smallest principal strains and acts on the plane that bisects the angle between the directions corresponding to largest and smallest principal directions.

1.14 Plane Strain and Mohr's Circles for Strain

13. At a point of a body, find the directions for which the shear strains attain the stationary values.
14. At a point of a body, find the directions for which the area change is stationary.
15. Express the ratio of the deformed area vector to the undeformed area vector in terms of Eulerian strain measures.
16. Prove that at each point of a deformable body there exists at least one direction which is preserved upon the deformation of the body.
17. For the case of two-dimensional state of an infinitesimal strain field, obtain the explicit forms of the compatibility conditions by elimination of displacement components.

Numerical

1. If the motion of a continuous medium is given by
$$x_1 = X_1 e^t - X_3(e^t - 1); x_2 = X_2 e^{-t} + X_3(1 - e^{-t}); x_3 = X_3$$
determine the displacement field in both material and spatial descriptions.

   Answer:
$$u_1 = (X_1 - X_3)(e^t - 1) = (x_1 - x_3)(e^t - 1);$$
$$u_2 = (X_2 - X_3)(e^{-t} - 1) = (x_2 - x_3)(1 - e^t);$$

2. The displacement field in a body is given by
$$x_1 = (2AX_1 + B)^{1/2}; x_2 = CX_2; x_3 = DX_3$$
where (X_1, X_2, X_3) and (x_1, x_2, x_3) are rectangular coordinates, and A, B, C and D are constants.
   (*i*) Determine the deformation tensors.
   (*ii*) Determine the lagrangian strain tensor and find the change in the square of arc length.
   (*iii*) The infinitesimal strains and rotations.
   If the coordinates (x_1, x_2, x_3) are taken to be cylindrical coordinates having the same origin as X_k. Find a geometrical meaning for the displacement field.

3. The cylindrical coordinates X_k are related to rectangular coordinates by
$$x_1 = X_1 \cos X_2, x_2 = X_1 \sin X_2; x_3 = X_3$$
   (*i*) Find the expression of the Green deformation tensor in terms of the components of the green deformation tensor in rectangular coordinates.

(ii) Express the components of a vector in cylindrical coordinates in terms of the components of the same vector in rectangular coordinates.

4. Given the displacement field $x_1 = X_1 + 2X_3$, $x_2 = X_2 - 2X_3$, $x_3 = X_3 - 2X_1 + 2X_2$, determine the Lagrangian and Eulerian finite strain tensors. Determine the principal-axes form of the two tensors.

Ans: $\begin{pmatrix} 2 & -2 & 0 \\ -2 & 2 & 0 \\ 0 & 0 & 4 \end{pmatrix}$, $\dfrac{1}{9}\begin{pmatrix} 2 & -2 & 0 \\ -2 & 2 & 0 \\ 0 & 0 & 4 \end{pmatrix}$

5. Determine principal strains and corresponding direction from the strain tensors

(i) $(E_{ij}) = \begin{pmatrix} 1 & 1 & 3 \\ 1 & 5 & 1 \\ 3 & 1 & 1 \end{pmatrix}$.

Ans: 6, 3, −2; $\dfrac{1}{\sqrt{6}}(1,2,1), \dfrac{1}{\sqrt{3}}(1,1,1), \dfrac{1}{\sqrt{2}}(1,0,-1)$

(ii) $(E_{ij}) = \begin{pmatrix} 5 & 2 & 2 \\ 2 & 2 & -4 \\ 2 & -4 & 2 \end{pmatrix}$.

Ans: 6, 6, −3; $\dfrac{1}{3}(2,2,-1), \dfrac{1}{3}(2,-1,2), \dfrac{1}{3}(-1,2,2)$

(iii) $(E_{ij}) = \begin{pmatrix} 5 & -1 & -1 \\ -1 & 4 & 0 \\ -1 & 0 & 4 \end{pmatrix}$.

Ans: 6, 6, −3; $\dfrac{1}{3}(2,2,-1), \dfrac{1}{3}(2,-1,2), \dfrac{1}{3}(-1,2,2)$

and obtain strain invariants from them. Calculate the strain invariants from the strain tensors. Show the equivalence of strain invariants.

6. The strain tensor at a point is given by $(E_{ij}) = \begin{pmatrix} 1 & -3 & \sqrt{2} \\ -3 & 1 & -\sqrt{2} \\ \sqrt{2} & -\sqrt{2} & 4 \end{pmatrix}$. Determine

(i) the extension of a line element in the direction of $\left(\dfrac{1}{2}, -\dfrac{1}{2}, \dfrac{1}{\sqrt{2}}\right)$.

1.14 Plane Strain and Mohr's Circles for Strain

(ii) Shear between the directions $\left(\frac{1}{2}, -\frac{1}{2}, \frac{1}{\sqrt{2}}\right)$ and $\left(-\frac{1}{2}, \frac{1}{2}, \frac{1}{\sqrt{2}}\right)$.

(iii) Principal strains, maximum normal strain, maximum shearing strain and strain invariants.

7. Let the deformation field of a continuum be given by the equations $x_2 = X_2 - \alpha X_1^2$, $x_1 = X_1 + \alpha X_2^2$, and $x_3 = X_3$, where α is a constant, determine the Lagrangian finite strain tensor E, and from it, assuming α is very small, deduce the infinitesimal strain tensor e. Verify this by calculating the displacement field and using the definition $e_{ij} = \frac{1}{2}(u_{i,j} + u_{j,i})$ for the infinitesimal theory.

8. A deformation field is given by the equations $x_1 = X_1/(X_1^2 + X_2^2)$, $x_2 = X_2/(X_1^2 + X_2^2)$, and $x_3 = X_3$. Determine the deformation tensor C with its principal values.

   Ans: $C_{(1)} = C_{(2)} = (X_1^2 + X_2^2)^{-2}, C_{(3)} = 1$

9. For the deformation field given by the equations $x_1 = X_1 + \alpha X_2$, $x_2 = X_2 - \alpha X_1$, and $x_3 = X_3$, where α is a constant, determine the matrix form of the Lagrangian finite strain tensor and the infinitesimal strain tensor and show that the circle of particles $X_1^2 + X_2^2 = 1$ deforms into the circle $x_1^2 + x_2^2 = 1 + \alpha^2$.

   Ans: $(E_{ij}) = \frac{1}{2}\begin{pmatrix} \alpha^2 & 0 & 0 \\ 0 & \alpha^2 & 0 \\ 0 & 0 & 0 \end{pmatrix}$ and $(e_{ij}) = \frac{1}{2}(1+\alpha^2)^{-2}\begin{pmatrix} -\alpha^2 & 0 & 0 \\ 0 & -\alpha^2 & 0 \\ 0 & 0 & 0 \end{pmatrix}$.

10. The uniform twist of a circular cylinder along its length may be represented by the deformation
    $X_1 = r\cos(\theta - Kz), X_2 = r\sin(\theta - Kz), X_3 = z,$
    where, K is a constant, (X_1, X_2, X_3) are rectangular coordinates of a point before deformation, and (r, θ, z) are cylindrical coordinates of the same point after deformation.
    (i) Find the deformation tensor C.
    (ii) Determine the invariants.
    (iii) Find the principal axes of C and principal strains.
    (iv) Draw the Cauchy strain ellipsoid.
    Show that the twist of a circular cylinder described is isochoric. Find the area changes and the magnitude of the element of the surface.

11. The displacement field of a body is described by

$$u_1 = A\frac{x_1 x_3}{r^3}, u_2 = A\frac{x_2 x_3}{r^3}, u_3 = A\left[\frac{x_3^2}{r^3} + \frac{\lambda + 3\mu}{\lambda + \mu}\frac{1}{r}\right]; r^2 = x_1^2 + x_2^2 + x_3^2$$

where, A, λ, μ are constants.

(i) Determine the infinitesimal strains and infinitesimal rotations.

(ii) Sketch the deformed shape of the spherical cavity $r = r_0$

(iii) Find the differences between finite strain components and the corresponding infinitesimal strains.

For this displacement field, determine the principal strains and principal axes.

12. The rectangular frame of reference x_k is rotated about the x_3 axis by an angle measured from the x_1 axis. Find the relationships between the strains and rotations in the two reference frames.

13. A parabolic cylinder coordinates X_k are related to the rectangular coordinates x_k by

$$x_1 = k(X_2 - X_1), x_2 = 2k\sqrt{X_1 X_2}, x_3 = X_3$$

(i) Sketch the coordinate surfaces and lines

(ii) Express the deformation tensor referred to parabolic coordinates in terms of the deformation tensors, referred to rectangular coordinates.

14. In a hollow circular cylinder whose axis is the axis, a possible two dimensional infinitesimal strain field is given by

$$e_{11} = \frac{A}{2\mu}\frac{x}{r^2}\left[1 - \frac{\lambda + \mu}{\lambda + 2\mu}\frac{2y^2}{r^2}\right], e_{22} = \frac{A}{2(\lambda + 2\mu)}\frac{x}{r^2}\left[1 - \frac{\lambda + \mu}{\mu}\frac{x^2 - y^2}{r^2}\right],$$

$$e_{12} = \frac{A(\lambda + \mu)}{\mu(\lambda + 2\mu)}\frac{yx^2}{r^4}$$

where, A, λ and μ are constants and (x, y, z) are rectangular coordinates.

(i) Does the stain field satisfy the compatibility conditions?

(ii) Determine the general displacement field corresponding to these strain fields, subject to compatibility conditions?

(iii) Is the displacement field so obtained single-valued and continuous? If not, under what condition can it be made single-valued?

15. The strain components are given by $E_{11} = E_{22} = E_{33} = \alpha F(X_1, X_2, X_3)$; $E_{12} = E_{13} = E_{31} = 0$ where, α is a constant. Show that in order to be compatible, F must be a linear function of (X_1, X_2, X_3).

16. The strain components are given by

1.14 Plane Strain and Mohr's Circles for Strain

$$E_{11} = \frac{1}{\alpha}f(X_2, X_3), \ E_{22} = E_{33} = -\frac{\nu}{\alpha}f(X_2, X_3), \ E_{12} = E_{23} = E_{31} = 0$$

Show that in order to be compatible, $f(X_2, X_3)$ must be a linear.

17. The Lagrangian description of a continuum motion is given by

$$x_1 = X_1 e^{-t} + X_3(e^{-t} - 1); x_2 = X_2 e^t - X_3(1 - e^{-t}); x_3 = X_3 e^t$$

Show that these equations are invertible and determine the Eulerian description of the motion.

Answer: $X_1 = x_1 e^t - x_3(e^t - 1); X_2 = x_2 e^{-t} + x_3(e^{-2t} - e^{-3t}); x_3 = X_3 e^{-t}$

18. For the motion given by the equations

$$x_1 = X_1 \cos\omega t + X_2 \sin\omega t; x_2 = -X_1 \sin\omega t + X_2 \cos\omega t; x_3 = (1 + kt)X_3$$

where, ω and k are constants, determine the displacement field in Eulerian form.

Answer: $u_1 = x_1(1 - \cos\omega t) + x_2 \sin\omega t$; $u_2 = -x_1 \sin\omega t + x_2(1 - \cos\omega t)$; $u_3 = kx_3 t / (1 + kt)$.

19. The state of strain throughout a continuum is specified by

$$\begin{pmatrix} x_1^2 & x_2^2 & x_1 x_3 \\ x_2^2 & x_3 & x_3^2 \\ x_1 x_3 & x_3^2 & 5 \end{pmatrix}.$$ Are the compatibility equations for strain satisfied?

20. A displacement field is given by $u_1 = 3x_1 x_2^2$, $u_2 = 2x_3 x_1$, $u_3 = x_3^2 - x_1 x_2$. Determine the strain tensor and check whether or not the compatibility conditions are satisfied.

Ans: $\begin{pmatrix} 3x_2^2 & 3x_1 x_2 + x_3 & -x_2/2 \\ 3x_1 x_2 + x_3 & 0 & x_1/2 \\ -x_2/2 & x_1/2 & 2x_3 \end{pmatrix}$, Yes

21. Show that the displacement field $u_1 = Ax_1 + 3x_2$, $u_2 = 3x_1 - Bx_2$, $u_3 = 5$ gives a state of plane strain and determine the relationship between A and B for which the deformation is isochoric (constant volume deformation).

22. For the shear displacement of Problem $x_1 = X_1$, $x_2 = X_2$, $x_3 = X_3 + \frac{2}{\sqrt{3}} X_2$

(i) determine the direction of the line element in the $X_2 X_3$ plane for which the normal strain is zero.

(ii) determine the shear angle γ_{23} for the deformation of the Problem.

Ans. $\gamma_{23} = \sin^{-1}\dfrac{2}{\sqrt{7}}$

(iii) determine the equation of the ellipse into which the circle $X_2^2 + X_3^2 = 1$ is deformed.

Ans: $x_2^2 + 9x_3^2 = 1$

23. For the displacement field $x_1 = X_1 + 2X_3$, $x_2 = X_2 - 2X_3$, $x_3 = X_3 - 2X_1 + 2X_2$ determine the deformation gradient F, and by a polar decomposition of F find the rotation tensor R and the right stretch tensor U.

Ans: $F = \begin{pmatrix} 1 & 0 & 2 \\ 0 & 1 & -2 \\ -2 & 2 & 1 \end{pmatrix}$, $R = \dfrac{1}{3}\begin{pmatrix} 2 & 1 & 2 \\ 1 & 2 & -2 \\ -2 & 2 & 1 \end{pmatrix}$, $U = \begin{pmatrix} 2 & -1 & 0 \\ -1 & 2 & 0 \\ 0 & 0 & 3 \end{pmatrix}$

24. For the displacement field $x_1 = X_1 + AX_3$, $x_2 = X_2$, $x_3 = X_3 - AX_1$ calculate the volume change and show that it is zero if A is a very small constant.

■ ■ ■

2

Theory of Motion

This chapter deals with the kinematics and global kinetics of continuous media. In continuum mechanics we consider *material bodies* in the form of solids, liquids, and gases. Let us begin by describing the model we use to represent such bodies. For this purpose, we define a material body B as the set of elements X, called *particles* or *material points,* which can be put into a one-to-one correspondence with the points of a regular region of physical space. Note that whereas a particle of classical mechanics has an assigned mass, a continuum particle is essentially a material point for which a density is defined.

2.1 Motion

Motion and flow are terms used to describe the instantaneous or continuing change in configuration of a continuum. The motion of a continuum body is a continuous time sequence of displacements. Thus, the material body will occupy different configurations at different times so that a particle occupies a series of points in space which describe a path-line. There is continuity during deformation or motion of a continuum body in the sense that:

(i) The material points forming a closed curve at any instant will always form a closed curve at any subsequent time.

(ii) The material points forming a closed surface at any instant will always form a closed surface at any subsequent time and the matter within the closed surface will always remain within.

The motion is said to be *kinematically admissible* if:

(*i*) Continuity of particle positions is preserved so that no gaps or interpenetration occurs.

(*ii*) Kinematic constraints on the motion (for example, support conditions) are preserved.

A kinematically admissible motion along a stage will be called a *stage motion*.

2.2 Material and Spatial Description

In the Lagrangian description the position and physical properties of the particles are described in terms of the material or referential coordinates and time. In this case the reference configuration is the configuration at $t = 0$. An observer standing in the referential frame of reference observes the changes in the position and physical properties as the material body moves in space as time progresses. The results obtained are independent of the choice of initial time and reference configuration, $K_0(B)$. This description is normally used in solid mechanics.

The motion of a continuum may be expressed in terms of material coordinates (Lagrangian description) by three equations

$$\boldsymbol{x} = \boldsymbol{x}(\boldsymbol{X},t); \; x_i = x_i(X_1, X_2, X_3, t), i = 1,2,3, \quad (2.1)$$

where t is real. Equation (2.1) describes motion of the material point completely in material method giving the subsequent position at time t. The co-ordinates $\boldsymbol{X} = (X_1, X_2, X_3)$ are independent coordinates called material coordinates or Lagrangian coordinates where as x_1, x_2, x_3, are dependent coordinates called spatial coordinates.

Axiom of continuity allows for the single-valued inverse of equation (2.1) to trace backwards where the particle currently located at $\boldsymbol{x} = (x_1, x_2, x_3)$ was located in the initial or referenced configuration $K_0(B)$. In this case the description of motion is made in terms of the spatial coordinates, in which case is called the spatial description or Eulerian description, *i.e.* the current configuration is taken as the reference configuration.

The Eulerian description, introduced by d'Alembert, focuses on the current configuration $K_t(B)$, giving attention to what is occurring at a fixed point in space as time progresses, instead of giving attention to individual particles as they move through space and time. By the inverse of the equations (2.1) in terms of the spatial coordinates (Eulerian description)

$$\boldsymbol{X} = \boldsymbol{X}(\boldsymbol{x},t); \quad X_i = X_i(x_1, x_2, x_3, t), i = 1,2,3, \quad (2.2)$$

A necessary and sufficient condition for this inverse function to exist is that the determinant of the Jacobian Matrix, often referred to simply as the Jacobian, should be different from zero. Thus,

2.3 Material and Spatial Time Derivatives

$$J = \left|\frac{\partial X_i}{\partial X_j}\right| = \left|\frac{\partial x_i}{\partial X_j}\right| \neq 0$$

Physically, the Lagrangian description fixes attention on specific particles of the continuum, whereas the Eulerian description concerns itself with a particular region of the space occupied by the continuum.

It is assumed that both (2.1) and (2.2) possess continuous partial derivatives with respect to their arguments to whatever order needed, except possibly at some singular points, lines, or surfaces. Generally, we shall not need more than second- or third-order derivatives. Single-valuedness of (2.1) and (2.2) is essential for the axiom of impenetrability of matter. Otherwise the body may split, or two disjoint bodies may intermingle.

Since equations (2.1) and (2.2) are the inverses of one another, any physical property of the continuum that is expressed with respect to a specific particle (Lagrangian, or material description) may also be expressed with respect to the particular location in space occupied by the particle (Eulerian, or spatial description). For example, if the material description of the density ρ is given by

$$\rho = \rho(\mathbf{X}, t); \quad \rho = \rho(X_i, t) \tag{2.3}$$

the spatial description is obtained by replacing X in this equation by the function given in (2.2). Thus the spatial description of the density is

$$\rho = \rho(\mathbf{X}(\mathbf{x},t),t) = \rho^*(\mathbf{x},t); \quad \rho = \rho(X_i(\mathbf{x},t),t) = \rho^*(x_i,t) \tag{2.4}$$

where the symbol ρ^* is used here to emphasize that the functional form of the Eulerian description is not necessarily the same as the Lagrangian form.

In the kinematics of continuous media, time rates of vectors and tensors are encountered. The most important among various time rates is the concept of the material derivative.

2.3 Material and Spatial Time Derivatives

The most important among various time rates is the concept of the material derivative. The time rate of change of any property of a continuum (such as temperature or velocity or stress tensor) with respect to specific particles of the moving continuum is called the material derivative of that property. The material derivative (also known as the substantial, or convective derivative) may be thought of as the time rate of change that would be measured by an observer traveling with the specific particles under study. The instantaneous position x_i of a particle is itself a property of the particle.

Let $\vec{F}$ be any vector property associated with a material point which happens to occupy the spatial position (x_1, x_2, x_3) at time t so that in spatial description we can write

$$\vec{F} = \vec{G}(x_1, x_2, x_3, t) = \text{a vector function of } (x_1, x_2, x_3, t) \quad (2.5)$$

After an interval of infinitesimal time δt, let this material point move on to a neighbouring position $(x_1 + \delta x_1, \delta x_2 + x_2, x_3 + \delta x_3)$. Let $\vec{F} + \delta \vec{F}$ be the value of the vector property there. Then, using Taylors series expansion, we write

$$\vec{F} + \delta \vec{F} = \vec{G}(x_1 + \delta x_1, x_2 + \delta x_2, x_3 + \delta x_3, t + \delta t)$$

$$= \vec{G}(x_1, x_2, x_3, t) + \frac{\partial \vec{G}}{\partial t} \delta t + \frac{\partial \vec{G}}{\partial x_k} \delta x_k + \ldots$$

$$= \vec{F} + \frac{\partial \vec{F}}{\partial t} \delta t + \frac{\partial \vec{F}}{\partial x_k} \delta x_k + \ldots \text{using Eq. (2.5)}$$

The change in $\vec{F}$ in time interval δt is given by

$$\delta \vec{F} = \frac{\partial \vec{F}}{\partial t} \delta t + \frac{\partial \vec{F}}{\partial x_k} \delta x_k + \ldots$$

Therefore, the time rate of change of $\vec{F}$ at the instant t is given by

$$\dot{\vec{F}} = \frac{d\vec{F}}{dt} = \lim_{\delta t \to 0} \frac{\delta \vec{F}}{\delta t} = \frac{\partial \vec{F}}{\partial t} + \frac{\partial \vec{F}}{\partial x_k} \frac{\partial x_k}{\partial t}$$

other terms vanish as $\delta t \to 0$. Therefore

$$\dot{\vec{F}} = \frac{d\vec{F}}{dt} = \left(\frac{\partial}{\partial t} + \frac{\partial}{\partial x_k} \frac{\partial x_k}{\partial t} \right) \vec{F} \quad (2.6)$$

which immediately suggests the introduction of the material derivative operator

$$\frac{d}{dt} \equiv \frac{\partial}{\partial t} + v_k \frac{\partial}{\partial x_k} \equiv \frac{\partial}{\partial t} + v.\nabla$$

which is used in taking the material derivatives of quantities expressed in spatial coordinates.

The first term on right hand side of equation (2.6) represents the rate of change of $\vec{F}$ at a particular location in space, regarded as fixed due to change in time only noted by a fixed observer and is called local rate of change of $\vec{F}$.

The second term $\frac{\partial \vec{F}}{\partial x_k} \frac{\partial x_k}{\partial t}$ gives the rate of change of $\vec{F}$ at a particular time due to the movement or convection from one location to another and is known as the convective rate of change of $\vec{F}$ and $\dot{\vec{F}}$, as in Eq. (2.6)

2.3 Material and Spatial Time Derivatives

is called the total time rate of change of $\vec{F}$ or substantial rate of change of $\vec{F}$ noted by an observer moving with the continuum. The above expression can also be written as

$$\frac{d\vec{F}}{dt} \equiv \dot{\vec{F}} = \frac{DF_k}{Dt} i_k = \hat{F}_k\, i_k$$

where,

$$\frac{DF_k}{Dt} = \frac{\partial F_k}{\partial t} + \frac{\partial F_k}{\partial x_l}\frac{\partial x_l}{\partial t} = \frac{\partial F_k}{\partial t} + F_{k,l}\frac{\partial x_l}{\partial t}$$

is called the material derivative of $\vec{F}$. We shall denote the material derivative by $\dfrac{D}{Dt}$. The first term $F_{k,l}$ on the extreme right of the expression $\dfrac{DF_k}{Dt}$ is sometimes called the local or nonstationary rate and the second term $\dfrac{\partial x_l}{\partial t}$ is known as the convective time rate.

In material description when vector property $\vec{F}$ associated with a material point is expressed in terms of material coordinates (X_1, X_2, X_3) where (X_1, X_2, X_3) are initial Cartesian coordinates of the material point, then

$$\vec{F} = \vec{F}(X_1, X_2, X_3, t) = \text{a vector function of } (X_1, X_2, X_3, t) \qquad (2.7)$$

The material time rate of change of a vector (or tensor) $\vec{F}$ is given by

$$\frac{d\vec{F}}{dt} = \lim_{\delta t \to 0}\frac{\vec{F}(X_1, X_2, X_3, t+\delta t) - \vec{F}(X_1, X_2, X_3, t)}{\delta t} = \frac{\partial \vec{F}}{\partial t} \qquad (2.8)$$

Similarly, for a scalar property $f = f(x_1, x_2, x_3, t)$,

$$\frac{df}{dt} = \left(\frac{\partial}{\partial t} + \frac{\partial x_k}{\partial t}\frac{\partial}{\partial x_k}\right) f$$

and for a scalar property, $f = f(X_1, X_2, X_3, t)$, $\dfrac{df}{dt} = \dfrac{\partial f}{\partial t}$. If we wish, we may employ the common notation $\dfrac{D}{Dt}$ to indicate the material derivatives of both material and spatial vectors and tensors with no ambiguity.

Let us now calculate the material time rate of change of a vector property associated with a specific material point of a continuum. When the vector property is expressed in terms of spatial coordinates in spatial description, in the determination of this time rate, one must take account not only of the change at a spatial point but also of the change in the field as observed by the material point due solely to its motion.

2.4 Velocity and Acceleration

Because the material coordinates are constant, the current spatial position is uniquely determined by the displacement vector u, pointing from the reference position to the current position. The global Cartesian components of this displacement vector in the spatial frame, by default called u, v, and w, are the primary dependent variables in the Solid Mechanics interface.

Velocity: The material derivative of the particle's position is the instantaneous velocity of the particle. The velocity vector $\vec{v}$ of a material point is defined to be the material time rate of change of the position vector of a particle. If $\vec{v}$ be the velocity vector of a material point whose position vector is $\vec{r}$ at time t, then

$$\vec{v} = \frac{d\vec{r}}{dt} = \dot{\vec{r}} \tag{2.9}$$

Let v_i be the components of the velocity of a material point which happens to occupy the spatial position (x_1, x_2, x_3) at time t. If X_i be the initial coordinates of this material point and u_i are components of displacement, then $x_i = X_i + u_i(X_1, X_2, X_3, t)$ and

$$v_i = \frac{dx_i}{dt} = \frac{d}{dt}(X_i + u_i) = \frac{du_i}{dt} \text{ or, } \mathbf{v} = \frac{d\mathbf{x}}{dt} = \frac{d}{dt}(\mathbf{X} + \mathbf{u}) = \frac{d\mathbf{u}}{dt} \tag{2.10}$$

as X_i is independent of time. In (2.9), if the displacement is expressed in the Lagrangian form $u_i = u_i(X,t)$ then

$$v_i \equiv \dot{u}_i = \frac{du_i(X,t)}{dt} = \frac{\partial u_i(X,t)}{\partial t} \text{ or, } \mathbf{v} = \dot{\mathbf{u}} = \frac{d\mathbf{u}(X,t)}{dt} = \frac{\partial \mathbf{u}(X,t)}{\partial t} \tag{2.11}$$

If, on the other hand, the displacement is in the Eulerian form $u_i = u_i(X,t)$, then

$$v_i(x,t) = \dot{u}_i(x,t) = \frac{du_i(x,t)}{dt} = \frac{\partial u_i(x,t)}{\partial t} + v_k(x,t)\frac{\partial u_i(x,t)}{\partial x_k}$$

or, $$v(x,t) = \dot{u}(x,t) = \frac{du(x,t)}{dt} = \frac{\partial u(x,t)}{\partial t} + v(x,t)\cdot\nabla_x u(x,t) \tag{2.12}$$

The operator $\frac{d}{dt} \equiv \frac{\partial}{\partial t} + \vec{v}\cdot\vec{\nabla}$ is known as differentiation following the motion. In equation (2.12) the velocity is given implicitly since it appears as a factor of the second term on the right. The function

$$v_i = v_i(x,t) \text{ or } v = v(x,t) \tag{2.13}$$

is said to specify the instantaneous velocity field.

2.4 Velocity and Acceleration

(a) Velocity components in polar co-ordinates: In polar co-ordinates, the matrix form of the components of ∇u can be written as

$$[\nabla u] = \begin{pmatrix} \dfrac{\partial u_r}{\partial r} & \dfrac{1}{r}\dfrac{\partial u_r}{\partial \theta} - \dfrac{u_\theta}{r} \\ \dfrac{\partial u_\theta}{\partial r} & \dfrac{1}{r}\dfrac{\partial u_\theta}{\partial \theta} + \dfrac{u_r}{r} \end{pmatrix}$$

In polar co-ordinates (r,θ), let the displacement vector $\vec{u}$ be given by $\vec{u} = u_r \hat{e}_r + u_\theta \hat{e}_\theta$. Therefore, the velocity vector $\vec{v}$ is defined as the time rate of change of the displacement vector for a given particle, that is

$$\vec{v} = \dfrac{d\vec{u}}{dt} = \dfrac{\partial \vec{u}}{\partial t} + (\vec{u} \cdot \vec{\nabla})\vec{u} = \begin{pmatrix} \dfrac{\partial u_r}{\partial t} \\ \dfrac{\partial u_\theta}{\partial t} \end{pmatrix} + \begin{pmatrix} \dfrac{\partial u_r}{\partial r} & \dfrac{1}{r}\dfrac{\partial u_r}{\partial \theta} - \dfrac{u_\theta}{r} \\ \dfrac{\partial u_\theta}{\partial r} & \dfrac{1}{r}\dfrac{\partial u_\theta}{\partial \theta} + \dfrac{u_r}{r} \end{pmatrix} \begin{pmatrix} u_r \\ u_\theta \end{pmatrix}$$

Therefore, the velocity components are given by

$$v_r = \dfrac{\partial u_r}{\partial t} + u_r \dfrac{\partial u_r}{\partial r} + \left(\dfrac{1}{r}\dfrac{\partial u_r}{\partial \theta} - \dfrac{u_\theta}{r} \right) u_\theta; \quad v_\theta = \dfrac{\partial u_\theta}{\partial t} + u_r \dfrac{\partial u_\theta}{\partial r} + \left(\dfrac{1}{r}\dfrac{\partial u_\theta}{\partial \theta} + \dfrac{u_r}{r} \right) u_\theta$$

(b) Velocity components in cylindrical co-ordinates: In cylindrical co-ordinates the matrix form the components of $\nabla \mathbf{u}$ can be written as

$$[\nabla u] = \begin{pmatrix} \dfrac{\partial u_r}{\partial r} & \dfrac{1}{r}\dfrac{\partial u_r}{\partial \theta} - \dfrac{u_\theta}{r} & \dfrac{\partial u_r}{\partial z} \\ \dfrac{\partial u_\theta}{\partial r} & \dfrac{1}{r}\dfrac{\partial u_\theta}{\partial \theta} + \dfrac{u_r}{r} & \dfrac{\partial u_\theta}{\partial z} \\ \dfrac{\partial u_z}{\partial r} & \dfrac{1}{r}\dfrac{\partial u_z}{\partial \theta} & \dfrac{\partial u_z}{\partial z} \end{pmatrix}$$

In cylindrical co-ordinates (r,θ,z), let the displacement vector $\vec{u}$ be given by $\vec{u} = u_r \hat{e}_r + u_\theta \hat{e}_\theta + u_z \hat{e}_z$. Therefore, the velocity vector $\vec{v}$ is given by $\vec{v} = v_r \hat{e}_r + v_\theta \hat{e}_\theta + v_z \hat{e}_z$. Thus, the velocity vector $\vec{v}$ is defined as the time rate of change of the displace vector for a given particle, that is

$$\vec{v} = \dfrac{d\vec{u}}{dt} = \dfrac{\partial \vec{u}}{\partial t} + (\vec{u} \cdot \vec{\nabla})\vec{u} = \begin{pmatrix} \dfrac{\partial u_r}{\partial t} \\ \dfrac{\partial u_\theta}{\partial t} \\ \dfrac{\partial u_z}{\partial t} \end{pmatrix} + \begin{pmatrix} \dfrac{\partial u_r}{\partial r} & \dfrac{1}{r}\dfrac{\partial u_r}{\partial \theta} - \dfrac{u_\theta}{r} & \dfrac{\partial u_r}{\partial z} \\ \dfrac{\partial u_\theta}{\partial r} & \dfrac{1}{r}\dfrac{\partial u_\theta}{\partial \theta} + \dfrac{u_r}{r} & \dfrac{\partial u_\theta}{\partial z} \\ \dfrac{\partial u_z}{\partial r} & \dfrac{1}{r}\dfrac{\partial u_z}{\partial \theta} & \dfrac{\partial u_z}{\partial z} \end{pmatrix} \begin{pmatrix} u_r \\ u_\theta \\ u_z \end{pmatrix}.$$

Therefore, the velocity components are given by

$$v_r = \frac{\partial u_r}{\partial t} + u_r \frac{\partial u_r}{\partial r} + \left(\frac{1}{r}\frac{\partial u_r}{\partial \theta} - \frac{u_\theta}{r}\right) u_\theta + \frac{\partial u_r}{\partial z} u_z$$

$$v_\theta = \frac{\partial u_\theta}{\partial t} + u_r \frac{\partial u_\theta}{\partial r} + \left(\frac{1}{r}\frac{\partial u_\theta}{\partial \theta} + \frac{u_r}{r}\right) u_\theta + \frac{\partial u_\theta}{\partial z} u_z$$

$$v_z = \frac{\partial u_z}{\partial t} + u_r \frac{\partial u_z}{\partial r} + \frac{u_\theta}{r}\frac{\partial u_z}{\partial \theta} + \frac{\partial u_z}{\partial z} u_z$$

(c) Velocity components in spherical polar co-ordinates: In spherical polar co-ordinates, the matrix form of the components of ∇u can be written as

$$[\nabla u] = \begin{pmatrix} \dfrac{\partial u_r}{\partial r} & \dfrac{1}{r}\dfrac{\partial u_r}{\partial \theta} - \dfrac{u_\theta}{r} & \dfrac{1}{r\sin\theta}\dfrac{\partial u_r}{\partial \phi} - \dfrac{u_\phi}{r} \\ \dfrac{\partial u_\theta}{\partial r} & \dfrac{1}{r}\dfrac{\partial u_\theta}{\partial \theta} + \dfrac{u_r}{r} & \dfrac{1}{r\sin\theta}\dfrac{\partial u_\theta}{\partial \phi} - \dfrac{u_\phi \cot\theta}{r} \\ \dfrac{\partial u_\phi}{\partial r} & \dfrac{1}{r}\dfrac{\partial u_\phi}{\partial \theta} & \dfrac{1}{r\sin\theta}\dfrac{\partial u_\phi}{\partial \phi} + \dfrac{u_r}{r} - \dfrac{u_\theta \cot\theta}{r} \end{pmatrix}$$

In spherical polar co-ordinates (r, θ, ϕ), let the displacement vector be given by $\vec{u} = u_r \hat{e}_r + u_\theta \hat{e}_\theta + u_\phi \hat{e}_\phi$. Then, the velocity vector is given by $\vec{v} = v_r \hat{e}_r + v_\theta \hat{e}_\theta + v_\phi \hat{e}_\phi$. Therefore, the velocity vector is defined as the time rate of change of the displacement vector for a given particle, that is

$$\vec{v} = \frac{d\vec{u}}{dt} = \frac{\partial \vec{u}}{\partial t} + (\vec{u}\cdot\vec{\nabla})\vec{u} = \begin{pmatrix} \dfrac{\partial u_r}{\partial t} \\ \dfrac{\partial u_\theta}{\partial t} \\ \dfrac{\partial u_\phi}{\partial t} \end{pmatrix}$$

$$+ \begin{pmatrix} \dfrac{\partial u_r}{\partial r} & \dfrac{1}{r}\dfrac{\partial u_r}{\partial \theta} - \dfrac{u_\theta}{r} & \dfrac{1}{r\sin\theta}\dfrac{\partial u_r}{\partial \phi} - \dfrac{u_\phi}{r} \\ \dfrac{\partial u_\theta}{\partial r} & \dfrac{1}{r}\dfrac{\partial u_\theta}{\partial \theta} + \dfrac{u_r}{r} & \dfrac{1}{r\sin\theta}\dfrac{\partial u_\theta}{\partial \phi} - \dfrac{u_\phi \cot\theta}{r} \\ \dfrac{\partial u_\phi}{\partial r} & \dfrac{1}{r}\dfrac{\partial u_\phi}{\partial \theta} & \dfrac{1}{r\sin\theta}\dfrac{\partial u_\phi}{\partial \phi} + \dfrac{u_r}{r} - \dfrac{u_\theta \cot\theta}{r} \end{pmatrix} \begin{pmatrix} u_r \\ u_\theta \\ u_\phi \end{pmatrix}.$$

2.4 Velocity and Acceleration

Therefore, the velocity components are given by

$$v_r = \frac{\partial u_r}{\partial t} + u_r \frac{\partial u_r}{\partial r} + \left(\frac{1}{r}\frac{\partial u_r}{\partial \theta} - \frac{u_\theta}{r}\right)u_\theta + \left(\frac{1}{r\sin\theta}\frac{\partial u_r}{\partial \phi} - \frac{u_\phi}{r}\right)u_r$$

$$v_\theta = \frac{\partial u_\theta}{\partial t} + u_r \frac{\partial u_\theta}{\partial r} + \left(\frac{1}{r}\frac{\partial u_\theta}{\partial \theta} + \frac{u_r}{r}\right)u_\theta + \left(\frac{1}{r\sin\theta}\frac{\partial u_\theta}{\partial \phi} - \frac{u_\phi \cot\theta}{r}\right)u_\phi$$

$$v_\phi = \frac{\partial u_\phi}{\partial t} + u_r \frac{\partial u_\phi}{\partial r} + \frac{u_\theta}{r}\frac{\partial u_\phi}{\partial \theta} + \left(\frac{1}{r\sin\theta}\frac{\partial u_\phi}{\partial \phi} + \frac{u_r}{r} - \frac{u_\theta \cot\theta}{r}\right)u_\phi$$

Acceleration: The acceleration of a particle is the rate of change of velocity of the particle. It is, therefore, the material derivative of velocity. The acceleration vector $\vec{Q}$ of a material point is defined as the time rate of change of the velocity vector for a given particle, that is, the *acceleration field*

$$\vec{a} = \frac{d\vec{v}}{dt} = \frac{d}{dt}\left(\frac{d\vec{r}}{dt}\right) \tag{2.13}$$

If $v_i = \frac{dx_i}{dt} = \dot{x}_i$ be the components of velocity and a_i be the components of acceleration, then

$$a_i = \frac{dv_i}{dt} = \frac{d}{dt}\left(\frac{dx_i}{dt}\right) \tag{2.14}$$

If the velocity is given in the Lagrangian form or material form, $v_i = v_i(X_1, X_2, X_3, t)$ then

$$a_i \equiv \dot{v}_i = \frac{dv_i(X,t)}{dt} = \frac{\partial v_i(X,t)}{\partial t} \quad \text{or} \quad a = \dot{v} = \frac{dv(X,t)}{dt} = \frac{\partial v(X,t)}{\partial t} \tag{2.15}$$

In the Lagrangian representation, the material particle with a given velocity or acceleration is identifiable. This scheme is an immediate extension of that of particle mechanics. If on the other hand, the velocity is given in spatial form, then $v_i = v_i(X_1, X_2, X_3, t)$ and so

$$a_i(x,t) = \dot{v}_i(x,t) = \frac{dv_i(x,t)}{dt} = \frac{\partial v_i(x,t)}{\partial t} + v_k(x,t)\frac{\partial v_i(x,t)}{\partial x_k}$$

or,

$$a(x,t) = \dot{v}(x,t) = \frac{dv(x,t)}{dt} = \frac{\partial v(x,t)}{\partial t} + v(x,t).\nabla_x v(x,t) \tag{2.16}$$

In the Eulerian description, the velocity and acceleration at time t at a spatial point are known, but the particle occupying this point is not known. As each particle moves through a spatial point, it acquires the velocity and acceleration associated with that point at that time. Also

$$\frac{d\vec{v}}{dt} = \frac{\partial \vec{v}}{\partial t} + \frac{\partial \vec{v}}{\partial x_k}\frac{\partial x_k}{\partial t} = \frac{\partial \vec{v}}{\partial t} + \vec{v}_{i,j}v_j$$

$$= \frac{\partial \vec{v}}{\partial t} + v_1\frac{\partial \vec{v}}{\partial x_1} + v_2\frac{\partial \vec{v}}{\partial x_2} + v_3\frac{\partial \vec{v}}{\partial x_3}$$

$$= \frac{\partial \vec{v}}{\partial t} + (\vec{v}\cdot\vec{\nabla})\vec{v} = \frac{\partial \vec{v}}{\partial t} + \vec{\nabla}(\frac{v^2}{2}) - \vec{v}\times rot\vec{v}$$

(a) Acceleration components in polar co-ordinates: In polar co-ordinates, the matrix form of the components of Δv can be written as

$$[\nabla v] = \begin{pmatrix} \frac{\partial v_r}{\partial r} & \frac{1}{r}\frac{\partial v_r}{\partial \theta} - \frac{v_\theta}{r} \\ \frac{\partial v_\theta}{\partial r} & \frac{1}{r}\frac{\partial v_\theta}{\partial \theta} + \frac{v_r}{r} \end{pmatrix}$$

In polar co-ordinates (r,θ), let the velocity vector $\vec{v}$ be given by $\vec{v} = v_r\hat{e}_r + v_\theta\hat{e}_\theta$ and $\vec{a} = a_r\hat{e}_r + a_\theta\hat{e}_\theta$. Therefore, the acceleration vector $\vec{a}$ is defined as the time rate of change of the velocity vector for a given particle, that is

$$\vec{a} = \frac{d\vec{v}}{dt} = \frac{\partial \vec{v}}{\partial t} + (\vec{v}\cdot\vec{\nabla})\vec{v} = \begin{pmatrix} \frac{\partial v_r}{\partial t} \\ \frac{\partial v_\theta}{\partial t} \end{pmatrix} + \begin{pmatrix} \frac{\partial v_r}{\partial r} & \frac{1}{r}\frac{\partial v_r}{\partial \theta} - \frac{v_\theta}{r} \\ \frac{\partial v_\theta}{\partial r} & \frac{1}{r}\frac{\partial v_\theta}{\partial \theta} + \frac{v_r}{r} \end{pmatrix} \begin{pmatrix} v_r \\ v_\theta \end{pmatrix}$$

Therefore, the acceleration components are given by

$$a_r = \frac{\partial v_r}{\partial t} + v_r\frac{\partial v_r}{\partial r} + \left(\frac{1}{r}\frac{\partial v_r}{\partial_,} - \frac{v_\theta}{r}\right)v_\theta; \quad a_\theta = \frac{\partial v_\theta}{\partial t} + v_r\frac{\partial v_\theta}{\partial r} + \left(\frac{1}{r}\frac{\partial v_\theta}{\partial_,} + \frac{v_r}{r}\right)v_\theta.$$

(b) Acceleration components in cylindrical co-ordinates: In cylindrical co-ordinates the matrix form of the components of ∇v can be written as

$$[\nabla v] = \begin{pmatrix} \frac{\partial v_r}{\partial r} & \frac{1}{r}\frac{\partial v_r}{\partial \theta} - \frac{v_\theta}{r} & \frac{\partial v_r}{\partial z} \\ \frac{\partial v_\theta}{\partial r} & \frac{1}{r}\frac{\partial v_\theta}{\partial \theta} + \frac{v_r}{r} & \frac{\partial v_\theta}{\partial z} \\ \frac{\partial v_z}{\partial r} & \frac{1}{r}\frac{\partial v_z}{\partial \theta} & \frac{\partial v_z}{\partial z} \end{pmatrix}$$

In cylindrical co-ordinates (r,θ,z), let the velocity vector $\vec{v}$ and acceleration $\vec{\theta}$ be $\vec{v} = v_r\hat{e}_r + v_\theta\hat{e}_\theta + v_z\hat{e}_z$. In cylindrical co-ordinates (r,θ,z), the acceleration $\vec{a}$ is given by $\vec{a} = a_r\hat{e}_r + a_\theta\hat{e}_\theta + a_z\hat{e}_z$. Therefore, the accelera-

2.4 Velocity and Acceleration

tion vector $\vec{a}$ is defined as the time rate of change of the velocity vector for a given particle, that is

$$\vec{a} = \frac{d\vec{v}}{dt} = \frac{\partial \vec{v}}{\partial t} + (\vec{v}\cdot\vec{\nabla})\vec{v} = \begin{pmatrix} \frac{\partial v_r}{\partial t} \\ \frac{\partial v_\theta}{\partial t} \\ \frac{\partial v_z}{\partial t} \end{pmatrix} + \begin{pmatrix} \frac{\partial v_r}{\partial r} & \frac{1}{r}\frac{\partial v_r}{\partial \theta} - \frac{v_\theta}{r} & \frac{\partial v_r}{\partial z} \\ \frac{\partial v_\theta}{\partial r} & \frac{1}{r}\frac{\partial v_\theta}{\partial \theta} + \frac{v_r}{r} & \frac{\partial v_\theta}{\partial z} \\ \frac{\partial v_z}{\partial r} & \frac{1}{r}\frac{\partial v_z}{\partial \theta} & \frac{\partial v_z}{\partial z} \end{pmatrix} \begin{pmatrix} v_r \\ v_\theta \\ v_z \end{pmatrix}.$$

Therefore, the acceleration components are given by

$$a_r = \frac{\partial v_r}{\partial t} + v_r \frac{\partial v_r}{\partial r} + \left(\frac{1}{r}\frac{\partial v_r}{\partial \theta} - \frac{v_\theta}{r}\right)v_\theta + \frac{\partial v_r}{\partial z}v_z$$

$$a_\theta = \frac{\partial v_\theta}{\partial t} + v_r \frac{\partial v_\theta}{\partial r} + \left(\frac{1}{r}\frac{\partial v_\theta}{\partial \theta} + \frac{v_r}{r}\right)v_\theta + \frac{\partial v_\theta}{\partial z}v_z$$

$$a_z = \frac{\partial v_z}{\partial t} + v_r \frac{\partial v_\theta}{\partial r} + \frac{v_\theta}{r}\frac{\partial v_z}{\partial \theta} + \frac{\partial v_z}{\partial z}v_z.$$

(c) Acceleration components in spherical co-ordinates: In spherical polar co-ordinates, the matrix form the components of can be written as

$$[\nabla v] = \begin{pmatrix} \frac{\partial v_r}{\partial r} & \frac{1}{r}\frac{\partial v_r}{\partial \theta} - \frac{v_\theta}{r} & \frac{1}{r\sin\theta}\frac{\partial v_r}{\partial \phi} - \frac{v_\phi}{r} \\ \frac{\partial v_\theta}{\partial r} & \frac{1}{r}\frac{\partial v_\theta}{\partial \theta} + \frac{v_r}{r} & \frac{1}{r\sin\theta}\frac{\partial v_\theta}{\partial \phi} - \frac{v_\phi \cot\theta}{r} \\ \frac{\partial v_\phi}{\partial r} & \frac{1}{r}\frac{\partial v_\phi}{\partial \theta} & \frac{1}{r\sin\theta}\frac{\partial v_\phi}{\partial \phi} + \frac{v_r}{r} - \frac{v_\theta \cot\theta}{r} \end{pmatrix}$$

In spherical polar co-ordinates (r,θ,ϕ), let the velocity vector $\vec{v}$ and the acceleration $\vec{a}$ be given by $\vec{v} = v_r \hat{e}_r + v_\theta \hat{e}_\theta + v_\phi \hat{e}_\phi$ and $\vec{a} = a_r \hat{e}_r + a_\theta \hat{e}_\theta + a_\phi \hat{e}_\phi$. Therefore, the acceleration vector $\vec{a}$ is defined as the time rate of change of the velocity vector for a given particle, that is

$$\vec{a} = \frac{d\vec{v}}{dt} = \frac{\partial \vec{v}}{\partial t} + (\vec{v}\cdot\vec{\nabla})\vec{v}$$

$$= \begin{pmatrix} \frac{\partial v_r}{\partial t} \\ \frac{\partial v_\theta}{\partial t} \\ \frac{\partial v_\phi}{\partial t} \end{pmatrix} + \begin{pmatrix} \frac{\partial v_r}{\partial r} & \frac{1}{r}\frac{\partial v_r}{\partial \theta} - \frac{v_\theta}{r} & \frac{1}{r\sin\theta}\frac{\partial v_r}{\partial \phi} - \frac{v_\phi}{r} \\ \frac{\partial v_\theta}{\partial r} & \frac{1}{r}\frac{\partial v_\theta}{\partial \theta} + \frac{v_r}{r} & \frac{1}{r\sin\theta}\frac{\partial v_\theta}{\partial \phi} - \frac{v_\phi \cot\theta}{r} \\ \frac{\partial v_\phi}{\partial r} & \frac{1}{r}\frac{\partial v_\phi}{\partial \theta} & \frac{1}{r\sin\theta}\frac{\partial v_\phi}{\partial \phi} + \frac{v_r}{r} - \frac{v_\theta \cot\theta}{r} \end{pmatrix} \begin{pmatrix} v_r \\ v_\theta \\ v_\phi \end{pmatrix}.$$

Therefore, the acceleration components are given by

$$a_r = \frac{\partial v_r}{\partial t} + v_r \frac{\partial v_r}{\partial r} + \left(\frac{1}{r}\frac{\partial v_r}{\partial \theta} - \frac{v_\theta}{r}\right)v_\theta + \left(\frac{1}{r\sin\theta}\frac{\partial v_r}{\partial \phi} - \frac{v_\phi}{r}\right)v_\phi$$

$$a_\theta = \frac{\partial v_\theta}{\partial t} + v_r \frac{\partial v_\theta}{\partial r} + \left(\frac{1}{r}\frac{\partial v_\theta}{\partial \theta} + \frac{v_r}{r}\right)v_\theta + \left(\frac{1}{r\sin\theta}\frac{\partial v_\theta}{\partial \phi} - \frac{v_\phi \cot\theta}{r}\right)v_\phi$$

$$a_\phi = \frac{\partial v_\phi}{\partial t} + v_r \frac{\partial v_\phi}{\partial r} + \frac{v_\theta}{r}\frac{1}{r}\frac{\partial v_\phi}{\partial \theta} + \left(\frac{1}{r\sin\theta}\frac{\partial v_\phi}{\partial \phi} + \frac{v_r}{r} - \frac{v_\theta \cot\theta}{r}\right)v_\phi.$$

Example: 2.1: Motion of a particle is given by

$$x_1 = X_1 + X_2 t + X_3 t^2, \quad x_2 = X_2 + X_3 t + X_1 t^2; \quad x_3 = X_3 + X_1 t + X_2 t^2$$

(i) Find at time t, the velocity and acceleration of a particle which was at (1,1,1) at $t = 0$.

(ii) Find at time t, the velocity and acceleration of a particle which is at (1,1,1) at time t.

Solution: (i) Here, given that $x_i = x_i(X_1, X_2, X_3, t)$, therefore $v_i = \frac{\partial x_i}{\partial t}$ and $a_i = \frac{\partial^2 x_i}{\partial t^2}$. Therefore, the velocities of a particle which was at (1,1,1) at $t = 0$ are given by

$$v_1 = \frac{\partial x_1}{\partial t} = X_2 + 2X_3 t = 1 + 2t \text{ at } (X_1, X_2, X_3) = (1,1,1)$$

$$v_2 = \frac{\partial x_2}{\partial t} = X_3 + 2X_1 t = 1 + 2t \text{ at } (X_1, X_2, X_3) = (1,1,1)$$

$$v_3 = \frac{\partial x_3}{\partial t} = X_1 + 2X_2 t = 1 + 2t \text{ at } (X_1, X_2, X_3) = (1,1,1)$$

and the acceleration of a particle which was at (1,1,1) at $t = 0$ are given by

$$a_1 = \frac{\partial^2 x_1}{\partial t^2} = 2X_3 = 2 \text{ at } (X_1, X_2, X_3) = (1,1,1)$$

$$a_2 = \frac{\partial^2 x_2}{\partial t^2} = 2X_1 = 2 \text{ at } (X_1, X_2, X_3) = (1,1,1)$$

$$a_2 = \frac{\partial^2 x_2}{\partial t^2} = 2X_1 = 2 \text{ at } (X_1, X_2, X_3) = (1,1,1)$$

(ii) Now, using the given relations, we have,

$$x_1 - x_2 t = X_1(1 - t^3) \text{ i.e., } X_1 = \frac{x_1 - x_2 t}{1 - t^3}$$

$$X_2 = \frac{x_2 - x_3 t}{1 - t^3}; \quad X_3 = \frac{x_3 - x_1 t}{1 - t^3}$$

2.4 Velocity and Acceleration

Therefore, v_i are given by

$$v_1 = X_2 + 2X_3t = \frac{x_2 + x_3t - 2x_1t^2}{1-t^3} = \frac{1+t-2t^2}{1-t^3} \text{ at } (x_1,x_2,x_3) = (1,1,1)$$

$$v_2 = X_3 + 2X_1t = \frac{x_3 + x_1t - 2x_2t^2}{1-t^3} = \frac{1+t-2t^2}{1-t^3} \text{ at } (x_1,x_2,x_3) = (1,1,1)$$

$$v_3 = X_1 + 2X_2t = \frac{x_1 + x_2t - 2x_3t^2}{1-t^3} = \frac{1+t-2t^2}{1-t^3} \text{ at } (x_1,x_2,x_3) = (1,1,1)$$

and a_i are given by

$$a_1 = \frac{\partial v_1}{\partial t} + v_1\frac{\partial v_1}{\partial x_1} + v_2\frac{\partial v_1}{\partial x_2} + v_3\frac{\partial v_1}{\partial x_3}$$

$$= \frac{\partial}{\partial t}\left(\frac{x_2 + x_3t - 2x_1t^2}{1-t^3}\right) + v_1\frac{\partial}{\partial x_1}(\frac{x_2 + x_3t - 2x_1t^2}{1-t^3})c + v_3\frac{\partial}{\partial x_3}(\frac{x_2 + x_3t - 2x_1t^2}{1-t^3})$$

$$= \frac{-4x_1 t + x_3t}{1-t^3} + (x_2 + x_3t - 2x_1t^2)c + \frac{1}{1-t^3}[v_1(-2t^2) + v_2 + v_3t]$$

$$= \frac{2-2t}{1-t^3}\left[\text{at }(x_1,x_2,x_3) = (1,1,1)\right]$$

Similarly, $a_2 = \dfrac{2-2t}{1-t^3} = a_3$ at $(x_1,x_2,x_3) = (1,1,1)$.

Example: 2.2: A continuum motion is expressed by

$$x_1 = X_1, \ x_2 = e^t\frac{X_2 + X_3}{2} + e^{-t}\frac{X_2 - X_3}{2}; \ x_3 = e^t\frac{X_2 + X_3}{2} - e^{-t}\frac{X_2 - X_3}{2}$$

Determine the velocity components in both their material and spatial forms.

Solution: From the second and third equations,

$$x_2 + x_3 = e^t\frac{X_2 + X_3}{2}, x_2 - x_3 = e^{-t}\frac{X_2 - X_3}{2}. \ x_2 - x_3 = e^{-t}\frac{X_2 - X_3}{2}.$$

Solving these simultaneously the inverse equations become

$$X_1 = x_1; X_2 = e^{-t}\frac{x_2 + x_3}{2} + e^t\frac{x_2 - x_3}{2}; X_3 = e^{-t}\frac{x_2 + x_3}{2} - e^t\frac{x_2 - x_3}{2}$$

Accordingly, the displacement components $u_i = x_i - x_i$ may be written in either the Lagrangian form

$$u_1 = x_1 - X_1 = x_1 - x_1 = 0$$
$$u_2 = x_2 - X_2 = x_2 - e^{-t}\frac{x_2 + x_3}{2} + e^{t}\frac{x_2 - x_3}{2}$$
$$u_3 = x_3 - X_3 = x_3 - e^{-t}\frac{x_2 + x_3}{2} - e^{t}\frac{x_2 - x_3}{2}$$

By using the inverse equations, we obtain the spatial description of the displacement field in component form

$$u_1 = x_1 - X_1 = x_1 - x_1 = 0$$
$$u_2 = x_2 - X_2 = x_2 - e^{-t}\frac{x_2 + x_3}{2} + e^{t}\frac{x_2 - x_3}{2}$$
$$u_3 = x_3 - X_3 = x_3 - e^{-t}\frac{x_2 + x_3}{2} - e^{t}\frac{x_2 - x_3}{2}$$

Therefore, the velocities of a particle are given by

$$v_1 = \frac{\partial u_1}{\partial t} = \frac{\partial X_1}{\partial t} = 0$$
$$v_2 = \frac{\partial u_2}{\partial t} = \frac{\partial X_2}{\partial t} = -e^{-t}\frac{x_2 + x_3}{2} + e^{t}\frac{x_2 - x_3}{2}$$
$$v_3 = \frac{\partial u_3}{\partial t} = \frac{\partial X_3}{\partial t} = -e^{-t}\frac{x_2 + x_3}{2} - e^{t}\frac{x_2 - x_3}{2}.$$

Relationship between spatial and material method of description: The primary quantity in spatial method is the velocity of the material point. If v_1, v_2, v_3 be the components of the velocity of the material point which happens to occupy the spatial point (x_1, x_2, x_3) at time t,

$$v_i = f_i(x_1, x_2, x_3, t)$$

We can now change from spatial method to material method by writing $\frac{dx_i}{dt}$ for v_i, the above equation reduces to

$$\frac{dx_1}{dt} = f_1(x_1, x_2, x_3, t); \quad \frac{dx_2}{dt} = f_2(x_1, x_2, x_3, t); \quad \frac{dx_3}{dt} = f_3(x_1, x_2, x_3, t) \quad (2.17)$$

The integration of the above equation lead to three constants of integration which can be considered as initial coordinates (X_1, X_2, X_3) identifying the material particle. Hence solution of Eq. (2.17) gives

$$x_1 = F_1(X_1, X_2, X_3, t); x_2 = F_2(X_1, X_2, X_3, t); x_3 = F_3(X_1, X_2, X_3, t)$$

These equations determine the motion completely in material method.

2.4 Velocity and Acceleration

Example 2.3: A velocity field is specified in Eulerian form by
$v_1 = x_1 + x_2 + x_3 + t,$ $v_2 = 2(x_1 + x_2 + x_3) + t;$ $v_3 = 3(x_1 + x_2 + x_3) + t.$
Determine in Lagrangian form position x_i of the particle at a point (X_1, X_2, X_3) at time $t = 0$.

Solution: Writing $\frac{dx_i}{dt}$ for v_i we change from Eulerian to Lagrangian form

$$v_1 = \frac{dx_1}{dt} = x_1 + x_2 + x_3 + t; v_2 = \frac{dx_2}{dt} = 2(x_1 + x_2 + x_3) + t;$$

$$v_3 = \frac{dx_3}{dt} = 3(x_1 + x_2 + x_3) + t$$

Let, $z = x_1 + x_2 + x_3$ then adding the above equations we get,

$$\frac{d}{dt}(x_1 + x_2 + x_3) = 6(x_1 + x_2 + x_3) + 3t$$

or,
$$\frac{dz}{dt} - 6z = 3t$$

or,
$$ze^{-6t} = \int 3te^{-6t} dt + C = -\frac{t}{2}e^{-6t} - \frac{1}{12}e^{-6t} + C$$

Since at $t = 0, z_0 = X_1 + X_2 + X_3$, therefore, $C = X_1 + X_2 + X_3 + \frac{1}{12}$. Hence,

$$z = x_1 + x_2 + x_3 = -\frac{t}{2} - \frac{1}{12} + \left(X_1 + X_2 + X_3 + \frac{1}{12}\right)e^{6t}$$

Now the velocity components are given by

$$\frac{dx_1}{dt} = x_1 + x_2 + x_3 + t = z + t$$

$$= -\frac{t}{2} - \frac{1}{12} + \left(X_1 + X_2 + X_3 + \frac{1}{12}\right)e^{6t} + t$$

$$= \frac{t}{2} - \frac{1}{12} + \left(X_1 + X_2 + X_3 + \frac{1}{12}\right)e^{6t}$$

$$\frac{dx_2}{dt} = 2(x_1 + x_2 + x_3) + t = 2z + t$$

$$= -2 \cdot \frac{t}{2} - 2 \cdot \frac{1}{12} + 2\left(X_1 + X_2 + X_3 + \frac{1}{12}\right)e^{6t} + t$$

$$= -\frac{1}{6} + 2\left(X_1 + X_2 + X_3 + \frac{1}{12}\right)e^{6t}$$

$$\frac{dx_3}{dt} = 3(x_1 + x_2 + x_3) + t = 3z + t$$

$$= -3 \cdot \frac{t}{2} - 3 \cdot \frac{1}{12} + 3\left(X_1 + X_2 + X_3 + \frac{1}{12}\right)e^{6t} + t$$

$$= -\frac{t}{2} - \frac{1}{4} + 3\left(X_1 + X_2 + X_3 + \frac{1}{12}\right)e^{6t}.$$

Example 2.4: A velocity field is specified in Lagrangian form by $v_1 = -X_2 e^{-t}, v_2 = -X_3, v_3 = 2t$. Determine acceleration components in Eulerian form.

Solution: In Lagrangian form $a_i = \frac{\partial v_i}{\partial t}$, so

$$a_1 = X_2 e^{-t}, a_2 = 0, a_3 = 2$$

To obtain components in Eulerian form,

$$\frac{dx_1}{dt} = v_1 = -X_2 e^{-t}, \frac{dx_2}{dt} = v_2 = -X_3,$$

$$\frac{dx_3}{dt} = v_3 = 2t$$

On integration, we get

or, $\qquad x_1 = X_1 + X_2 e^{-t}, x_2 = X_2 - X_3 t, x_3 = X_3 + t^2$

or, $\qquad x_2 + x_3 t = X_2 + t^3 \text{ i.e., } X_2 = x_2 + x_3 t - t^3$

Thus the acceleration components are

$$a_1 = X_2 e^{-t} = \left(x_2 + x_3 t - t^3\right)e^{-t}, a_2 = 0, a_3 = 2$$

Example 2.5: Let a certain motion of a continuum be given by the component equations $x_1 = X_1 e^{-t}, x_2 = X_2 e^{t}; x_3 = X_3 + X_2 \left(e^{-t} - 1\right)$ and let the temperature field of the body be given by the spatial description, $\theta = e^{-t}(x_1 - 2x_2 + 3x_3)$. Determine the velocity field in spatial form, and using that, compute the material derivative $\frac{d\theta}{dt}$ of the temperature field.

2.4 Velocity and Acceleration

Solution: Note that the initial configuration serves as the reference configuration. The velocity components in material form are readily determined to be

$$v_1 = -X_1 e^{-t}, v_2 = X_2 e^{t}; v_3 = -X_2 e^{-t}.$$

Also, the motion equations may be inverted directly to give

$$X_1 = x_1 e^{t}, X_2 = x_2 e^{-t}; X_3 = x_3 - x_2(e^{-2t} - e^{-t}).$$

which upon substitution into the above velocity expressions yields the spatial components,

$$v_1 = -x_1, \quad v_2 = x_2; \quad v_3 = -x_2 e^{-2t}.$$

We may determine the displacement components in material form directly from the motion equations

$$u_1 = x_1 - X_1 = X_1 e^{-t} - X_1 = X_1(e^{-t} - 1)$$
$$u_2 = x_2 - X_2 = X_2 e^{t} - X_2 = X_2(e^{t} - 1)$$
$$u_3 = x_3 - X_3 = X_3 + X_2 e^{-t} - X_3 = X_2(e^{-t} - 1)$$

and, using the inverse equations computed before, we obtain the spatial displacements

$$u_1 = x_1 - X_1 = X_1 e^{-t} - X_1 = -x_1(e^{t} - 1)$$
$$u_2 = x_2 - X_2 = X_2 e^{t} - X_2 = -x_2(e^{-t} - 1)$$
$$u_3 = x_3 - X_3 = X_3 + X_2 e^{-t} - X_3 = x_2(e^{-2t} - e^{-t}).$$

Therefore, by differentiating the above displacement components, we get

$$v_1 = -x_1 e^{t} + v_1(1 - e^{t}); v_2 = x_2 e^{-t} + v_2(1 - e^{-t}); v_3 = -x_2(2e^{-2t} - e^{-t}) + v_2(e^{-2t} - e^{-t})$$

which results in a set of equations having the desired velocity components on both sides of the equations. In general, this set of equations must be solved simultaneously. In this case, the solution is quite easily obtained, yielding

$$v_1 = -x_1; \quad v_2 = x_2; \quad v_3 = -x_2 e^{-2t}.$$

Therefore, the material derivative $\dfrac{d\theta}{dt}$ in spatial form is given by

$$\frac{d\theta}{dt} = \frac{\partial\theta}{\partial t} + \frac{\partial x_k}{\partial t}\cdot\frac{\partial\theta}{\partial x_k} = \frac{\partial\theta}{\partial t} + \frac{\partial x_1}{\partial t}\cdot\frac{\partial\theta}{\partial x_1} + \frac{\partial x_2}{\partial t}\cdot\frac{\partial\theta}{\partial x_2} + \frac{\partial x_3}{\partial t}\cdot\frac{\partial\theta}{\partial x_3}$$

$$= -e^{-t}(x_1 - 2x_2 + 3x_3) + v_1.e^{-t} + v_2\cdot(-2e^{-t}) + v_3\cdot(-3e^{-t})$$

$$= -e^{-t}(x_1 - 2x_2 + 3x_3) - e^{-t}x_1 - 2x_2 e^{-t} - 3x_3 e^{-t}$$

which may be converted to its material form using the original motion equations, resulting in

$$\frac{d\theta}{dt} = -e^{-t}(x_1 - 2x_2 + 3x_3) - 2X_1 e^{-2t} - 3X_2(2e^{-2t} - e^{-t}) - 3X_3 e^{-t}$$

$$= -e^{-t}\left[X_1 e^{-t} - 2X_2 e^{t} + 3X_3 + 3X_2(e^{t} - 1)\right] - 2X_1 e^{-2t} - 3X_2 v$$

2.5 Flow

Flow sometimes carries the connotation of a motion leading to a permanent deformation as for example, in plasticity studies. In fluid flow, however, the word denotes continuing motion.

A deformable body is called a fluid if the deformation increases indefinitely with the continued application of any force, how small it may be, on the body and cannot return to its original configuration when the force is withdrawn. This continuous shear deformation is called the flow of the fluid.

Steady and unsteady motion:

In general, the continuum physical properties including the velocity field is dependent on time. The motion is then called 'unsteady'. When the physical properties and the velocity field does not change in time so that its local time rate of change is zero, the motion is said to be 'steady'. In particular,

$$\frac{\partial\vec{v}}{\partial t} = 0, i.e., \vec{v} = \vec{v}(x_1, x_2, x_3)$$

so that $\vec{v}$ is a function of spatial coordinate alone.

Equilibrium point: A continuum is said to be in equilibrium if its velocity at every point does not seem to change with time with respect to an observer moving with it so that $\frac{d\vec{v}}{dt} = 0$ throughout the continuum.

Stagnation point: A spatial point where, the velocity $v = (v_1, v_2, v_3) = 0$ is called a stagnation point. If at a given place $x = (x_1, x_2, x_3)$ the velocity does not change in time, we say that the motion is steady at that point. For a steady motion, therefore,

$$v = v(x) = v(x_1, x_2, x_3).$$

2.5 Flow

A motion is one dimensional (or lineal) if two components of the velocity vanish, and the third depends only on one space variable, for example,

$$v_1 = v_1(x_1,t), \quad v_2 = 0, \quad v_3 = 0.$$

A motion is a plane motion if the velocity component perpendicular to a set of parallel planes is zero and the two velocity components in these planes are functions of two plane coordinates, for example,

$$v_1 = v_1(x_1,x_2,t), v_2 = v_2(x_1,x_2,t), v_3 = 0.$$

Path lines, stream lines and streak lines:

A *path line* is the curve or path followed by a particle during motion or flow. A path line is the trace of a particle $X = (X_1, X_2, X_3)$ as t varies. Thus

$$x_i = x_i(X,t) = x_i(X_1, X_2, X_3, t), \quad X = \text{fixed}$$

gives the path line of a particle initially located at X. The path line is also obtained as the integral curve of that $dx_i = v_i dt$ passes through X at $t = 0$. The differential equation of the path line is $\vec{v} = \dfrac{d\vec{r}}{dt}$, or

$$\frac{dx_1}{v_1(x_1,x_2,x_3,t)} = \frac{dx_2}{v_2(x_1,x_2,x_3,t)} = \frac{dx_3}{v_3(x_1,x_2,x_3,t)} = dt. \quad (2.18)$$

The *stream lines* at a given time are the curves drawn in a continuum that are tangent at every point of it is in the instantaneous direction of the velocity at that point. Hence, the differential equation of the stream line at a given instant t is $\vec{v} = k d\vec{r}$ or

$$\frac{dx_1}{v_1(x_1,x_2,x_3,t)} = \frac{dx_2}{v_2(x_1,x_2,x_3,t)} = \frac{dx_3}{v_3(x_1,x_2,x_3,t)}. \quad (2.19)$$

Experimentally, by dropping a small visible floating particle and taking a long time exposure, we see the trace of the particle revealing its path line. A short time exposure of a fluid onto which many floating particles are dropped will show the instantaneous direction of the velocity field, thus revealing the stream lines.

The stream sheets and stream tubes, respectively, are the collection of stream lines that intersect an open and a closed curve. Since the material particles move in the direction of the velocity field, they cannot cross the stream sheets and stream tubes.

By drawing tangent curves to any vector field at a spatial point at time t, one obtains other types of lines, for example, vortex lines. The sheets and tubes associated with such lines are similarly defined.

A *streak line* at a given time is the curve that connects the temporary location of all the particles that have past to a fixed spatial point in the flow field. Let $x = x(X,t)$, represents the path line for a material point X at time t. We put it as $X = X(x,t)$. Then the streak line through a spatial point x at the time t' are the locus of

$$x = x\left[X(x,t'), t\right]; 0 \leq t' \leq t$$

as t' varies.

Example 2.6: **Find the stream line, path line and streak line of a continuum particle for the velocity field**

$$v_1 = \frac{x_1}{t}, v_2 = x_2, v_3 = 0.$$

Solution: It is a two dimensional unsteady motion. The differential equation of the stream line at a given instant is given by

$$\frac{dx_1}{v_1} = \frac{dx_2}{v_2} = \frac{dx_3}{v_3} \Rightarrow \frac{dx_1}{\frac{x_1}{t}} = \frac{dx_2}{x_2} = \frac{dx_3}{0}$$

Where t is a constant. Keeping t constant, equation of streamline at t is given by

$$t \log x_1 = \log \frac{x_2}{c_1} \text{ and } x_3 = c_2 \Rightarrow x_2 = c_1 (x_1)^t \text{ and } x_3 = c_2$$

where, c_1, c_2 are constants. This is the required equation of the stream lines as plane curves at a given instant t. For the stream lines at time t passing through point $\left(x_1^{(1)}, x_2^{(1)}, x_3^{(1)}\right)$, we have $c_1 = \frac{x_2^{(1)}}{x_1^{(1)t}}, c_2 = x_3^{(1)}$. Therefore, we have

$$x_2 = \frac{x_2^{(1)}}{x_1^{(1)t}} x_1^t \text{ and } x_3 = c_2 = x_3^{(1)},$$

which gives the equation of the stream lines.

The differential equation of the path line is given by

$$\frac{dx_1}{dt} = v_1 = \frac{x_1}{t}, \frac{dx_2}{dt} = v_2 = x_2, \frac{dx_3}{dt} = v_3 = 0$$

where t is a variable.

Now

$$\frac{dx_1}{dt} = \frac{x_1}{t} \Rightarrow \log \frac{x_1}{c_3} = \log t \Rightarrow x_1 = c_3 t$$

2.5 Flow

$$\frac{dx_2}{dt} = x_2 \Rightarrow \log\frac{x_2}{c_4} = t \Rightarrow x_2 = c_4 e^t$$

$$\frac{dx_3}{dt} = 0 \Rightarrow x_3 = \text{constant} = c_5$$

where, c_3, c_4 and c_5 are constants. Let $\left(x_1^{(0)}, x_2^{(0)}, x_3^{(0)}\right)$ be the co-ordinates of the chosen fluid particle at time $t = t_0$, then

$$c_3 = \frac{x_1^{(0)}}{t_0}, c_4 = x_2^{(0)} e^{-t_0}, c_5 = x_3^{(0)}.$$

For such values of c_3, c_4 and c_5, the solution becomes,

$$x_1 = \frac{x_1^{(0)}}{t_0} t, x_2 = x_2^{(0)} e^{t-t_0}, x_3 = x_3^{(0)}.$$

Eliminating t, equation of the path line is

$$x_2 = c_4 \exp\left(\frac{x_1}{c_3}\right), x_3 = c_5 \Rightarrow x_2 = x_2^{(0)} \exp\left(\frac{x_1}{x_1^{(0)}} - 1\right) t_0, x_3 = x_3^{(0)}$$

Let the fluid particle $\left(x_1^{(0)}, x_2^{(0)}, x_3^{(0)}\right)$ passes through a fixed point $\left(x_1^{(1)}, x_2^{(1)}, x_3^{(1)}\right)$ at an instant of time $t = t'$, then

$$x_1^{(1)} = \frac{x_1^{(0)}}{t_0} t', x_2^{(1)} = x_2^{(0)} e^{t'-t_0}, x_3^{(1)} = x_3^{(0)}$$

$$\Rightarrow x_1^{(0)} = \frac{t_0}{t'} x_1^{(1)}; x_2^{(0)} = x_2^{(1)} e^{t_0-t'}; x_3^{(0)} = x_3^{(1)}$$

where, t' is the parameter. Substituting, we get

$$x_1 = \frac{t_0}{t'} x_1^{(1)}, \quad x_2 = x_2^{(1)} e^{t-t'}, \quad x_3 = x_3^{(0)} = x_3^{(1)},$$

which gives the equation of streak line at the point $\left(x_1^{(1)}, x_2^{(1)}, x_3^{(1)}\right)$.

Example 2.7: Find the stream line, path line and streak line of a continuum particle for the velocity field $v_1 = x_1, v_2 = -x_2, v_3 = 0$.

Solution: It is a two dimensional unsteady motion as $\frac{d}{dt}\vec{v} = \vec{0}$. The differential equation of the stream line at a given instant t is given by

$$\frac{dx_1}{v_1} = \frac{dx_2}{v_2} = \frac{dx_3}{v_3} \Rightarrow \frac{dx_1}{x_1} = \frac{dx_2}{x_2} = \frac{dx_3}{0}$$

where t is a constant. Keeping t constant, equation of streamline at t is given by

$$\log x_1 + \log x_2 = \log c_1 \text{ and } x_3 = c_2 \Rightarrow x_1 x_2 = c_1 \text{ and } x_3 = c_2$$

where, c_1, c_2 and c_3 are constants. This is the required equation of the stream lines as plane curves at a given instant t and is the rectangular hyperbola.

The differential equation of the path line is given by

$$\frac{x_1}{dt} = v_1 = x_1, \frac{dx_2}{dt} = v_2 = -x_2, \frac{dx_3}{dt} = v_3 = 0$$

where t is a variable. Now

$$\frac{dx_1}{dt} = x_1 \Rightarrow \log \frac{x_1}{c_3} = t \Rightarrow x_1 = c_3 e^t$$

$$\frac{dx_2}{dt} = -x_2 \Rightarrow \log \frac{x_2}{c_4} = -t \Rightarrow x_2 = c_4 e^{-t}$$

$$\frac{dx_3}{dt} = 0 \Rightarrow x_3 = \text{constant} = c_5$$

where, c_3, c_4 and c_5 are constants. Let $\left(x_1^{(0)}, x_2^{(0)}, x_3^{(0)}\right)$ be the co-ordinates of the chosen fluid particle at time $t = t_0$, then

$$c_3 = x_1^{(0)} e^{-t_0}, c_4 = x_2^{(0)} e^{t_0}, c_5 = x_3^{(0)}.$$

For such values of c_3, c_4 and c_5, the solution becomes,

$$x_1 = x_1^{(0)} e^{t-t_0}, x_2 = x_2^{(0)} e^{t_0-t}, x_3 = x_3^{(0)}.$$

Eliminating t, equation of the path line is $x_1 x_2 = x_1^{(0)} x_2^{(0)}, x_3 = x_3^{(0)}$.

Let the fluid particle $\left(x_1^{(0)}, x_2^{(0)}, x_3^{(0)}\right)$ passes through a fixed point $\left(x_1^{(1)}, x_2^{(1)}, x_3^{(1)}\right)$ at an instant of time $t = t'$, then

$$x_1^{(1)} = x_1^{(0)} e^{t'-t_0}, \quad x_2^{(1)} = x_2^{(0)} e^{t_0-t'}, \quad x_3^{(1)} = x_3^{(0)}$$
$$\Rightarrow x_1^{(0)} = x_1^{(1)} e^{t_0-t'}, \quad x_2^{(0)} = x_2^{(1)} e^{t'-t_0}, \quad x_3^{(0)} = x_3^{(1)}$$

where, t' is the parameter. Substituting, we get

$$x_1 = x_1^{(1)} e^{t-t'}, \quad x_2 = x_2^{(1)} e^{t'-t}, \quad x_3 = x_3^{(0)} = x_3^{(1)},$$

Eliminating t', we get the equation of the streak line as $x_1 x_2 = x_1^{(1)} x_2^{(1)}, x_3 = x_3^{(1)}$. Hence in this case path lines, stream lines and streak lines coincide(or which are the same curves when the motion is steady).

2.5 Flow

Example 2.8: Find the stream line, path line and streak line of a continuum particle for the velocity field

$$v_1 = \frac{x_1}{1+t}, \quad v_2 = x_2, \quad v_3 = 0$$

Solution: The differential equation of the stream line at a given instant is given by

$$\frac{dx_1}{v_1} = \frac{dx_2}{v_2} = \frac{dx_3}{v_3} \Rightarrow \frac{dx_1}{\frac{x_1}{1+t}} = \frac{dx_2}{x_2} = \frac{dx_3}{0}$$

where t is a constant. Keeping t constant, equation of streamline at t is given by

$$(1+t)\log\frac{x_1}{c_1} = \log x_2 \text{ and } x_3 = c_2 \Rightarrow x_2 = \left(\frac{x_1}{c_1}\right)^{1+t} \text{ and } x_3 = c_2$$

which is the required equation of the stream line at a given instant t.

The differential equation of the path line is given by

$$\frac{dx_1}{dt} = v_1 = \frac{x_1}{1+t}, \frac{dx_2}{dt} = v_2 = x_2, \frac{dx_3}{dt} = v_3 = 0$$

where t is a variable. Now

$$\frac{dx_1}{dt} = \frac{x_1}{1+t} \Rightarrow \log\frac{x_1}{c_3} = \log(1+t) \Rightarrow x_1 = c_3(1+t)$$

$$\frac{dx_2}{dt} = x_2 \Rightarrow \log\frac{x_2}{c_4} = t \Rightarrow x_2 = c_4 e^t$$

$$\frac{dx_3}{dt} = 0 \Rightarrow x_3 = \text{constant} = c_5$$

Eliminating t, equation of the path line is $x_2 = c_4 \exp\left(\frac{x_1}{c_3} - 1\right), x_3 = c_5$.

Example 2.9: For the velocity field, $v_1 = \frac{3x_1^2 - r^2}{r^5}, v_2 = \frac{3x_1 x_2}{r^5}, v_3 = \frac{3x_1 x_3}{r^5}$

where, $r^2 = x_1^2 + x_2^2 + x_3^2$ show that motion is irrotational. Find the velocity potential and stream lines.

Solution: For the given velocity field $\vec{v} = (v_1, v_2, v_3)$, we have

$$\text{rot } \vec{v} = \begin{vmatrix} \hat{i} & \hat{j} & \hat{k} \\ \dfrac{\partial}{\partial x_1} & \dfrac{\partial}{\partial x_2} & \dfrac{\partial}{\partial x_3} \\ v_1 & v_2 & v_3 \end{vmatrix} = \begin{vmatrix} \hat{i} & \hat{j} & \hat{k} \\ \dfrac{\partial}{\partial x_1} & \dfrac{\partial}{\partial x_2} & \dfrac{\partial}{\partial x_3} \\ \dfrac{3x_1^2 - r^2}{r^5} & \dfrac{3x_1 x_2}{r^5} & \dfrac{3x_1 x_3}{r^5} \end{vmatrix}$$

$$= \left[\dfrac{\partial}{\partial x_2}\left(\dfrac{3x_1 x_3}{r^5}\right) - \dfrac{\partial}{\partial x_3}\left(\dfrac{3x_1 x_2}{r^5}\right) \right] \hat{i} + \left[\dfrac{\partial}{\partial x_3}\left(\dfrac{3x_1^2 - r^2}{r^5}\right) - \dfrac{\partial}{\partial x_1}\left(\dfrac{3x_1 x_3}{r^5}\right) \right] \hat{j}$$

$$+ \left[\dfrac{\partial}{\partial x_1}\left(\dfrac{3x_1 x_2}{r^5}\right) - \dfrac{\partial}{\partial x_2}\left(\dfrac{3x_1^2 - r^2}{r^5}\right) \right] \hat{k}$$

$$= \left[3x_1 x_3 \left(\dfrac{-5}{r^6}\right)\dfrac{\partial r}{\partial x_2} - 3x_1 x_2 \left(\dfrac{-5}{r^6}\right)\dfrac{\partial r}{\partial x_3} \right] \hat{i} + \left[\dfrac{-2r}{r^5}\dfrac{\partial r}{\partial x_3} + (3x_1^2 - r^2)\left(\dfrac{-5}{r^6}\right)\dfrac{\partial r}{\partial x_3} \right] \hat{j}$$

$$- \left[\dfrac{3x_3}{r^5} + 3x_1 x_3 \left(\dfrac{-5}{r^6}\right)\dfrac{\partial r}{\partial x_1} \right] \hat{j} - \left[\dfrac{-2r}{r^5}\dfrac{\partial r}{\partial x_2} + (3x_1^2 - r^2)\left(\dfrac{-5}{r^6}\right)\dfrac{\partial r}{\partial x_2} \right] \hat{k}$$

$$= \left[-\dfrac{15 x_1 x_3 \, x_2}{r^6 \, r} + \dfrac{15 x_1 x_2 \, x_3}{r^6 \, r} \right] \hat{i} + \left[\dfrac{-2r}{r^5}\dfrac{x_3}{r} + (3x_1^2 - r^2)\left(\dfrac{-5}{r^6}\right)\dfrac{x_3}{r} \right] \hat{j}$$

$$- \left[\dfrac{3x_3}{r^5} + 3x_1 x_3 \left(\dfrac{-5}{r^6}\right)\dfrac{x_1}{r} \right] \hat{j} - \left[\dfrac{-2r}{r^5}\dfrac{x_2}{r} + (3x_1^2 - r^2)\left(\dfrac{-5}{r^6}\right)\dfrac{x_2}{r} \right] \hat{k} = \vec{0}$$

Since rot rot $\vec{v} = \vec{0}$, so the given velocity field is irrotational and there exists a scalar potential function $\varnothing$ such that $\vec{v} = \vec{\nabla}\varnothing$, i.e., $v_i = -\dfrac{\partial \varnothing}{\partial x_i}$. If $\varnothing$ be the velocity potential, then

$$v_i . dx_i = -\dfrac{\partial \varnothing}{\partial x_i} . dx_i = -d\varnothing.$$

Therefore,

$$d\varnothing = -\left\{ \dfrac{3x_1^2 - r^2}{r^5} dx_1 + \dfrac{3x_1 x_2}{r^5} dx_2 + \dfrac{3x_1 x_3}{r^5} dx_3 \right\}$$

$$= -\dfrac{(3x_1^2 - r^2) dx_1 + 3x_1 x_2 dx_2 + 3x_1 x_3 dx_3}{r^5}$$

$$= -\dfrac{3x_1 (x_1 dx_1 + x_2 dx_2 + x_3 dx_3)}{r^5} + \dfrac{r^2 dx_1}{r^5} = -3x_1 \dfrac{r.dr}{r^5} + \dfrac{dx_1}{r^3}$$

2.5 Flow

$$= -3x_1 \frac{dr}{r^4} + \frac{dx_1}{r^3} = d\left(\frac{x_1}{r^3}\right) \Rightarrow \emptyset = \frac{x_1}{r^3} = \frac{x_1}{(x_1^2 + x_2^2 + x_3^2)^{3/2}}$$

The differential equation of the stream line is given by

$$\frac{dx_1}{v_1} = \frac{dx_2}{v_2} = \frac{dx_3}{v_3} \Rightarrow \frac{dx_1}{3x_1^2 - r^2} = \frac{dx_2}{3x_1 x_2} = \frac{dx_3}{3x_1 x_3}$$

From the last pair of equations

$$\frac{dx_2}{x_2} = \frac{dx_3}{x_3} \Rightarrow \log \frac{x_2}{c_2} = \log x_3 \Rightarrow x_2 = c_2 x_3$$

We write differential equation of stream line as

$$\frac{x_1 dx_1 + x_2 dx_2 + x_3 dx_3}{x_1(3x_1^2 - r^2) + 3x_1 x_2^2 + 3x_1 x_3^2} = \frac{x_2 dx_2 + x_3 dx_3}{3x_1 x_2^2 + 3x_1 x_3^2}$$

$$\Rightarrow \frac{x_1 dx_1 + x_2 dx_2 + x_3 dx_3}{3x_1(x_1^2 + x_2^2 + x_3^2) - x_1(x_1^2 + x_2^2 + x_3^2)} = \frac{x_2 dx_2 + x_3 dx_3}{3x_1(x_2^2 + x_3^2)}$$

$$\Rightarrow \frac{3(x_1 dx_1 + x_2 dx_2 + x_3 dx_3)}{x_1^2 + x_2^2 + x_3^2} = \frac{2(x_2 dx_2 + x_3 dx_3)}{x_2^2 + x_3^2}$$

$$\Rightarrow 3\log(x_1^2 + x_2^2 + x_3^2) - \log c_3 = 2\log(x_2^2 + x_3^2) \text{ on integration}$$

$$\Rightarrow (x_1^2 + x_2^2 + x_3^2)^3 = c_3(x_2^2 + x_3^2)^2$$

Example 2.10: A velocity field is given by $v_1 = \frac{x_1}{1+t}, v_2 = \frac{2x_2}{1+t}, v_3 = \frac{3x_3}{1+t}$. Determine the acceleration components for this motion in Eulerian form. Integrate the velocity equation to obtain the relations $x_i = x_i(X_1, X_2, X_3, t)$. From these, determine the acceleration components in Lagrangian form. Determine the stream lines and path lines.

Solution: The acceleration components $a_i = \frac{\partial v_i}{\partial t} + v_j v_{i,j}$ in spatial coordinates form are given by

$$a_1 = \frac{\partial v_1}{\partial t} + v_1 \frac{\partial v_1}{\partial x_1} + v_2 \frac{\partial v_1}{\partial x_2} + v_3 \frac{\partial v_1}{\partial x_3} = -\frac{x_1}{(1+t)^2} + \frac{x_1}{1+t} \cdot \frac{1}{1+t} + \frac{2x_2}{1+t} \cdot 0 + \frac{3x_3}{1+t} \cdot 0 = 0.$$

$$a_2 = \frac{\partial v_2}{\partial t} + v_1 \frac{\partial v_2}{\partial x_1} + v_2 \frac{\partial v_2}{\partial x_2} + v_3 \frac{\partial v_2}{\partial x_3} = -\frac{2x_2}{(1+t)^2} + \frac{x_1}{1+t} \cdot 0 + \frac{2x_2}{1+t} \cdot \frac{2}{1+t} + \frac{3x_3}{1+t} \cdot 0 = \frac{2x_2}{(1+t)^2}.$$

$$a_3 = \frac{\partial v_3}{\partial t} + v_1 \frac{\partial v_3}{\partial x_1} + v_2 \frac{\partial v_3}{\partial x_2} + v_3 \frac{\partial v_3}{\partial x_3} = -\frac{3x_3}{(1+t)^2} + \frac{x_1}{1+t} \cdot 0 + \frac{2x_2}{1+t} \cdot 0 + \frac{3x_3}{1+t} \cdot \frac{3}{1+t} = \frac{6x_3}{(1+t)^2}.$$

On integration of the given velocity components $v_1 = \dfrac{x_1}{1+t}, v_2 = \dfrac{2x_2}{1+t}, v_3 = \dfrac{3x_3}{1+t}$, we get

$$\dfrac{dx_1}{x_1} = \dfrac{dt}{1+t}, \quad \dfrac{dx_2}{x_2} = \dfrac{2dt}{1+t}, \quad \dfrac{dx_3}{x_3} = \dfrac{3dt}{1+t}$$

$$\Rightarrow \int_{X_1}^{x_1} \dfrac{dx_1}{x_1} = \int_0^t \dfrac{dt}{1+t}, \quad \int_{X_2}^{x_2} \dfrac{dx_2}{x_2} = \int_0^t \dfrac{2dt}{1+t}, \quad \int_{X_3}^{x_3} \dfrac{dx_3}{x_3} = \int_0^t \dfrac{3dt}{1+t}$$

$$\Rightarrow x_1 = X_1(1+t), \quad x_2 = X_2(1+t)^2, \quad x_3 = X_3(1+t)^3$$

which are the required relations like $x_i = x_i(X_1, X_2, X_3, t)$. In Lagrangian form the velocity components are given by

$$v_1 = \dfrac{\partial x_1}{\partial t} = X_1, \quad v_2 = \dfrac{\partial x_2}{\partial t} = 2X_2(1+t), \quad v_3 = \dfrac{\partial x_3}{\partial t} = 3X_3(1+t)^2$$

The differential equation of the stream line at a given instant t is given by

$$\dfrac{dx_1}{v_1} = \dfrac{dx_2}{v_2} = \dfrac{dx_3}{v_3} \Rightarrow \dfrac{dx_1}{x_1} = \dfrac{dx_2}{2x_2} = \dfrac{dx_3}{3x_3}$$

where t is a constant. Solution of these equations gives

$$\int_{X_1}^{x_1} \dfrac{dx_1}{x_1} = \dfrac{1}{2} \int_{X_2}^{x_2} \dfrac{dx_2}{x_2} \Rightarrow 2\log(\dfrac{x_1}{X_1}) = \log(\dfrac{x_2}{X_2}) \Rightarrow \left(\dfrac{x_1}{X_1}\right)^2 = \dfrac{x_2}{X_2}$$

$$\int_{X_1}^{x_1} \dfrac{dx_1}{x_1} = \dfrac{1}{3} \int_{X_3}^{x_3} \dfrac{dx_3}{x_3} \Rightarrow 3\log(\dfrac{x_1}{X_1}) = \log(\dfrac{x_3}{X_3}) \Rightarrow \left(\dfrac{x_1}{X_1}\right)^3 = \dfrac{x_3}{X_3}$$

$$\dfrac{1}{2}\int_{X_2}^{x_2} \dfrac{dx_2}{x_2} = \dfrac{1}{3}\int_{X_3}^{x_3} \dfrac{dx_3}{x_3} \Rightarrow 3\log(\dfrac{x_2}{X_2}) = 2\log(\dfrac{x_3}{X_3}) \Rightarrow \left(\dfrac{x_2}{X_2}\right)^3 = (\dfrac{x_3}{X_3})^2$$

Thus the equation of the stream line at a given instant t is given by

$$\left(\dfrac{x_1}{X_1}\right)^2 = \dfrac{x_2}{X_2}, \quad \left(\dfrac{x_1}{X_1}\right)^3 = \dfrac{x_3}{X_3}, \quad \left(\dfrac{x_2}{X_2}\right)^3 = \left(\dfrac{x_3}{X_3}\right)^2.$$

The differential equation of the path line is given by

$$\dfrac{dx_1}{x_1} = \dfrac{dx_2}{2x_2} = \dfrac{dx_3}{3x_3} = \dfrac{dt}{1+t}$$

where t is a variable.

Solution of these equations gives

2.5 Flow

$$\int_{X_1}^{x_1} \frac{dx_1}{x_1} = \int_0^t \frac{dt}{1+t} \Rightarrow \log\left(\frac{x_1}{X_1}\right) = \log(1+t) \Rightarrow x_1 = X_1(1+t)$$

$$\int_{X_2}^{x_2} \frac{dx_2}{x_2} = 2\int_0^t \frac{dt}{1+t} \Rightarrow \log\left(\frac{x_2}{X_2}\right) = 2\log(1+t) \Rightarrow x_2 = X_2(1+t)^2$$

$$\int_{X_3}^{x_3} \frac{dx_3}{x_3} = 3\int_0^t \frac{dt}{1+t} \Rightarrow \log\left(\frac{x_3}{X_3}\right) = 3\log(1+t) \Rightarrow x_3 = X_3(1+t)^3.$$

Thus the equation of the path line is given by

$$x_1 = X_1(1+t), x_2 = X_2(1+t)^2, x_3 = X_3(1+t)^3.$$

Eliminating t, we see that the two equations are identical, *i.e.*, the stream line and path line coincide.

Example 2.11: **Show that $\Phi = (x-t)(y-t)$ be the velocity potential of the two dimensional irrotational motion of a continuum and then show that stream lines at time t are $(x-t)^2 + (y-t)^2$ = constant and that paths of particles have the equations**

$$\log|x-y| = \frac{1}{2}\left[(x+y) - a(x-y)^{-1}\right] + b,$$

where a and b are constants.

Solution: Since $\Phi = (x-t)(y-t)$, so

$$\nabla^2 \Phi = \frac{\partial^2 \Phi}{\partial x^2} + \frac{\partial^2 \Phi}{\partial y^2} + \frac{\partial^2 \Phi}{\partial z^2} = \frac{\partial}{\partial x}(y-t) + \frac{\partial}{\partial y}(x-t) + 0 = 0$$

Thus $\Phi = (x-t)(y-t)$ represents the possible velocity potential of an incompressible fluid. Here

$$v_1 = -\frac{\partial \Phi}{\partial x} = -(y-t); \quad v_2 = -\frac{\partial \Phi}{\partial y} = -(x-t).$$

The differential equation of the stream line at a given instant t is given by

$$\frac{dx}{v_1} = \frac{dy}{v_2} \Rightarrow \frac{dx}{-(y-t)} = \frac{dy}{-(x-t)}$$

where t is a constant. Solution of these equations gives

$$(x-t)dx = (y-t)dy \Rightarrow (x-t)d(x-t) - (y-t)d(y-t) = 0$$
$$\Rightarrow (x-t)^2 + (y-t)^2 = \text{constant}.$$

Also, $v_1 = \dfrac{dx}{dt} = -(y-t); v_2 = \dfrac{dy}{dt} = -(x-t)$.

Therefore,

$$\dfrac{d}{dt}(x+y) = -(x+y) + 2t \Rightarrow e^t \dfrac{d}{dt}(x+y) + e^t(x+y) = 2te^t$$

$$\Rightarrow e^t(x+y) = 2\int t\, e^t dt + c = 2\left[te^t - e^t + c_1\right]; c_1 = \text{constant}$$

$$\Rightarrow (x+y) = 2\left[t - 1 + c_1 e^{-t}\right].$$

Again, using $v_1 = \dfrac{dx}{dt} = -(y-t); v_2 = \dfrac{dy}{dt} = -(x-t)$, we get

$$\dfrac{d}{dt}(x-y) = (x-y) \Rightarrow (x-y) = c_2 e^t ; c_2 = \text{constant}$$

$$\Rightarrow \log(x-y) = \log c_2 + t.$$

Using this result, we get

$$x + y = 2t - 2 + 2c_1 c_2 (x-y)^{-1} \Rightarrow \dfrac{1}{2}(x+y) = t - 1 + c_1 c_2 (x-y)^{-1}$$

$$\Rightarrow \dfrac{1}{2}(x+y) - c_1 c_2 (x-y)^{-1} = t - 1 = \log(x-y) - \log c_2 - 1$$

$$\log(x-y) = \dfrac{1}{2}(x+y) - \dfrac{c_1 c_2}{x-y} + \log c_2 + 1 = \dfrac{1}{2}\left[(x+y) - a(x-y)^{-1}\right] + b$$

where, $a = c_1 c_2$ and $b = \log \log c_2 + 1$.

2.6 Boundary Surface

A surface is a boundary surface or material surface when it always consists of the same set of particles as it moves.

Theorem 2.1: (**Lagrange's criterion**) **A necessary and sufficient condition for a given surface $F(x_1, x_2, x_3, t) = 0$ to be a boundary (material) surface is that**

$$\dot{F} = \dfrac{dF}{dt} = \dfrac{\partial F}{\partial t} + v_k F_{,k} = 0. \qquad (2.20)$$

Proof: *Necessary part:* Let $F(x_1, x_2, x_3, t) = 0$ be a boundary surface. The necessary condition implies that the component of the velocity of a particle on the surface along the normal to the surface must be equal to the normal velocity of the surface itself at the point.

2.6 Boundary Surface

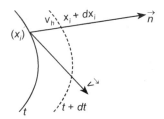

Fig. 2.1: Boundary surface

If v_n be the normal velocity of the surface at a point (x_1, x_2, x_3) and (v_1, v_2, v_3) is a component of velocity of a particle at the point P and (n_1, n_2, n_3) is the direction cosine of the normal to the surface at the point P. Then the above condition implies that

$$v_n = n_i v_i$$

Since v_n is the normal component of velocity, the displacement components (dx_1, dx_2, dx_3) of the point $P(x_1, x_2, x_3)$ of the surface are given by

$$dx_1 = n_1 v_1 dt, d_2 = n_2 v_2 dt, dx_3 = n_3 v_3 dt, i.e., dx_i = n_i v_i dt$$

At time t, let $F(x_1, x_2, x_3, t) = 0$ be the position of the particle of the surface $F(x_1, x_2, x_3, t) = 0$. At time $t + dt$, the point moves to the position $Q(x_1 + dx_1, x_2 + dx_2, x_3 + dx_3)$. Since the particle continues to lie on the surface $F(x_1, x_2, x_3, t) = 0$, we must have

$$F(x_1 + dx_1, x_2 + dx_2, x_3 + dx_3) = 0$$

Expanding in a Taylor's series and retaining only the first order terms of the expansion, we have

$$F(x_1, x_2, x_3, t) + \frac{\partial F}{\partial x_1} dx_1 + \frac{\partial F}{\partial x_2} dx_2 + \frac{\partial F}{\partial x_3} dx_3 + \frac{\partial F}{\partial t} dt = 0$$

$$\Rightarrow \frac{\partial F}{\partial x_1} dx_1 + \frac{\partial F}{\partial x_2} dx_2 + \frac{\partial F}{\partial x_3} dx_3 + \frac{\partial F}{\partial t} dt = 0, \text{ as } F(x_1, x_2, x_3, t) = 0$$

$$\Rightarrow \frac{\partial F}{\partial x_i} dx_i + \frac{\partial F}{\partial t} dt = 0 \Rightarrow \left(\frac{\partial F}{\partial x_i} n_i \right) v_i dt + \frac{\partial F}{\partial t} dt = 0$$

Since $dt \neq 0$, we must have

$$\left(\frac{\partial F}{\partial x_i} n_i \right) v_i + \frac{\partial F}{\partial t} = 0 \Rightarrow v_i = -\frac{\frac{\partial F}{\partial t}}{\frac{\partial F}{\partial x_i} n_i}$$

But the direction cosine n_i of the normal to the surface $F(x_1,x_2,x_3,t)=0$ is given by

$$n_i = \frac{1}{R}\frac{\partial F}{\partial x_i}; R = \sqrt{\left\{\left(\frac{\partial F}{\partial x_1}\right)^2 + \left(\frac{\partial F}{\partial x_2}\right)^2 + \left(\frac{\partial F}{\partial x_3}\right)^2\right\}}$$

$$\Rightarrow v_n = -\frac{\dfrac{\partial F}{\partial t}}{\dfrac{\partial F}{\partial x_i}\dfrac{1}{R}\dfrac{\partial F}{\partial x_i}} = -\frac{\dfrac{\partial F}{\partial t}}{\dfrac{R^2}{R}} = -\frac{1}{R}\frac{\partial F}{\partial t}$$

$$\Rightarrow -\frac{1}{R}\frac{\partial F}{\partial t} = v_n = n_i v_i = \frac{1}{R}\left\{v_i \frac{\partial F}{\partial x_i}\right\}$$

$$\Rightarrow \frac{\partial F}{\partial t} + v_i \frac{\partial F}{\partial x_i} = 0 \Rightarrow \frac{dF}{dt} = 0; \frac{1}{R} \neq 0.$$

Hence the condition is necessary. The relation is the condition to be satisfied by the surface $F(x_1,x_2,x_3,t)=0$ in order that surface may be a boundary surface.

Sufficient condition: Let $F(x_1,x_2,x_3,t)=0$ satisfies the condition $\dfrac{dF}{dt}=0$, i.e.,

$$\frac{\partial F}{\partial t} + v_1 \frac{\partial F}{\partial x_1} + v_2 \frac{\partial F}{\partial x_2} + v_3 \frac{\partial F}{\partial x_3} = 0 \qquad (2.21)$$

which is a first order partial differential equation with differential equation of the paths of the particle given by

$$\frac{dx_1}{v_1} = \frac{dx_2}{v_2} = \frac{dx_3}{v_3} = \frac{dt}{1} = \frac{dF}{0}$$

Integrals of this equation have the form $x_i = x_i(X_1, X_2, X_3, t)$, where X_1, X_2, X_3 are three arbitrary constants which specify the particle. The general solution of equation (2.21) is given by

$$F = \varnothing(X_1, X_2, X_3)$$

where, $\varnothing$ is arbitrary function. This shows that when $F=0$, a particle once on the surface remains on the surface throughout the motion.

Example 2.12: Find the condition that $\dfrac{x^2}{a^2}f_1(t) + \dfrac{y^2}{b^2}f_2(t) + \dfrac{z^2}{c^2}f_3(t) = 1$, is a possible form of a boundary surface for an incompressible flow.

Solution: The given surface can be written as

$$F(x,y,z,t) = \frac{x^2}{a^2}f_1(t) + \frac{y^2}{b^2}f_2(t) + \frac{z^2}{c^2}f_3(t) - 1 = 0$$

2.6 Boundary Surface

Now the surface $F(x,y,z,t) = 0$ can be a possible boundary surface, if it satisfies the boundary condition

$$\frac{\partial F}{\partial t} + u\frac{\partial F}{\partial x} + v\frac{\partial F}{\partial y} + w\frac{\partial F}{\partial z} = 0$$

$$\Rightarrow \frac{x^2}{a^2}f_1'^{(t)} + \frac{y^2}{b^2}f_2'^{(t)} + \frac{z^2}{c^2}f_3'^{(t)} + u\frac{2x}{a^2}f_1 + v\frac{2y}{b^2}f_2 + w\frac{2z}{c^2}f_3 = 0$$

$$\Rightarrow \frac{2x}{a^2}f_1\left\{u + \frac{x}{2}\frac{f_1'}{f_1}\right\} + \frac{2y}{b^2}f_2\left\{v + \frac{y}{2}\frac{f_2'}{f_2}\right\} + \frac{2z}{c^2}f_3\left\{w + \frac{z}{2}\frac{f_3'}{f_3}\right\} = 0$$

This equation is identically satisfies if we take

$$u + \frac{x}{2}\frac{f_1'}{f_1} = 0, \; v + \frac{y}{2}\frac{f_2'}{f_2} = 0, \; w + \frac{z}{2}\frac{f_3'}{f_3} = 0$$

$$\Rightarrow \quad u = -\frac{x}{2}\frac{f_1'}{f_1}, \; v = -\frac{y}{2}\frac{f_2'}{f_2}, \; w = -\frac{z}{2}\frac{f_3'}{f_3}$$

which are the expressions for velocity components. Since the flow is incompressible, so u, v, w satisfy the equation of continuity

$$\frac{\partial u}{\partial x} + \frac{\partial v}{\partial y} + \frac{\partial w}{\partial z} = 0 \Rightarrow -\frac{1}{2}\left[\frac{f_1'}{f_1} + \frac{f_2'}{f_2} + \frac{f_3'}{f_3}\right] = 0$$

$$\Rightarrow \log\log(f_1 f_2 f_3) = \text{constant} \Rightarrow f_1 f_2 f_3 = \text{constant} = k$$

Example 2.13: Show that the variable ellipsoid $\frac{x^2}{a^2 k^2 t^{2n}} + kt^n + \left\{\frac{y^2}{b^2} + \frac{z^2}{c^2}\right\} = 1$ is a possible form for the bounding surface of a liquid at any time t.

Solution: The given surface can be written as

$$F(x, y, z, t) = \frac{x^2}{a^2 k^2 t^{2n}} + kt^n \left\{\frac{y^2}{b^2} + \frac{z^2}{c^2}\right\} - 1 = 0$$

Now the surface $F(x, y, z, t) = 0$ can be a possible boundary surface, if it satisfies the boundary condition

$$\frac{\partial F}{\partial t} + u\frac{\partial F}{\partial x} + v\frac{\partial F}{\partial y} + w\frac{\partial F}{\partial z} = 0$$

$$\Rightarrow -\frac{x^2 \cdot 2n}{a^2 k^2 t^{2n+1}} + nkt^{n-1}\left\{\frac{y^2}{b^2} + \frac{z^2}{c^2}\right\} + u\frac{2x}{a^2 k^2 t^{2n}} + v\frac{2kyt^n}{b^2} + w\frac{2kzt^n}{c^2} = 0$$

$$\Rightarrow \left(u - \frac{nx}{t}\right)\frac{2x}{a^2k^2t^{2n}} + \left(v + \frac{ny}{2t}\right)\frac{2kyt^n}{b^2} + \left(w + \frac{nz}{2t}\right)\frac{2kzt^n}{c^2} = 0$$

This equation is identically satisfied if we take

$$u - \frac{nx}{t} = 0, v + \frac{ny}{2t} = 0, w + \frac{nz}{2t} = 0$$

$$\Rightarrow u = \frac{nx}{t}, v = -\frac{ny}{2t}, w = -\frac{nz}{2t}$$

which are the expressions for velocity components u, v, w.
    Also

$$\frac{\partial u}{\partial x} + \frac{\partial v}{\partial y} + \frac{\partial w}{\partial z} = \frac{n}{t} - \frac{n}{2t} - \frac{n}{2t} = 0$$

which shows that the velocity components satisfy the continuity equation. Thus the given equation $F(x, y, z, t) = 0$ is a possible form for the boundary surface of a liquid with velocity components

$$u = \frac{nx}{t}, v = -\frac{ny}{2t}, w = -\frac{nz}{2t}.$$

Example 2.14: **Show that all necessary conditions can be satisfied by a velocity potential of the form $\emptyset(x, y, z) = \alpha x^2 + \beta y^2 + \gamma z^2$ and a boundary surface of the form $F(x, y, z, t) = ax^4 + by^4 + cz^4 - \mu(t) = 0$, where, $\mu(t)$ is a given function of the time and $\alpha, \beta, \gamma, a, b, c$ are the suitable functions of the time.**

Solution: The necessary conditions are:
(i) The function $\emptyset(x, y, z)$ satisfies Laplace's equation, i.e., $\nabla^2 \emptyset(x, y, z) = 0$ for incompressible fluid flow.
(ii) The given equation $F(x, y, z, t) = 0$ satisfies the conditions for bounding surface

$$\frac{dF}{dt} = \frac{\partial F}{\partial t} + u\frac{\partial F}{\partial x} + v\frac{\partial F}{\partial y} + w\frac{\partial F}{\partial z} = 0$$

The velocity potential $\emptyset(x, y, z)$ is of the form $\emptyset(x, y, z) = \alpha x^2 + \beta y^2 + \gamma z^2$. The Laplace's equation $\nabla^2 \emptyset(x, y, z) = 0$ will be satisfied if

$$2\alpha + 2\beta + 2\gamma = 0 \Rightarrow \alpha + \beta + \gamma = 0$$

where, α, β, γ are the suitable functions of the time. The velocity components u, v, w are given by

$$u = -\frac{\partial \varnothing}{\partial x} = -2\alpha\, x, \quad v = -\frac{\partial \varnothing}{\partial y} = -2\beta\, y, \quad w = -\frac{\partial \varnothing}{\partial x} = -2\gamma\, z$$

Again, $F(x,y,z,t) = 0$ is a possible form for the bounding surface of a liquid, if

$$\frac{\partial F}{\partial t} + u\frac{\partial F}{\partial x} + v\frac{\partial F}{\partial y} + w\frac{\partial F}{\partial z} = 0$$

$$\Rightarrow x^4 \frac{\partial a}{\partial t} + y^4 \frac{\partial b}{\partial t} + z^4 \frac{\partial c}{\partial t} - \mu'(t) + 4aux^3 + 4bvy^3 + 4cwz^3 = 0$$

$$\Rightarrow x^4 \frac{\partial a}{\partial t} + y^4 \frac{\partial b}{\partial t} + z^4 \frac{\partial c}{\partial t} - \mu'(t) - 8\alpha a x^4 - 8\beta b y^4 - 8\gamma c z^4 = 0$$

$$\Rightarrow x^4 \left(\frac{\partial a}{\partial t} - 8\alpha a\right) + y^4 \left(\frac{\partial b}{\partial t} - 8\beta b\right) + z^4 \left(\frac{\partial c}{\partial t} - 8\gamma c\right) - \mu'(t) = 0$$

Comparing this with the equation of the bounding surface of the form $F(x,y,z,t) = ax^4 + by^4 + cz^4 - \mu(t) = 0$, we have

$$\frac{\left(\frac{\partial a}{\partial t} - 8\alpha a\right)}{a} = \frac{\left(\frac{\partial b}{\partial t} - 8\beta b\right)}{b} = \frac{\left(\frac{\partial c}{\partial t} - 8\gamma c\right)}{c} = \frac{\mu'(t)}{\mu(t)}.$$

This condition will hold if $\alpha, \beta, \gamma, a, b, c$ are the suitable functions of the time. Hence the velocity potential $\varnothing(x,y,z)$ and the boundary surface $F(x,y,z,t) = ax^4 + by^4 + cz^4 - \mu(t) = 0$ satisfy the necessary condition for the velocity potential and boundary surface if $\alpha, \beta, \gamma, a, b, c$ are the suitable functions of the time.

2.7 Material Derivative of the Element of Arc, Surface, and Volume

In many problems of continuum mechanics, we are not concerned with the change of shape of a continuum body but with the rate at which this change is taking place. Thus in fluid mechanics it is usually required to find the rate of change of velocity. We therefore begin by investigating the time rate of change a line element. The time rates of elements of arc, surface, and volume are often required in the calculations of time rates of integrals taken over material lines, surfaces, and volumes.

2.7.1 Material derivatives of the displacement gradient

Consider a line element $P_0 Q_0$ at $P_0(X_1, X_2, X_3)$ of a continuum body occupying initial configuration B_0 at time $t = 0$. When the body undergoes deformation, the same particles which initially lie on the line P_0Q_0 will now lie on a new line element PQ of length dl oriented in the direction (n_1, n_2, n_3) occupying current configuration B at time t. Let P has coordinates (x_1, x_2, x_3) and Q has coordinates $(x_1 + dx_1, x_2 + dx_2, x_3 + dx_3)$ then the differential position dx_i can be written as $dx_i = \dfrac{\partial x_i}{\partial X_k} dX_k$. Consider

$$\frac{d}{dt}(dx_i) = \frac{d}{dt}\left[\frac{\partial x_i}{\partial X_k} dX_k\right] = \frac{d}{dt}\left(\frac{\partial x_i}{\partial X_k}\right) dX_k = \frac{\partial}{\partial X_k}\left(\frac{dx_i}{dt}\right) dX_k$$

Let v_i be the velocity of the typical particle which was at $P_0(X_1, X_2, X_3)$ subsequently occupies the position $P(x_1, x_2, x_3)$ at time t, then $v_i = \dfrac{dx_i}{dt}$. Therefore,

$$\frac{d}{dt}(dx_i) = \frac{\partial v_i}{\partial X_k} dX_k = \frac{\partial v_i}{\partial x_l} \cdot \frac{\partial x_l}{\partial X_k} dX_k$$

$$= \frac{\partial v_i}{\partial x_l} dx_l = v_{i,l} dx_l; \text{ as } dx_i = \frac{\partial x_i}{\partial X_k} dX_k$$

which is the expression for the material derivative of the displacement gradients. Thus

(i) The material derivative of the displacement gradients $dx_i \equiv \dfrac{\partial x_i}{\partial X_k} dX_k$ is given by

$$\frac{d}{dt}\left(\frac{\partial x_i}{\partial X_k}\right) = v_{i,l} \frac{\partial x_l}{\partial X_k}; \quad \frac{d}{dt}(dx_i) = v_{i,l} dx_l$$

(ii) Let us take the relation $\dfrac{\partial x_l}{\partial X_p} \dfrac{\partial X_p}{\partial x_k} = \delta_{kl}$, the material derivative gives

$$\frac{d}{dt}\left(\frac{\partial x_l}{\partial X_p} \frac{\partial X_p}{\partial x_k}\right) = \frac{d}{dt}(\delta_{kl}) \Rightarrow \frac{d}{dt}\left(\frac{\partial x_l}{\partial X_p}\right)\frac{\partial X_p}{\partial x_k} + \frac{\partial x_l}{\partial X_p}\frac{d}{dt}\left(\frac{\partial X_p}{\partial x_k}\right) = 0$$

$$\Rightarrow \frac{\partial x_l}{\partial X_p} \frac{d}{dt}\left(\frac{\partial X_p}{\partial x_k}\right) = -\frac{d}{dt}\left(\frac{\partial x_l}{\partial X_p}\right)\frac{\partial X_p}{\partial x_k} = -\frac{\partial}{\partial X_p}\left(\frac{dx_l}{dt}\right)\frac{\partial X_p}{\partial x_k}$$

$$= -\frac{\partial v_l}{\partial X_p}\frac{\partial X_p}{\partial x_k} = -\frac{\partial v_l}{\partial x_k}\frac{\partial x_k}{\partial X_p}\frac{\partial X_p}{\partial x_k} = -\frac{\partial v_l}{\partial x_k} = -v_{l,k}$$

Multiplying both sides by $\dfrac{\partial X_p}{\partial x_l}$, we obtain

$$\frac{\partial X_p}{\partial x_l} \frac{\partial x_l}{\partial X_p} \frac{d}{dt}\left(\frac{\partial X_p}{\partial x_k}\right) = -v_{l,k} \frac{\partial X_p}{\partial x_l} \Rightarrow \frac{d}{dt}\left(\frac{\partial X_p}{\partial x_k}\right) = -v_{l,k} \frac{\partial X_p}{\partial x_l}.$$

2.7.2 Material derivatives of the square of the arc length

Consider a line element P_0Q_0 at $P_0(X_1, X_2, X_3)$ of a continuum body occupying initial configuration B_0 at time $t = 0$. When the body undergoes deformation, the same particles which initially lie on the line P_0Q_0 will now lie on a new line element PQ of length dl oriented in the direction (n_1, n_2, n_3) occupying current configuration B at time t. If P has coordinates (x_1, x_2, x_3) and Q has coordinates $(x_1 + dx_1, x_2 + dx_2, x_3 + dx_3)$ then

$$dl^2 = dx_1^2 + dx_2^2 + dx_3^2 = dx_i dx_i = \delta_{ij} dx_i dx_j$$

and $n_i = \dfrac{dx_i}{dl}$. Now the Material derivatives of the square of the arc length is given by

$$\frac{d}{dt}(dl^2) = \delta_{ij}\left[dx_j \frac{d}{dt}(dx_i) + dx_i \frac{d}{dt}(dx_j)\right]$$

$$= \delta_{ij}\left[dx_j \frac{d}{dt}\left(\frac{\partial x_i}{\partial X_k} dX_k\right) + dx_i \frac{d}{dt}\left(\frac{\partial x_j}{\partial X_l} dX_l\right)\right]$$

$$= \delta_{ij}\left[dx_j \frac{d}{dt}\left(\frac{\partial x_i}{\partial X_k}\right) dX_k + dx_i \frac{d}{dt}\left(\frac{\partial x_j}{\partial X_l}\right) dX_l\right]$$

$$= \delta_{ij}\left[dx_j \frac{\partial}{\partial X_k}\left(\frac{dx_i}{dt}\right) dX_k + dx_i \frac{\partial}{\partial X_l}\left(\frac{dx_j}{dt}\right) dX_l\right]$$

$$= \delta_{ij}\left[dx_j \frac{\partial v_i}{\partial X_k} dX_k + dx_i \frac{\partial v_j}{\partial X_l} dX_l\right] \text{ as } v_i = \frac{dx_i}{dt}$$

$$= \delta_{ij}\left[dx_j \frac{\partial v_i}{\partial x_l} \frac{\partial x_l}{\partial X_k} dX_k + dx_i \frac{\partial v_j}{\partial x_l} \frac{\partial x_l}{\partial X_l} dX_l\right]$$

$$= \frac{\partial v_j}{\partial x_l} dx_l dx_j + \frac{\partial v_i}{\partial x_l} dx_l dx_i = \frac{\partial v_j}{\partial x_i} dx_i dx_j + \frac{\partial v_i}{\partial x_j} dx_j dx_i$$

$$= \left(\frac{\partial v_i}{\partial x_j} + \frac{\partial v_j}{\partial x_i}\right) dx_i dx_j = 2d_{ij} dx_i dx_j \text{ where } d_{ij} = \frac{1}{2}\left(\frac{\partial v_i}{\partial x_j} + \frac{\partial v_j}{\partial x_i}\right) \quad (2.22)$$

The equation (2.22) gives the time rate of change of square of the length of a line element. From the equation (2.22), it follows that $2d_{ij}dx_i dx_j$ is a scalar but the product $dx_i dx_j$ is a tensor of order two. Therefore, by quotient law of tensor, d_{ij} is a second order tensor. Furthermore

$$d_{ji} = \frac{1}{2}\left(\frac{\partial v_j}{\partial x_i} + \frac{\partial v_i}{\partial x_j}\right) = \frac{1}{2}\left(\frac{\partial v_i}{\partial x_j} + \frac{\partial v_j}{\partial x_i}\right) = d_{ij}$$

So the tensor d_{ij} is symmetric. d_{ij} is called symmetric second order strain-rate tensor or the deformation rate tensor of Euler. Now

(i) For an arbitrary pair of points (arbitrary dx) it is clear from equation (2.22) that $\frac{d}{dt}(dl^2) = 0$ if and only if $d_{ij} = 0$. But this implies that the element of arc length for any pair is not changing in time. This, of course, is the definition of rigid body motion.

(ii) The necessary and sufficient condition for the motion of a body to be rigid is $d_{ij} = 0$.

2.7.3 Material Derivative of Angle Between Two Line Elements

Consider a line element $P_0 Q_0$ and $P_0 R_0$ at $P_0(X_1, X_2, X_3)$ of a continuum body occupying initial configuration B_0 at time $t = 0$. When continuum body undergoes deformation these two line elements PQ and PR at $P(x_1, x_2, x_3)$ of respective lengths dl and δl oriented in the respective directions (n_1, n_2, n_3) and (m_1, m_2, m_3) and inclined at an angle θ in the current deformed state. If Q has coordinates $(x_1 + dx_1, x_2 + dx_2, x_3 + dx_3)$, R has coordinates $(x_1 + \delta x_1, x_2 + \delta x_2, x_3 + \delta x_3)$ then

$$m_i = \frac{\delta x_i}{\delta l}, \quad n_i = \frac{dx_i}{dl}; \quad \cos\theta = m_i \cdot n_i = \frac{dx_i}{dl} \cdot \frac{\delta x_i}{\delta l} \qquad (2.23)$$

Let v_i be the velocity of the typical particle which was at $P_0(X_1, X_2, X_3)$ subsequently occupies the position $P(x_1, x_2, x_3)$ at time t. Let $v_i + dv_i$ be the velocity of another particle occupying the position $Q(x_1 + dx_1, x_2 + dx_2, x_3 + dx_3)$ and $v_i + \delta v_i$ be the velocity of the third particle occupying the position $R(x_1 + \delta x_1, x_2 + \delta x_2, x_3 + \delta x_3)$ at time t, then $v_i = f_i(x_1, x_2, x_3, t)$ and

$$v_i + dv_i = f_i(x_1 + dx_1, x_2 + dx_2, x_3 + dx_3, t)$$

$$= f_i(x_1, x_2, x_3, t) + \frac{\partial f_1}{\partial x_1}dx_1 + \frac{\partial f_2}{\partial x_2}dx_2 + \frac{\partial f_3}{\partial x_3}dx_3 + \ldots$$

$$= f_i(x_1, x_2, x_3, t) + \frac{\partial f_i}{\partial x_j}dx_j = v_i + \frac{\partial v_i}{\partial x_j}dx_j$$

2.7 Material Derivative of the Element of Arc, Surface, and Volume

$$dv_i = \frac{\partial v_i}{\partial x_j} dx_j \text{ and similarly } \delta v_i = \frac{\partial v_i}{\partial x_j} \delta x_j \qquad (2.24)$$

Now dv_i, the relative velocity of $Q(x_1 + dx_1, x_2 + dx_2, x_3 + dx_3)$ with respect to the position $P(x_1, x_2, x_3)$, will be time rate of the change of dx_i, the relative position of Q with respect to P so that we can write

$$dv_i = \frac{d}{dt}(dx_i) \text{ and similarly, } \delta v_i = \frac{d}{dt}(\delta x_i) \qquad (2.25)$$

Now we shall find the rate of change of angle θ. From equation (2.22) we get

$$\frac{d}{dt}(dl^2) = 2dl\frac{d}{dt}(dl) = 2d_{ij}dx_i dx_j \text{ where } d_{ij} = \frac{1}{2}\left(\frac{\partial v_i}{\partial x_j} + \frac{\partial v_j}{\partial x_i}\right)$$

or,

$$\frac{\frac{d}{dt}(dl)}{dl} = d_{ij}\frac{dx_i}{dl}\frac{dx_j}{dl} = d_{ij} n_i . n_j \qquad (2.26)$$

Similarly,

$$\frac{\frac{d}{dt}(\delta l)}{\delta l} = d_{ij}\frac{\delta x_i}{\delta l}\frac{\delta x_j}{\delta l} = d_{ij} m_i . m_j$$

From equation (2.23) we get $dx_i \delta x_i = dl\, \delta l \cos\theta$, differentiating with respect to t, we get

$$\frac{d}{dt}(dx_i \delta x_i) = \cos\theta\left[\frac{d}{dt}(dl)\delta l + dl\frac{d}{dt}(\delta l)\right] - dl\,\delta l \sin\theta \cdot \dot\theta$$

$$\frac{\frac{d}{dt}(dx\,\delta x_i)}{dl\,\delta l} = \cos\theta\left[\frac{\frac{d}{dt}(dl)}{dl} + \frac{\frac{d}{dt}(\delta l)}{\delta l}\right] - \sin\theta \cdot \dot\theta$$

or,

$$\frac{\frac{d}{dt}(\delta x_i)dx_i + \frac{d}{dt}(dx_i)\delta x_i}{dl\,\delta l} = \cos\theta\left[d_{ij}n_i \cdot n_j + d_{ij}m_i \cdot m_j\right] - \sin\theta \cdot \dot\theta$$

or,

$$\frac{\frac{\partial v_i}{\partial x_j}\delta x_j dx_i + \frac{\partial v_i}{\partial x_j}dx_j \delta x_i}{dl\,\delta l} = \cos\theta\left[d_{ij}n_i \cdot n_j + d_{ij}m_i \cdot m_j\right] - \sin\theta \cdot \dot\theta$$

or,

$$\frac{\delta v_i dx_i + dv_i \delta x_i}{dl\,\delta l} = \cos\theta\left[d_{ij}n_i \cdot n_j + d_{ij}m_i \cdot m_j\right] - \sin\theta \cdot \dot\theta\text{ ; from Eq.(2.24)}$$

$$\text{or,}\quad \frac{\dfrac{\partial v_i}{\partial x_j}\delta x_j dx_i + \dfrac{\partial v_i}{\partial x_j}dx_j \delta x_i}{dl\ \delta l} = \cos\theta\left[d_{ij}n_i\cdot n_j + d_{ij}m_i\cdot m_j\right] - \sin\theta\cdot\dot{\theta}$$

(interchanging the dummy indices)

$$\text{or,}\quad \left(\frac{\partial v_i}{\partial x_j} + \frac{\partial v_j}{\partial x_i}\right)\frac{dx_i}{dl}\cdot\frac{\delta x_j}{\delta l} = \cos\theta\left[d_{ij}n_i\cdot n_j + d_{ij}m_i\cdot m_j\right] - \sin\theta\cdot\dot{\theta}$$

$$\text{or,}\quad 2d_{ij}n_i\cdot m_j = \cos\theta\left(n_i\cdot n_j + m_i\cdot m_j\right)d_{ij} - \sin\theta\cdot\dot{\theta} \qquad (2.27)$$

If $d_{ij} = 0$, we have from equation (2.22)

$$\frac{d}{dt}\left(dl^2\right) = 2dl\frac{d}{dt}(dl) = 0$$

and from equation (2.27), we get

$$\dot{\theta} = \frac{d}{dt}(\theta) = 0.$$

This implies that length of a line element and angle between two linear elements does not change with time and the body has undergone only rigid body motion. Thus the necessary and sufficient condition for the motion of the body to be rigid is symmetric second order strain-rate tensor $d_{ij} = 0$.

In other words d_{ij} cause a time rate of change in the length and orientation of a line element, it may be taken as exact measure of rate of strain.

Geometrical interpretation: From equation (2.26), we get

$$\frac{\dfrac{d}{dt}(dl)}{dl} = d_{ij}\frac{dx_i}{dl}\frac{dx_j}{dl} = d_{ij}n_i.n_j$$

$$\text{or,}\quad \frac{1}{dl}\cdot\frac{d}{dt}(dl - dL) = d_{ij}n_i.n_j \qquad (2.28)$$

where dL is original length of a line element, L.H.S. represents the time rate of change of extension of a line element per unit current length called the rate of extension and is denoted by $d_{(n)}$. Thus time rate of extension of a line element currently oriented in the direction (n_1, n_2, n_3) is given by

$$d_{(n)} = d_{ij}n_i.n_j \qquad (2.29)$$

where, d_{ij} is a symmetric second order strain-rate tensor. To give a geometrical interpretation of d_{11}, d_{22}, d_{33} consider a line element currently parallel to X_1 axis. Then we have

$$n_1 = 1, n_2 = 0, n_3 = 0; d_{(1)} = d_{11}.$$

2.7 Material Derivative of the Element of Arc, Surface, and Volume

Thus d_{11} is time rate of extension of line element currently parallel to X_1 axis. Similarly, d_{22}, d_{33} represent the time rate of extension of line element currently parallel to X_2 and X_3 axis respectively.

Theorem 2.2: (Time rate of change of shear between two orthogonal line elements) For small displacement gradients, strain-rate is equal to the time rate of change of small strain.

Proof: Consider a pair of orthogonal line elements currently oriented in the directions (n_1, n_2, n_3) and (m_1, m_2, m_3) then $\theta = \dfrac{\pi}{2}$. From equation (2.27), we get

$$2d_{ij} n_i \cdot m_j = \cos\frac{\pi}{2}(n_i \cdot n_j + m_i \cdot m_j) d_{ij} - \sin\frac{\pi}{2} \cdot \dot\theta = -\dot\theta = \frac{d}{dt}\left(\frac{\pi}{2} - \theta\right)$$

or
$$\frac{d}{dt}\left(\frac{\pi}{2} - \theta\right) = 2d_{ij} n_i . m_j \qquad (2.30)$$

where, $d_{ij} = \dfrac{1}{2}\left(\dfrac{\partial v_i}{\partial x_j} + \dfrac{\partial v_j}{\partial x_i}\right)$ = symmetric second order strain-rate tensor at $P_0(X_1, X_2, X_3)$. L.H.S. represents the rate at which right angle between two line elements decreases and is called the rate of shear. We denote it by $S_{(n,m)}$. Thus the rate of shear between two orthogonal line elements currently oriented in the directions (n_1, n_2, n_3) and (m_1, m_2, m_3) respectively is given by

$$S_{(n,m)} = 2d_{ij} n_i . m_j. \qquad (2.31)$$

To give the geometrical interpretation of d_{12}, we take a pair of line elements currently parallel to X_1 and X_2 axes, respectively. Then

$$n_1 = 1, n_2 = 0, n_3 = 0; \quad m_1 = 0, m_2 = 1,$$
$$m_3 = -\dot\theta = 2d_{ij} n_i . m_j = 2d_{12}; \quad i.e., d_{12} = -\frac{1}{2}\cdot\dot\theta. \qquad (2.32)$$

Thus d_{12} represents half the time rate of shear between two line elements currently parallel to X_1 axis and X_2 axis axes, respectively. Similarly, for d_{23} and d_{31}. Strain-rate tensor d_{ij} has properties which in almost every respect are analogous to those of small strain tensor E_{ij} or e_{ij}.

The tensor d_{ij} differs from E_{ij} in the sense that it is exact measure of strain-rate whereas E_{ij} can never be an exact measure of strain, it is an approximate measure.

Now, $\dfrac{dE_{ij}}{dt} = \dfrac{1}{2}\left[\dfrac{d}{dt}\left(\dfrac{\partial u_i}{\partial X_j}\right) + \dfrac{d}{dt}\left(\dfrac{\partial u_j}{\partial X_i}\right)\right]$

or,
$$\frac{dE_{ij}}{dt} = \frac{1}{2}\left[\frac{\partial}{\partial X_j}\left(\frac{du_i}{dt}\right) + \frac{\partial}{\partial X_i}\left(\frac{du_j}{dt}\right)\right] = \frac{1}{2}\left[\frac{\partial v_i}{\partial X_j} + \frac{\partial v_j}{\partial X_i}\right].$$

In case of small displacement gradients $\frac{\partial v_i}{\partial X_j} = \frac{\partial v_i}{\partial x_j}; \frac{\partial v_j}{\partial X_i} = \frac{\partial v_j}{\partial x_i}$ and therefore

$$\frac{dE_{ij}}{dt} = \frac{1}{2}\left[\frac{\partial v_i}{\partial X_j} + \frac{\partial v_j}{\partial X_i}\right] = \frac{1}{2}\left[\frac{\partial v_i}{\partial x_j} + \frac{\partial v_j}{\partial x_i}\right] = d_{ij} = \frac{de_{ij}}{dt}. \qquad (2.33)$$

Therefore, for small displacement gradients, strain-rate is equal to the time rate of change of small strain.

Example 2.15: A steady field is given by $v_1 = 3x_1^2 x_2, v_2 = 2x_2^2 x_3, v_3 = x_1 x_2 x_3^2$ determine the rate of extension at $P(1,1,1)$ in the direction of $\left(\frac{3}{5}, 0, -\frac{4}{5}\right)$. Determine the rate of shear between orthogonal direction $\left(\frac{3}{5}, 0, -\frac{4}{5}\right)$ and $\left(\frac{4}{5}, 0, \frac{3}{5}\right)$.

Solution: The spin rate tensors are given by

$$d_{11} = \frac{1}{2}\left[\frac{\partial v_1}{\partial x_1} + \frac{\partial v_1}{\partial x_1}\right] = \frac{\partial v_1}{\partial x_1} = 6x_1 x_2 = 6 \text{ at } P$$

$$d_{22} = \frac{1}{2}\left[\frac{\partial v_2}{\partial x_2} + \frac{\partial v_2}{\partial x_2}\right] = \frac{\partial v_2}{\partial x_2} = 4x_2 x_3 = 4 \text{ at } P$$

$$d_{33} = \frac{1}{2}\left[\frac{\partial v_3}{\partial x_3} + \frac{\partial v_3}{\partial x_3}\right] = \frac{\partial v_3}{\partial x_3} = 2x_1 x_2 x_3 = 2 \text{ at } P$$

$$d_{12} = d_{21} = \frac{1}{2}\left[\frac{\partial v_1}{\partial x_2} + \frac{\partial v_2}{\partial x_1}\right] = \frac{3}{2}x_1^2 = \frac{3}{2} \text{ at } P$$

$$d_{13} = d_{31} = \frac{1}{2}\left[\frac{\partial v_1}{\partial x_3} + \frac{\partial v_3}{\partial x_1}\right] = \frac{1}{2}x_2 x_3^2 = \frac{1}{2} \text{ at } P$$

$$d_{23} = d_{23} = \frac{1}{2}\left[\frac{\partial v_2}{\partial x_3} + \frac{\partial v_3}{\partial x_2}\right] = \frac{1}{2}\left[2x_2^2 + x_1 x_3^2\right] = \frac{3}{2} \text{ at } P$$

In matrix notation, the spin rate tensors at $P(1,1,1)$ are given by

2.7 Material Derivative of the Element of Arc, Surface, and Volume

$$(d_{ij}) = \begin{pmatrix} 6x_1x_2 & \frac{3}{2}x_1^2 & \frac{1}{2}x_2x_3^2 \\ \frac{3}{2}x_1^2 & 4x_2x_3 & \frac{1}{2}(2x_2^2 + x_1x_3^2) \\ \frac{1}{2}x_2x_3^2 & \frac{1}{2}(2x_2^2 + x_1x_3^2) & 2x_1x_2x_3 \end{pmatrix} = \begin{pmatrix} 6 & \frac{3}{2} & \frac{1}{2} \\ \frac{3}{2} & 4 & \frac{3}{2} \\ \frac{1}{2} & \frac{3}{2} & 2 \end{pmatrix}.$$

At the point $P(1,1,1)$, the extension rate in the direction of $n = \left(\frac{3}{5}, 0, -\frac{4}{5}\right)$ is given by

$$d_{ij}n_in_j = \left(\frac{3}{5}, 0, -\frac{4}{5}\right) \begin{pmatrix} 6 & \frac{3}{2} & \frac{1}{2} \\ \frac{3}{2} & 4 & \frac{3}{2} \\ \frac{1}{2} & \frac{3}{2} & 2 \end{pmatrix} \begin{pmatrix} \frac{3}{5} \\ 0 \\ -\frac{4}{5} \end{pmatrix} = \frac{74}{25}.$$

At the point $P(1,1,1)$, the rate of shear between the orthogonal direction $l = \left(\frac{3}{5}, 0, -\frac{4}{5}\right)$ and $m = \left(\frac{4}{5}, 0, \frac{3}{5}\right)$ is given by

$$2d_{ij}l_im_j = 2\left(\frac{3}{5}, 0, -\frac{4}{5}\right) \begin{pmatrix} 6 & \frac{3}{2} & \frac{1}{2} \\ \frac{3}{2} & 4 & \frac{3}{2} \\ \frac{1}{2} & \frac{3}{2} & 2 \end{pmatrix} \begin{pmatrix} \frac{4}{5} \\ 0 \\ \frac{3}{5} \end{pmatrix} = \frac{89}{25}.$$

2.7.4 Material derivatives of the Jacobian

The Jacobian of transformation $x_i = x_i(X_1, X_2, X_3, t)$ is given by

$$J = J\begin{pmatrix} x_1 & x_2 & x_3 \\ X_1 & X_2 & X_3 \end{pmatrix} = \frac{\partial(x_1, x_2, x_3)}{\partial(X_1, X_2, X_3)} = \begin{vmatrix} \frac{\partial x_1}{\partial X_1} & \frac{\partial x_1}{\partial X_2} & \frac{\partial x_1}{\partial X_3} \\ \frac{\partial x_2}{\partial X_1} & \frac{\partial x_2}{\partial X_2} & \frac{\partial x_2}{\partial X_3} \\ \frac{\partial x_3}{\partial X_1} & \frac{\partial x_3}{\partial X_2} & \frac{\partial x_3}{\partial X_3} \end{vmatrix}.$$

Hence the material derivative of the Jacobian of transformation $x_i = x_i(X_1, X_2, X_3, t)$ is given by

$$\frac{dJ}{dt} = \frac{d}{dt}\begin{vmatrix} \frac{\partial x_1}{\partial X_1} & \frac{\partial x_1}{\partial X_2} & \frac{\partial x_1}{\partial X_3} \\ \frac{\partial x_2}{\partial X_1} & \frac{\partial x_2}{\partial X_2} & \frac{\partial x_2}{\partial X_3} \\ \frac{\partial x_3}{\partial X_1} & \frac{\partial x_3}{\partial X_2} & \frac{\partial x_3}{\partial X_3} \end{vmatrix} = \begin{vmatrix} \frac{d}{dt}\left(\frac{\partial x_1}{\partial X_1}\right) & \frac{d}{dt}\left(\frac{\partial x_1}{\partial X_2}\right) & \frac{d}{dt}\left(\frac{\partial x_1}{\partial X_3}\right) \\ \frac{\partial x_2}{\partial X_1} & \frac{\partial x_2}{\partial X_2} & \frac{\partial x_2}{\partial X_3} \\ \frac{\partial x_3}{\partial X_1} & \frac{\partial x_3}{\partial X_2} & \frac{\partial x_3}{\partial X_3} \end{vmatrix} +$$

$$\begin{vmatrix} \frac{\partial x_1}{\partial X_1} & \frac{\partial x_1}{\partial X_2} & \frac{\partial x_1}{\partial X_3} \\ \frac{d}{dt}\left(\frac{\partial x_2}{\partial X_1}\right) & \frac{d}{dt}\left(\frac{\partial x_2}{\partial X_2}\right) & \frac{d}{dt}\left(\frac{\partial x_2}{\partial X_3}\right) \\ \frac{\partial x_3}{\partial X_1} & \frac{\partial x_3}{\partial X_2} & \frac{\partial x_3}{\partial X_3} \end{vmatrix} + \begin{vmatrix} \frac{\partial x_1}{\partial X_1} & \frac{\partial x_1}{\partial X_2} & \frac{\partial x_1}{\partial X_3} \\ \frac{\partial x_2}{\partial X_1} & \frac{\partial x_2}{\partial X_2} & \frac{\partial x_2}{\partial X_3} \\ \frac{d}{dt}\left(\frac{\partial x_3}{\partial X_1}\right) & \frac{d}{dt}\left(\frac{\partial x_3}{\partial X_2}\right) & \frac{d}{dt}\left(\frac{\partial x_3}{\partial X_3}\right) \end{vmatrix}$$

$$= \begin{vmatrix} \frac{\partial v_1}{\partial X_1} & \frac{\partial v_1}{\partial X_2} & \frac{\partial v_1}{\partial X_3} \\ \frac{\partial x_2}{\partial X_1} & \frac{\partial x_2}{\partial X_2} & \frac{\partial x_2}{\partial X_3} \\ \frac{\partial x_3}{\partial X_1} & \frac{\partial x_3}{\partial X_2} & \frac{\partial x_3}{\partial X_3} \end{vmatrix} + \begin{vmatrix} \frac{\partial x_1}{\partial X_1} & \frac{\partial x_1}{\partial X_2} & \frac{\partial x_1}{\partial X_3} \\ \frac{\partial v_2}{\partial X_1} & \frac{\partial v_2}{\partial X_2} & \frac{\partial v_2}{\partial X_3} \\ \frac{\partial x_3}{\partial X_1} & \frac{\partial x_3}{\partial X_2} & \frac{\partial x_3}{\partial X_3} \end{vmatrix} + \begin{vmatrix} \frac{\partial x_1}{\partial X_1} & \frac{\partial x_1}{\partial X_2} & \frac{\partial x_1}{\partial X_3} \\ \frac{\partial x_2}{\partial X_1} & \frac{\partial x_2}{\partial X_2} & \frac{\partial x_2}{\partial X_3} \\ \frac{\partial v_3}{\partial X_1} & \frac{\partial v_3}{\partial X_2} & \frac{\partial v_3}{\partial X_3} \end{vmatrix}$$

as $\quad \dfrac{d}{dt}\left(\dfrac{\partial x_i}{\partial X_j}\right) = \dfrac{\partial}{\partial X_j}\left(\dfrac{dx_i}{dt}\right) = \dfrac{\partial v_i}{\partial X_j}$

$$= \frac{\partial(v_1, x_2, x_3)}{\partial(X_1, X_2, X_3)} + \frac{\partial(x_1, v_2, x_3)}{\partial(X_1, X_2, X_3)} + \frac{\partial(x_1, x_2, v_3)}{\partial(X_1, X_2, X_3)}$$

$$= \frac{\partial(v_1, x_2, x_3)}{\partial(x_1, x_2, x_3)} \cdot \frac{\partial(x_1, x_2, x_3)}{\partial(X_1, X_2, X_3)} + \frac{\partial(v_2, x_2, x_3)}{\partial(x_3, x_2, x_3)} \cdot \frac{\partial(x_1, x_2, x_3)}{\partial(X_1, X_2, X_3)}$$
$$+ \frac{\partial(v_3, x_2, x_3)}{\partial(x_1, x_2, x_3)} \cdot \frac{\partial(x_1, x_2, x_3)}{\partial(X_1, X_2, X_3)}$$

$$= \begin{vmatrix} \frac{\partial v_1}{\partial x_1} & \frac{\partial v_1}{\partial x_2} & \frac{\partial v_1}{\partial x_3} \\ 0 & 1 & 0 \\ 0 & 0 & 1 \end{vmatrix} J + \begin{vmatrix} 1 & 0 & 0 \\ \frac{\partial v_2}{\partial X_1} & \frac{\partial v_2}{\partial X_2} & \frac{\partial v_2}{\partial X_3} \\ 0 & 0 & 1 \end{vmatrix} J + \begin{vmatrix} 1 & 0 & 0 \\ 0 & 1 & 0 \\ \frac{\partial v_3}{\partial X_1} & \frac{\partial v_3}{\partial X_2} & \frac{\partial v_3}{\partial X_3} \end{vmatrix} J$$

$$= J\frac{\partial v_1}{\partial X_1} + J\frac{\partial v_2}{\partial X_2} + J\frac{\partial v_3}{\partial X_3} = J\left(\frac{\partial v_1}{\partial X_1} + \frac{\partial v_2}{\partial X_2} + \frac{\partial v_3}{\partial X_3}\right) = J\frac{\partial v_i}{\partial X_i} = J\text{ div }\vec{v}.$$

2.7.5 Material derivatives of the element of area

In this section we investigate the change of area with the deformation. We have already found that an infinitesimal rectangular parallelepiped with edge vectors $I_1 dX_1, I_2 dX_2$ and $I_3 dX_3$, after deformation becomes a rectilinear parallelepiped with edge vectors $C_1 dX_1, C_2 dX_2$ and $C_3 dX_3$ where $C_j = \dfrac{\partial x_j}{\partial X_k}i_j$ (Fig. 2.2).

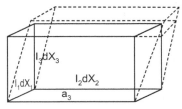

Fig. 2.2: Deformation of an infinitesimal rectangular parallelepiped

We calculate the area vector da_3 according as

$$da_3 = (C_1 dX_1)\times(C_2 dX_2) = \frac{\partial x_k}{\partial X_1}\frac{\partial x_l}{\partial X_2}dX_1 dX_2 (i_k \times i_l)$$

$$= \frac{\partial x_k}{\partial X_1}\frac{\partial x_l}{\partial X_2}dA_3 \varepsilon_{klm} i_m; dA_3 = dX_1 dX_2, \quad i_k \times i_l = \varepsilon_{klm} i_m.$$

where, the alternating symbol ε_{ijk} is the permutation symbol, defined by

$\varepsilon_{ijk} = 0$, if any two of i, j, k are equal
$\phantom{\varepsilon_{ijk}} = 1$, if i, j, k are even permutation of 1,2,3
$\phantom{\varepsilon_{ijk}} = -1$ if i, j, k are odd permutation of 1,2,3

This may be expressed in terms of Jacobian

$$J = \left|\frac{\partial x_k}{\partial X_k}\right| = \varepsilon_{klm}\frac{\partial x_k}{\partial X_1}\frac{\partial x_l}{\partial X_2}\frac{\partial x_m}{\partial X_3}.$$

Through the chain rule of partial differentiation it is clear that

$$\frac{\partial x_k}{\partial X_j}\frac{\partial X_j}{\partial x_l} = \delta_{kl}; \quad \frac{\partial X_j}{\partial x_k}\frac{\partial x_k}{\partial X_m} = \delta_{jm}.$$

Each one of these two sets is a set of nine linear equations for the nine unknowns $\dfrac{\partial x_k}{\partial X_j}$ or $\dfrac{\partial X_j}{\partial x_k}$. A unique solution exists, since the jacobian of the transformation is assumed not to vanish. Using Cramer's rule of determinants, the solution for $\dfrac{\partial X_j}{\partial x_k}$ may be obtained in terms of $\dfrac{\partial x_k}{\partial X_j}$. Thus

$$\frac{\partial X_i}{\partial x_p} = \frac{\text{cofactor } \dfrac{\partial x_p}{\partial X_i}}{J} = \frac{1}{2J}\varepsilon_{ijk}\varepsilon_{pqr}\frac{\partial x_q}{\partial X_j}\frac{\partial x_r}{\partial X_k} \Rightarrow J\frac{\partial X_3}{\partial x_m} = \varepsilon_{klm}\frac{\partial x_k}{\partial X_1}\frac{\partial x_l}{\partial X_2}.$$

Therefore, the area vector da_3 is given by

$$da_3 = \frac{\partial x_k}{\partial X_1}\frac{\partial x_l}{\partial X_2}dA_3\varepsilon_{klm}i_m = J\frac{\partial X_3}{\partial x_m}dA_3 i_m$$

Similar expressions are obtained for the vector area da_1 and da_2 by replacing the index 3 by 1 and 2, respectively as

$$da_1 = J\frac{\partial X_1}{\partial x_m}dA_1 i_m; \quad da_2 = J\frac{\partial X_2}{\partial x_m}dA_2$$

The area vector da is therefore given by

$$da = da_1 + da_2 + da_3 = J\left[\frac{\partial X_1}{\partial x_m}dA_1 + \frac{\partial X_2}{\partial x_m}dA_2 + \frac{\partial X_3}{\partial x_m}dA_3\right]i_m = J\frac{\partial X_k}{\partial x_m}dA_k i_m$$

Therefore, $da_m = J\dfrac{\partial X_k}{\partial_m}dA_k$. Now the material derivative of the element of area is given by

$$\frac{d}{dt}(da_k) = \frac{d}{dt}\left(J\frac{\partial X_m}{\partial x_k}dA_m\right) = \left[\frac{dJ}{dt}\frac{\partial X_m}{\partial x_k} + J\frac{d}{dt}\left(\frac{\partial X_m}{\partial x_k}\right)\right]dA_m$$

$$= J\frac{\partial v_i}{\partial X_i}dA_m - Jv_{l,k}\frac{\partial X_m}{\partial x_l}dA_m; \text{ as } \frac{dJ}{dt} = J\frac{\partial v_i}{\partial X_i}, \frac{d}{dt}\left(\frac{\partial X_m}{\partial x_k}\right) = -v_{l,k}\frac{\partial X_m}{\partial x_l}$$

$$= v_{m,m}da_k - v_{m,k}da_m.$$

2.7.6 Material Derivatives of the Volume Element

Consider a continuum body occupying, at time $t = 0$ an elementary rectangular parallelepiped of volume dV_0 with one of its vertices at a point $P_0(X_1, X_2, X_3)$ of a continuum body in the undeformed state. Due to deformation, the rectangular parallelepiped deforms into a skew parallelepiped of volume element dV with one of its vertices at a point $P(x_1, x_2, x_3)$ in the deformed state. Then

2.7 Material Derivative of the Element of Arc, Surface, and Volume

$$dV = JdV_0, \text{ where, } J = J\begin{pmatrix} x_1 & x_2 & x_3 \\ X_1 & X_2 & X_3 \end{pmatrix} = \frac{\partial(x_1, x_2, x_3)}{\partial(X_1, X_2, X_3)}$$

Hence the material derivative of the volume element is given by

$$\frac{d}{dt}(dV) = \frac{d}{dt}(JdV_0) = \frac{dJ}{dt}dV_0 + J\frac{d}{dt}(dV_0)$$

$$= \frac{dJ}{dt}dV_0, \text{ as } dV_0 \text{ is independent of } t$$

$$= \left[\frac{\partial v_1}{\partial X_1} + \frac{\partial v_2}{\partial X_2} + \frac{\partial v_3}{\partial X_3}\right]JdV_0 = \frac{\partial v_i}{\partial X_i}dV = v_{i,i}dV$$

which gives the rate of change of volume element. Now

$$\frac{d}{dt}(dV) = v_{i,i}dV \Rightarrow \frac{\frac{d}{dt}(dV)}{dV} = v_{i,i}$$

Thus $v_{i,i}$ is the rate of increase of volume per unit volume and is called time rate of dilation or the first deformation invariant.

Theorem 2.3: (Reynolds transport theorem): Let $\psi(x,t)$ denotes a material property per unit volume of a continuum. Here $\psi(x,t)$ may be a scalar, vector or a second order tensor. The total amount of $\psi(x,t)$ in the material volume at time t is given by $V(t)$

$$\bar{\psi}(t) = \int_{V(t)} \psi(x,t)dV = \int_{V(t)} \psi(x_1, x_2, x_3, t)dV \qquad (2.34)$$

The time rate of increase of the total amount $\bar{\psi}(t)$ as the volume $V(t)$ moves with the continuum is given by

$$\frac{d\bar{\psi}}{dt} = \frac{d}{dt}\int_{V(t)} \psi(x,t)dV = \frac{d}{dt}\int_{V(t)} \psi(x_1, x_2, x_3, t)dV = \int_{V(t)} \frac{d}{dt}[\psi(x,t)dV]$$

Since the volume always consist of same set of particles, the order of the integration and differentiation can be changed. Therefore,

$$\frac{d\bar{\psi}}{dt} = \int_{V(t)} \frac{d}{dt}[\psi(x,t)dV] = \int_{V(t)} \frac{d\psi}{dt}dV + \int_{V(t)} \psi(x,t)\frac{d}{dt}[dV]$$

$$= \int_{V}^{(t)}\left[\frac{d\psi}{dt} + \psi(x,t)\operatorname{div}\bar{v}\right]dV, \text{ as } \frac{d}{dt}[dV] = \operatorname{div}\bar{v}dV$$

$$= \int_{V(t)}\left[\frac{\partial\psi}{\partial t} + v_k\frac{\partial\psi}{\partial x_k} + \psi\frac{\partial v_k}{\partial x_k}\right]dV; \bar{v} = (v_1, v_2, v_3)$$

$$= \int_{V(t)} \left[\frac{\partial \psi}{\partial t} + \frac{\partial}{\partial x_k}(\psi v_k) \right] dV = \int_{V(t)} \frac{\partial \psi}{\partial t} dV + \iint_S \psi v_k n_k dS \qquad (2.35)$$

where we apply the Gauss theorem to the second integral and S is the boundary of $V(t)$ and is the direction cosine of the unit outward normal to the element dS at instant t. Equation (2.35) states that the time rate of change of the total property over a material volume V is equal to the rate at which ψ is increasing with time t in the volume V instantaneously coincide with the fixed spatial volume and the flux of ψ across the boundary of the fixed volume at the instant t. The equation given by (2.35) is called the Raynold's transport theorem.

(i) If $div\,\bar{v} = \bar{0}$, i.e., if the motion of the continuum is volume preserving then

$$\frac{d\bar{\psi}}{dt} = \int_{V(t)} \left[\frac{d\psi}{dt} + \psi(x,t) div\,\bar{v} \right] dV = \int_{V(t)} \frac{d\psi}{dt} dV$$

(ii) If we take $\psi(x,t) = \xi(x,t)$, ξ is the density of the fluid then from equation (2.34)

$$\bar{\psi}(t) = \int_{V(t)} \xi(x,t) dV = \int_{V(t)} \xi(x_1, x_2, x_3, t) dV$$

which is the total mass of fluid in the volume $V(t)$. Therefore, the theorem of conservation of mass

$$\frac{d\bar{\psi}}{dt} = \int_{V(t)} \left[\frac{d\xi}{dt} + \xi(x,t) div\,\bar{v} \right] dV = 0$$

Since the volume $V(t)$ is arbitrary

$$\frac{d\xi}{dt} + \xi\,div\,\bar{v} = 0, \text{ i.e., } \frac{\partial \xi}{\partial t} + div(\xi\bar{v}) = 0 \qquad (2.36)$$

Equation (2.36) represents the equation of continuity in Euler's description

(iii) Let us take $\psi(x,t) = F(x,t)\xi(x,t)$, then

$$\frac{d\bar{\psi}}{dt} = \frac{d}{dt} F(x,t)\xi(x,t) dV = \int_{V(t)} \left[\frac{d}{dt}\{F(x,t)\xi(x,t)\} + F(x,t)\xi(x,t) div\,\bar{v} \right] dV$$

$$= \int_{V(t)} \left[\frac{dF}{dt}\xi + F\{\frac{d\xi}{dt} + \xi\,div\,\bar{v}\} \right] dV = \int_{V(t)} \frac{dF}{dt}\xi dV \qquad (2.37)$$

Hence the theorem.

2.8. Kinematics of Line, Surface and Volume Integrals

We restricted ourselves to the properties of a continuum associated with specific particles and are functions of its coordinates. Now we consider the integrals extending over certain material volume. A simple example being mass

$$m = \int_V \rho(x_1, x_2, x_3, t) dV.$$

The general expression off such an integral is

$$\int_v P_{ij\ldots}(x_1, x_2, x_3, t) dv$$

where, $P_{ij\ldots}(x_1, x_2, x_3, t)$ is the tensor property of the continuum and V the volume that considered part of continuum occupies at time t. We find its time rate of change.

Theorem 2.4: (Material derivative of any field over material volume): If $P_{ij\ldots}(x_1, x_2, x_3, t)$ is a tensor property of a continuum and $V(t)$ is the material volume that considered portion of the continuum occupies at time, then the time rate of change of the integral of $P_{ij\ldots}(x_1, x_2, x_3, t)$ extended over a material volume V is given by

$$\frac{d}{dt}\int_V P_{ij\ldots}(x_1, x_2, x_3, t) dV = \int_V \frac{\partial}{\partial t} P_{ij\ldots}(x_1, x_2, x_3, t) dV + \int_V (v_k P_{ij\ldots}(x_1, x_2, x_3, t))_{,k} dV$$

Proof: Since the region of integration is a definite portion of the continuum which occupies of the same particles for all time and moving with the continuum, the operation of differentiation and operation of integration may be interchanged. Now

$$\frac{d}{dt}\int_V P_{ij\ldots}(x_1, x_2, x_3, t) dV = \int_V \frac{d}{dt}\left[P_{ij\ldots}(x_1, x_2, x_3, t) dV\right]$$

$$= \int_V \frac{d}{dt}\left[P_{ij\ldots}(x_1, x_2, x_3, t)\right] dV + \int_V P_{ij\ldots}(x_1, x_2, x_3, t) \frac{d}{dt}(dV)$$

$$= \int_V \frac{d}{dt}[P_{ij\ldots}(x_1, x_2, x_3, t)] dV + \int_V P_{ij\ldots}(x_1, x_2, x_3, t) v_{k,k} dV; \quad \frac{d}{dt}(dV) = v_{k,k} dV$$

$$= \int_V \left(\frac{\partial}{\partial t} + v_k \frac{\partial}{\partial x_k}\right) P_{ij\ldots}(x_1, x_2, x_3, t) dV + \int_V P_{ij\ldots}(x_1, x_2, x_3, t) v_{k,k} dV$$

$$= \int_V \frac{\partial}{\partial t} P_{ij\ldots}(x_1, x_2, x_3, t) dV + \int_V [v_k P_{ij\ldots,k}(x_1, x_2, x_3, t) + v_{k,k} P_{ij\ldots}(x_1, x_2, x_3, t)] dV$$

$$= \int_V \frac{\partial}{\partial t} P_{ij\ldots}(x_1,x_2,x_3,t) dV + \int_V (v_k P_{ij\ldots}(x_1,x_2,x_3,t))_{,k} dV$$

If $v_{k,k} = 0$, it follows that

$$\frac{d}{dt} \int_V P_{ij\ldots}(x_1,x_2,x_3,t) dV = \int_V \frac{d}{dt}[P_{ij\ldots}(x_1,x_2,x_3,t)] dV$$

i.e., if the motion of the continuum is volume preserving.

Theorem 2.5 (Reynolds transport theorem for volume property): The time rate of increase of total tensor property $P_{ij\ldots}(x_1,x_2,x_3,t)$ in that portion of the continuum instantaneously occupying a volume V is equal to the sum of the amount of the property created within V per unit time and the flux (rate of normal flow) $v_k P_{ij\ldots}(x_1,x_2,x_3,t)$ through the bounding surface S of V.

Proof: Let $P_{ij\ldots}(x_1, x_2, x_3, t)$ is a tensor property of a continuum and $V(t)$ is the material volume that considered portion of the continuum occupies at time, then the time rate of change of the integral of extended over a material volume is given by

$$\frac{d}{dt} \int_V P_{ij\ldots}(x_1,x_2,x_3,t) dV = \int_V \frac{\partial}{\partial t} P_{ij\ldots}(x_1,x_2,x_3,t) dV + \int_V (v_k P_{ij\ldots}(x_1,x_2,x_3,t))_{,k}$$

(2.38)

Since the region of integration is a definite portion of the continuum which consists of the same particles for all time and moving with the continuum, the operation of differentiation and operation of integration may be changed.

Applying Gauss's theorem

$$\frac{d}{dt} \int_V P_{ij\ldots}(x_1,x_2,x_3,t) dV = \int_V \frac{\partial}{\partial t} P_{ij\ldots}(x_1,x_2,x_3,t) dV + \int_S v_k P_{ij\ldots}(x_1,x_2,x_3,t) n_k dS$$

or, $$\frac{d}{dt} \int_V P_{ij\ldots}(x_1,x_2,x_3,t) dV = \int_V \frac{\partial}{\partial t} P_{ij\ldots}(x_1,x_2,x_3,t) dV + \int_S v_k P_{ij\ldots}(x_1,x_2,x_3,t) dS_k$$

(2.39)

If we select volume V and surface S to coincide instantaneously with fixed spatial volume V' and its surface S', then we have

$$\frac{d}{dt} \int_V P_{ij\ldots}(x_1,x_2,x_3,t) dV = \int_{V'} \frac{\partial}{\partial t} P_{ij\ldots}(x_1,x_2,x_3,t) dV + \int_{S'} v_k P_{ij\ldots}(x_1,x_2,x_3,t) dS_k$$

Thus the time rate of change of the total tensor property $P_{ij\ldots}(x_1,x_2,x_3,t)$ over a material volume V is

(i) identical to the rate of creation $P_{ij....}(x_1,x_2,x_3,t)$ in a fixed volume V' instantaneously coinciding with V and

(ii) the normal flux off $P_{ij...}(x_1,x_2,x_3,t)$ over the boundary S' of V'.

So far we have considered volume integrals. Integrals extended over curves can be considered similarly.

2.9 Spin Vector and Spin Tensor

Consider two neighbouring particles at positions $P(x_1,x_2,x_3)$ and $Q(x_1+dx_1,x_2+dx_2,x_3+dx_3)$ of the continuum in current deformed state with velocity v_i and v_i+dv respectively, at a fixed time t. For spatial method of description

$$v_i = f_i(x_1,x_2,x_3,t) \qquad (2.40)$$

Then $v_i + dv_i$ will be similar functions $x_1+dx_1, x_2+dx_2, x_3+dx_3$ so that we can write

$$v_i + dv_i = f_i(x_1+dx_1, x_2+dx_2, x_3+dx_3, t)$$

Since positions $P(x_1,x_2,x_3)$ and $Q(x_1+dx_1,x_2+dx_2,x_3+dx_3)$ are very closed together, dx_i are small. Using Taylors's series and neglecting terms containing powers of dX_i higher than first, we get

$$v_i + dv_i = f_i(x_1+dx_1, x_2+dx_2, x_3+dx_3, t)$$

$$= f_i(x_1,x_2,x_3,t) + \frac{\partial f_1}{\partial x_1}dx_1 + \frac{\partial f_2}{\partial x_2}dx_2 + \frac{\partial f_3}{\partial x_3}dx_3 + ...$$

$$= f_i(x_1,x_2,x_3,t) + \frac{\partial f_i}{\partial x_j}dx_j$$

$$= v_i + \frac{\partial v_i}{\partial x_j}dx_j$$

or
$$dv_i = \frac{\partial v_i}{\partial x_j}dx_j = v_{i,j}dx_j \qquad (2.41)$$

which is the relative velocity of the particle at $Q(x_1+dx_1,x_2+dx_2,x_3+dx_3)$ with respect to that at $P(x_1,x_2,x_3)$. The spatial gradient of the instantaneous velocity field defines the velocity gradient tensor $\frac{\partial v_i}{\partial x_j}$. Now equation (2.41) can be expressed in the form

$$dv_i = \frac{\partial v_i}{\partial x_j}.dx_j = \left(\frac{1}{2}\left[\frac{\partial v_i}{\partial x_j}+\frac{\partial v_j}{\partial x_i}\right]+\frac{1}{2}\left[\frac{\partial v_i}{\partial x_j}-\frac{\partial v_j}{\partial x_i}\right]\right).dx_j = (d_{ij}+w_{ij}).dx_j$$

$$= d_{ij}.dx_j + w_{ij}.dx_j = dv_i^{(1)} + dv_i^{(2)} \qquad (2.42)$$

where,

$$d_{ij} = \frac{1}{2}\left[\frac{\partial v_i}{\partial x_j} + \frac{\partial v_j}{\partial x_i}\right] = \text{Symmetric small strain tensor of order } 2 = d_{ji}$$

$$w_{ij} = \frac{1}{2}\left[\frac{\partial v_i}{\partial x_j} - \frac{\partial v_j}{\partial x_i}\right] = \text{Skew-symmetric tensor of order } 2 = -w_{ji}$$

$dv_i^{(1)} = d_{ij}.dx_j$ and $dv_i^{(2)} = w_{ij}.dx_j$.

This decomposition is valid even if v_i and $\frac{\partial v_i}{\partial x_j}$ are finite quantities. The symmetric tensor d_{ij} is called the *rate of deformation tensor*. Many other names are used for this tensor; among them rate of strain, stretching, strain rate and velocity strain tensor. The skew-symmetric tensor w_{ij} is called the *vorticity or spin tensor*.

Thus the relative displacement dv_i consists of two parts:

(i) $dv_i^{(1)} = d_{ij}.dx_j$ corresponding to symmetric strain-rate tensor d_{ij} of order 2

(ii) $dv_i^{(2)} = w_{ij}.dx_j$ corresponding to skew-symmetric tensor w_{ij} of order 2.

In order to study $dv_i^{(2)}$ we form a vector $w_i = e_{ijk}w_{kj}$, therefore

$$e_{ijk}w_i = e_{ijk}e_{ipq}w_{qp} = (\delta_{jp}\delta_{kq} - \delta_{jq}\delta_{kp})w_{qp}$$
$$= w_{kj} - w_{jk} = w_{kj} + w_{kj} = 2w_{kj}; \quad \text{as} \quad w_{jk} = -w_{kj}$$

or,
$$w_{kj} = \frac{1}{2}e_{ijk}w_i. \tag{2.43}$$

Therefore, $dv_i^{(2)}$ becomes

$$dv_i^{(2)} = w_{kj}.dx_j = \frac{1}{2}e_{ijk}w_i.dx_j$$

or,
$$\vec{dv^{(2)}} = \frac{1}{2}\vec{w} \times d\vec{x} \tag{2.44}$$

where $d\vec{x}$ is the vector connecting the positions $P(x_1,x_2,x_3)$ and $Q(x_1+dx_1,x_2+dx_2,x_3+dx_3)$ of the continuum. Now

$$w_i = e_{ijk}w_{kj} = \frac{1}{2}e_{ijk}\left[\frac{\partial v_i}{\partial x_j} - \frac{\partial v_j}{\partial x_i}\right] = \frac{1}{2}(e_{ijk}v_{k,j} - e_{ijk}v_{j,k})$$

$$= \frac{1}{2}(e_{ijk}v_{k,j} - e_{ikj}v_{k,j}) = \frac{1}{2}(e_{ijk}v_{k,j} + e_{ijk}v_{k,j})$$

2.9 Spin vector and Spin Tensor

$$w_i = e_{ijk} v_{k,j} = (rot\ \vec{v})_i, \text{ i.e., } \vec{w} = rot\vec{v}. \tag{2.45}$$

Therefore we have the following conclusions:

(i) In order to study $dv_i^{(1)} = d_{ij}.dx_j$, we observe that d_{ij} is a strain-rate tensor causing a rate of change in the relative position of any two neighbouring particles.

(ii) The physical significance of the strain-rate tensor is understood by studying the time rate of change of material lengths. For this, we first introduce the physical quantity $d_{(n)}$ called the stretching in the direction of a unit vector n at x, by

$$d_{(n)} = \frac{1}{ds}\frac{d}{dt}(ds) = d_{ij} n_i n_j;\ n_i = \frac{dx_i}{ds}.$$

Therefore the normal component of the deformation rate tensor is the stretching. This point is clarified further by selecting n in the direction of one of the coordinate axes, say x_1. Then $n_1 = 1, n_2 = n_3 = 0$, and we get $d_{(1)} = d_{11}$. Similarly, the mixed components are half the shearings in rectangular co-ordinates.

(iii) Therefore the relative velocity $dv_i^{(2)} = w_{ij}.dx_j$ represents a rigid body spin with angular velocity $\frac{1}{2}\vec{w} = \frac{1}{2}rot\vec{v}$ where $\vec{w} = rot\ \vec{v}$ is called spin vector. The spin tensor has properties analogous to those of rotation tensor except that no approximation is involved in its derivation. The physical significance of the spin tensor is understood if we consider the rate at which a material element x rotates about a fixed direction v.

The velocity of the neighbourhood of $P(x_1, x_2, x_3)$ is given by

$$v_i + dv_i = v_i + dv_i^{(1)} + dv_i^{(2)} = v_i + d_{ij}.dx_j + w_{ij}.dx_j$$

Therefore, the velocity of the neighbourhood of a point $P(x_1, x_2, x_3)$ now appears decomposed into three parts:

(i) The translational motion with the velocity of point $P(x_1, x_2, x_3)$,
(ii) Motion causing a rate of change in relative position
(iii) A spin about point $P(x_1, x_2, x_3)$ with angular velocity $\frac{1}{2}\vec{w} = \frac{1}{2}rot\vec{v}$.
(iv) The spin is the angular velocity of the principal axes of the deformation-rate tensor.

In case of small displacement gradients $w_{ij} = \frac{dR_{ij}}{dt}$, i.e., $\vec{w} = \frac{d\vec{R}}{dt}$. Since d_{ij} is a symmetric, second-order tensor, the concepts of principal axes, principal values, invariants, a rate of deformation quadric, and a rate of deformation deviator tensor may be associated with it. Also, equations of compatibility for the components of the rate of deformation tensor, analogous to those presented in the previous Chapter for the linear strain tensors may be developed.

Example 2.16: For the velocity field $v_1 = -Ux_2$, $v_2 = Ux_1$, $v_3 = -U$. Find spin tensor w_{ij} and strain-rate tensor d_{ij}.

Solution: The spin tensors are given by

$$w_{11} = \frac{1}{2}\left[\frac{\partial v_1}{\partial x_1} - \frac{\partial v_1}{\partial x_1}\right] = 0, \; w_{22} = \frac{1}{2}\left[\frac{\partial v_2}{\partial x_2} - \frac{\partial v_2}{\partial x_2}\right] = 0, \; w_{33} = \frac{1}{2}\left[\frac{\partial v_3}{\partial x_3} - \frac{\partial v_3}{\partial x_3}\right]$$

$$w_{12} = \frac{1}{2}\left[\frac{\partial v_1}{\partial x_2} - \frac{\partial v_2}{\partial x_1}\right] = \frac{1}{2}(-U-U) = -U = w_{21}$$

$$w_{13} = \frac{1}{2}\left[\frac{\partial v_1}{\partial x_3} - \frac{\partial v_3}{\partial x_1}\right] = \frac{1}{2}(0-0) = 0 = w_{31}$$

$$w_{23} = \frac{1}{2}\left[\frac{\partial v_2}{\partial x_3} - \frac{\partial v_3}{\partial x_2}\right] = \frac{1}{2}(0-0) = 0 = w_{23}$$

$$(w_{ij}) = \begin{pmatrix} w_{11} & w_{12} & w_{13} \\ w_{21} & w_{22} & w_{23} \\ w_{31} & w_{32} & w_{33} \end{pmatrix} = \begin{pmatrix} 0 & -U & 0 \\ -U & 0 & 0 \\ 0 & 0 & 0 \end{pmatrix}.$$

The spin rate tensors are given by

$$d_{11} = \frac{1}{2}\left[\frac{\partial v_1}{\partial x_1} + \frac{\partial v_1}{\partial x_1}\right] = 0, d_{22} = \frac{1}{2}\left[\frac{\partial v_2}{\partial x_2} + \frac{\partial v_2}{\partial x_2}\right] = 0, d_{33} = \frac{1}{2}\left[\frac{\partial v_3}{\partial x_3} + \frac{\partial v_3}{\partial x_3}\right] = 0$$

$$d_{12} = \frac{1}{2}\left[\frac{\partial v_1}{\partial x_2} + \frac{\partial v_2}{\partial x_1}\right] = \frac{1}{2}(-U+U) = 0 = d_{21}$$

$$d_{13} = \frac{1}{2}\left[\frac{\partial v_1}{\partial x_3} + \frac{\partial v_3}{\partial x_1}\right] = \frac{1}{2}(0-0) = 0 = d_{31}$$

$$d_{23} = \frac{1}{2}\left[\frac{\partial v_2}{\partial x_3} + \frac{\partial v_3}{\partial x_2}\right] = \frac{1}{2}(0-0) = 0 = d_{2:}$$

$$(d_{ij}) = \begin{pmatrix} d_{11} & d_{12} & d_{13} \\ d_{21} & d_{22} & d_{23} \\ d_{31} & d_{32} & d_{33} \end{pmatrix} = \begin{pmatrix} 0 & 0 & 0 \\ 0 & 0 & 0 \\ 0 & 0 & 0 \end{pmatrix}$$

2.9 Spin vector and Spin Tensor

Example 2.17: For velocity field given by $v_1 = x_1^2 x_2 + x_2^3$, $v_2 = -(x_1^3 + x_1 x_2^2), v_3 = 0$, determine the principal strain rates and principal axes at $P(1,2,3)$. Determine the rate of extension at $P(1,2,3)$. in the direction $\left(\frac{1}{3}, -\frac{2}{3}, \frac{2}{3}\right)$. What is the maximum shear rate?

Solution: The spin rate tensors are given by

$$d_{11} = \frac{1}{2}\left[\frac{\partial v_1}{\partial x_1} + \frac{\partial v_1}{\partial x_1}\right] = \frac{\partial v_1}{\partial x_1} = 2x_1 x_2 = 4 \text{ at } P(1,2,3)$$

$$d_{22} = \frac{1}{2}\left[\frac{\partial v_2}{\partial x_2} + \frac{\partial v_2}{\partial x_2}\right] = \frac{\partial v_2}{\partial x_2} = -2x_1 x_2 = -4 \text{ at } P(1,2,3)$$

$$d_{33} = \frac{1}{2}\left[\frac{\partial v_3}{\partial x_3} + \frac{\partial v_3}{\partial x_3}\right] = \frac{\partial v_3}{\partial x_3} = 0 \text{ at } P(1,2,3)$$

$$d_{12} = d_{21} = \frac{1}{2}\left[\frac{\partial v_1}{\partial x_2} + \frac{\partial v_2}{\partial x_1}\right] = -\frac{1}{2}\left(2x_1^2 + x_2^2\right) = -3 \text{ at } P(1,2,3)$$

$$d_{13} = d_{31} = \frac{1}{2}\left[\frac{\partial v_1}{\partial x_3} + \frac{\partial v_3}{\partial x_1}\right] = 0 \text{ at } P; d_{23} = d_{23} = \frac{1}{2}\left[\frac{\partial v_2}{\partial x_3} + \frac{\partial v_3}{\partial x_2}\right] = 0 \text{ at } P(1,2,3).$$

In matrix notation, the spin rate tensors at $P(1,2,3)$ are given by

$$(d_{ij}) = \begin{pmatrix} 2x_1 x_2 & -\frac{1}{2}(2x_1^2 + x_2^2) & 0 \\ -\frac{1}{2}(2x_1^2 + x_2^2) & -2x_1 x_2 & 0 \\ 0 & 0 & 0 \end{pmatrix} = \begin{pmatrix} 4 & -3 & 0 \\ -3 & -4 & 0 \\ 0 & 0 & 0 \end{pmatrix}.$$

The principal strain rates at the point P are the roots of the characteristic equation

$$\begin{vmatrix} d_{11} - d & d_{12} & d_{13} \\ d_{21} & d_{22} - d & d_{23} \\ d_{31} & d_{32} & d_{33} - d \end{vmatrix} = \begin{vmatrix} 4 - d & -3 & 0 \\ -3 & -4 - d & 0 \\ 0 & 0 & 0 - d \end{vmatrix} = 0$$

$$\Rightarrow d^3 - 25d = 0 \Rightarrow d = -5, 0, 5.$$

The principal directions of stress at P are given by the equation

$$(d_{11}-d)n_1 + d_{12}n_2 + d_{13}n_3 = 0$$
$$d_{21}n_1 + (d_{22}-d)n_2 + d_{23}n_3 = 0$$
$$d_{31}n_1 + d_{32}n_2 + (d_{33}-d)n_3 = 0.$$

For $d = d_1 = -5$, the above system of equations become

$$\begin{pmatrix} 4+5 & -3 & 0 \\ -3 & -4+5 & 0 \\ 0 & 0 & 0+5 \end{pmatrix} \begin{pmatrix} n_1^{(1)} \\ n_2^{(1)} \\ n_3^{(1)} \end{pmatrix} = \begin{pmatrix} 0 \\ 0 \\ 0 \end{pmatrix} \Rightarrow \begin{pmatrix} n_1^{(1)} \\ n_2^{(1)} \\ n_3^{(1)} \end{pmatrix} = \begin{pmatrix} 1 \\ 3 \\ 0 \end{pmatrix}.$$

For $d = d_2 = 0$, the above system of equations become

$$\begin{pmatrix} 4-0 & -3 & 0 \\ -3 & -4-0 & 0 \\ 0 & 0 & 0-0 \end{pmatrix} \begin{pmatrix} n_1^{(2)} \\ n_2^{(2)} \\ n_3^{(2)} \end{pmatrix} = \begin{pmatrix} 0 \\ 0 \\ 0 \end{pmatrix} \Rightarrow \begin{pmatrix} n_1^{(2)} \\ n_2^{(2)} \\ n_3^{(2)} \end{pmatrix} = \begin{pmatrix} 0 \\ 0 \\ 1 \end{pmatrix}.$$

For $d = d_3 = 5$, the above system of equations become

$$\begin{pmatrix} 4-5 & -3 & 0 \\ -3 & -4-5 & 0 \\ 0 & 0 & 0-5 \end{pmatrix} \begin{pmatrix} n_1^{(3)} \\ n_2^{(3)} \\ n_3^{(3)} \end{pmatrix} = \begin{pmatrix} 0 \\ 0 \\ 0 \end{pmatrix} \Rightarrow \begin{pmatrix} n_1^{(3)} \\ n_2^{(3)} \\ n_3^{(3)} \end{pmatrix} = \begin{pmatrix} -3 \\ 1 \\ 0 \end{pmatrix}.$$

Thus the required principal directions are

$$\frac{1}{\sqrt{10}}(1,3,0); (0,0,1); \frac{1}{\sqrt{10}}(-3,1,0).$$

At the point $P(1,2,3)$, the extension rate in the direction of $n = \left(\frac{1}{3}, -\frac{2}{3}, \frac{2}{3}\right)$ is given by

$$d_{ij}n_i n_j = \left(\frac{1}{3}, -\frac{2}{3}, \frac{2}{3}\right) \begin{pmatrix} 4 & -3 & 0 \\ -3 & -4 & 0 \\ 0 & 0 & 0 \end{pmatrix} \begin{pmatrix} \frac{1}{3} \\ -\frac{2}{3} \\ \frac{2}{3} \end{pmatrix} = -\frac{24}{25}.$$

The maximum shear rate is $= \dfrac{5-(-5)}{2} = 5.$

2.10 Irrotational Motion and Velocity Potential

We can classify the state of motion of the continuum into two distinct classes:

(i) When the state of motion does not involve the rotational part of the motion so that $rot\ \vec{v} = \vec{0}$ at every point of the continuum, the motion is said to be irrotational. In this case, there exists a scalar potential function Φ, called velocity potential, such that

$$\vec{v} = -\nabla\Phi, \quad i.e., \quad v_i = -\frac{\partial \Phi}{\partial x_i}.$$

(ii) When the state of motion involves the rotational part of the motion so that the vorticity vector $rot\ \vec{v} \neq \vec{0}$, at every point of the continuum, the motion is said to be rotational or vortex motion.

This division of of the motion into two distinct classes has the great practical value of reducing complicated problems to simple components each of which can be handled by the analytical method.

Example 2.18: For a velocity field is described by $v_1 = Ux_3$, $v_2 = Ux_3$, $v_3 = U(x_1 + x_2)$. Show that motion is irrotational. Find the velocity and the stream lines.

Solution: Let $\vec{v} = (v_1, v_2, v_3)$, then

$$rot\ \vec{v} = \vec{\nabla} \times \vec{v} = \begin{vmatrix} \hat{i} & \hat{j} & \hat{k} \\ \frac{\partial}{\partial x_1} & \frac{\partial}{\partial x_2} & \frac{\partial}{\partial x_3} \\ v_1 & v_2 & v_3 \end{vmatrix} = \begin{vmatrix} \hat{i} & \hat{j} & \hat{k} \\ \frac{\partial}{\partial x_1} & \frac{\partial}{\partial x_2} & \frac{\partial}{\partial x_3} \\ Ux_3 & Ux_3 & U(x_1+x_2) \end{vmatrix} = \vec{0}.$$

Since $rot\ \vec{v} = \vec{0}$, so at every point of the continuum, the given velocity field is irrotational. Therefore, there exists a scalar potential function $\varnothing$ such that $\vec{v} = \vec{\nabla}\varnothing$, i.e. $v_i = -\frac{\partial \varnothing}{\partial x_i}$. If $\varnothing$ be the velocity potential, then

$$v_i.dx_i = -\frac{\partial \varnothing}{\partial x_i}.dx_i = -d\varnothing.$$

Therefore,

$$d\varnothing = -U[x_3 dx_1 + x_3 dx_2 + (x_1+x_2)dx_3] = -U[x_3 d(x_1+x_2) + (x_1+x_2)dx_3]$$
$$= -U\ d[x_3(x_1+x_2)] \Rightarrow \varnothing = -Ux_3(x_1+x_2).$$

The differential equation of the stream line is given by

$$\frac{dx_1}{v_1} = \frac{dx_2}{v_2} = \frac{dx_3}{v_3} \Rightarrow \frac{dx_1}{Ux_3} = \frac{dx_2}{Ux_3} = \frac{dx_3}{U(x_1+x_2)}.$$

From the first pair of equations

$$dx_1 = dx_2 \Rightarrow (x_1 - x_2) = c_1.$$

Also,

$$\frac{d(x_1+x_2)}{2Ux_3} = \frac{dx_3}{U(x_1+x_2)} \Rightarrow (x_1+x_2)d(x_1+x_2) = 2x_3 dx_3$$

$$\Rightarrow (x_1+x_2)^2 - 2x_3^2 = c_2.$$

Example 2.19: A velocity field is described by $v_1 = -Ux_2$, $v_2 = Ux_1$, $v_3 = 0$. Show that motion is rotational.

Solution: Let $\vec{v} = (v_1, v_2, v_3)$, then

$$rot\,\vec{v} = \vec{\nabla} \times \vec{v} = \begin{vmatrix} \hat{i} & \hat{j} & \hat{k} \\ \frac{\partial}{\partial x_1} & \frac{\partial}{\partial x_2} & \frac{\partial}{\partial x_3} \\ v_1 & v_2 & v_3 \end{vmatrix} = \begin{vmatrix} \hat{i} & \hat{j} & \hat{k} \\ \frac{\partial}{\partial x_1} & \frac{\partial}{\partial x_2} & \frac{\partial}{\partial x_3} \\ -Ux_2 & Ux_1 & 0 \end{vmatrix} = 2U\hat{k} \neq \vec{0}$$

Therefore, at every point of the continuum, the motion is rotational.

2.10.1 Vortex Motion and Vortex Line

As already known that when the state of motion involves rotational part of the motion such that vorticity vector $\vec{w} = rot\,\vec{v} \neq \vec{0}$, it is vortex motion and represents motion in circles about the central point.

A vortex line is a curve drawn in the continuum at any given instant of time such that the tangent at any point of it is in the instantaneous direction of the vorticity vector $\vec{w}$ at that point. The differential equation of the vortex line at a given instant of time t is

$$\frac{dx_1}{w_1} = \frac{dx_2}{w_2} = \frac{dx_3}{w_3}$$

or,

$$\frac{dx_1}{\frac{\partial v_3}{\partial x_2} - \frac{\partial v_2}{\partial x_3}} = \frac{dx_2}{\frac{\partial v_1}{\partial x_3} - \frac{\partial v_3}{\partial x_1}} = \frac{dx_3}{\frac{\partial v_2}{\partial x_1} - \frac{\partial v_1}{\partial x_2}}.$$

2.11 Objective Tensor

In the formulation of physical laws, it is desirable to employ, as far as possible, quantities that are independent of the motion of the observer. Such quantities are called *objective* or *material frame-indifferent*. For example, the location of a point will appear different to observers located at different places. Similarly the velocity of a point is dependent on the velocity of the observer. Therefore, these quantities are not objective. On the other hand, the distance between two points and the angles between two directions are independent of the rigid motion of the frame of reference (the observer). Newton's laws of motion have long been known to be valid only in a special frame of reference called the Galilean frame. A Galilean frame differs from a fixed reference frame by a constant translatory velocity. Attempts to free the principles of mechanics from the motion of the observer were resolved by Einstein's theory of general relativity.

We wish to stay in the domain of classical mechanics with regard to basic axioms. However, we would like to employ the principle of objectivity in the description of material properties.

Let a rectangular reference frame $OX_1X_2X_3$ be in relative rigid motion with respect to another frame $O'X_1'X_2'X_3'$ i.e. rectangular frame of reference $O'X_1'X_2'X_3'$ is translated and rotate to obtain $OX_1X_2X_3$ (Fig. 2.3).

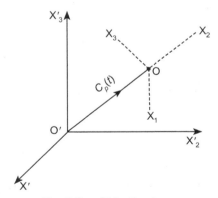

Fig. 2.3: Objective tensor

A point with rectangular coordinates (x_1, x_2, x_3) at time t with respect to the frame $OX_1X_2X_3$ will have rectangular coordinates $O'X_1'X_2'X_3'$ and are related to each other by

$$x_p'(t') = \alpha_{pi}(t)x_i(t) + c_p(t); t' = t - a \qquad (2.46)$$

where, a is a constant allowing us to select the origin of time different in $O'X_1'X_2'X_3'$ than in $OX_1X_2X_3$; $c_p(t)$ is a translation and $\alpha_{pi}(t)$ is the

Cosine of the angle between axes OX'_i and OX_j describing orthogonal rotation subject to

$$\alpha_{ij}\alpha_{ik} = \delta_{jk} = \alpha_{ji}\alpha_{ki} \qquad (2.47)$$

from which it follows that $\det \alpha = \pm 1$. For rigid motions $\det \alpha = 1$. Note that $\det \alpha = \pm 1$ include, beside the rigid motions, a reflection in a coordinate frame (mirror image) thus carrying a right-handed frame of reference to a left-handed one. The totality of all transformations $|\alpha|$ subject to $\det \alpha = \pm 1$ form a group called the *full group of orthogonal transformations*. At present, both forms are employed in continuum physics.

An objective vector $\vec{a}$, i.e. a_i in frame $OX_1X_2X_3$ will appear in frame $O'X'_1X'_2X'_3$ as

$$a'_p = \alpha_{pi}a_i \qquad (2.48)$$

An objective second order tensor a_{ij} in frame $OX_1X_2X_3$ will appear in the frame $O'X'_1X'_2X'_3$ as

$$a'_{pq} = \alpha_{pi}(t)\alpha_{qj}(t)a_{ij} \qquad (2.49)$$

Two motions $x_p(X,t)$ and $x'(X,t')$ are called objectively equivalent, if and only if

$$x'_p(X,t') = \alpha_{pi}(t)x_i(X,t) + c_p(t); \quad t' = t - a \qquad (2.50)$$

where, $\alpha_{ij}\alpha_{ik} = \delta_{jk} = \alpha_{ji}\alpha_{ki}$. For vectors and tensors that are independent of time t, objectivity readily applies. Two objectively equivalent motions differ only relative to the reference frame and time. For a fixed frame and time, the two motions may be made to coincide by the superposition of an arbitrary rigid motion of one and by a shift of time.

Any tensorial quantity is said to be *objective or material frame-indifferent*, if in any two objectively equivalent motions it obeys the appropriate tensor transformation law for all times.

Theorem 2.6: The distance between two points is objective.

Proof: Using the transformation equation, we have

$$dl'^2 = dx'_p\, dx'_p = \alpha_{pi}dx_i\alpha_{pj}dx_j = \alpha_{pi}\alpha_{pj}dx_i dx_j = \delta_{ij}dx_i dx_j$$

$$= dx_i dx_i = dl^2 \qquad (2.51)$$

Thus, we have to prove that the transformation (2.46) subject to Eq. (2.47) is the most general one in which the length are preserved. A similar proof is produced in a similar fashion for the preservation of angles.

Theorem 2.7: The velocity vector is not objective.

Proof: For vectors and tensors that are independent of time, objectivity readily applies. For time-dependent quantities, we have

2.11 Objective Tensor

$$v'_p = \frac{dx'_p}{dt} = \frac{dx_i}{dt}\alpha_{pi} + \frac{d\alpha_{pi}}{dt}x_i + \dot{c}_p(t)$$

or,
$$v'_p = \alpha_{pi}v_i + \dot{\alpha}_{pi}x_i + \dot{c}_p \tag{2.52}$$

This is not of the form on (2.48). Therefore, the velocity is not objective. Similarly, we can show that the acceleration is not objective either.

Theorem 2.8: The deformation rate tensor d_{ij} is objective but the spin tensor w_{ij} is not.

Proof: Differentiating the relation $\alpha_{pi}\alpha_{qi} = \delta_{pq}$ with respect to t, we get

$$\dot{\alpha}_{pi}\alpha_{qi} + \alpha_{pi}\dot{\alpha}_{qi} = 0$$

To calculate, $\dfrac{\partial x_k}{\partial x'_q}$ we first solve for x_k from Eq. (2.46).

Hence,

$$\alpha_{pi}x_i(t) = x'_p - c_p \Rightarrow \alpha_{pk}\alpha_{pi}x_i(t) = \alpha_{pk}(x'_p - c_p)$$
$$\Rightarrow \delta_{ik}x_i(t) = \alpha_{pk}(x'_p - c_p) \Rightarrow x_k = \alpha_{pk}(x'_p - c_p)$$
$$\Rightarrow \frac{\partial x_k}{\partial x'_q} = \alpha_{pk}\frac{\partial x'_p}{\partial x'_q} = \alpha_{pk}\delta_{pq} = \alpha_{pk}$$

Form Eq. (2.52), we have, $v'_p = \alpha_{pi}v_i + \dot{\alpha}_{pi}x_i + \dot{c}_p$. Using this expression, we have

$$v'_{p,q} = \frac{\partial v'_p}{\partial x'_q} = \alpha_{pi}v_{i,k}\frac{\partial x_k}{\partial x'_q} + \dot{\alpha}_{pi}\frac{\partial x_i}{\partial x'_q}$$
$$= \alpha_{pi}v_{i,k}\alpha_{qk} + \dot{\alpha}_{pi}\alpha_{qi} = \alpha_{pi}\alpha_{qj}v_{i,j} + \dot{\alpha}_{pi}\alpha_{qi}$$

Similarly, we find

$$v'_{q,p} = \alpha_{qi}\alpha_{pj}v_{i,j} + \dot{\alpha}_{qi}\alpha_{pi} = \alpha_{qj}\alpha_{pi}v_{j,i} + \dot{\alpha}_{qi}\alpha_{pi}$$

If we add these two expressions, we get

$$v'_{p,q} + v'_{q,p} = \alpha_{qj}\alpha_{pi}(v_{j,i} + v_{i,j}) + \dot{\alpha}_{pi}\alpha_{qi} + \dot{\alpha}_{qi}\alpha_{pi}$$
$$\Rightarrow 2d'_{pq} = 2d_{ij}\alpha_{pi}\alpha_{qj} \Rightarrow d'_{pq} = d_{ij}\alpha_{pi}\alpha_{qj}; \text{ as } \dot{\alpha}_{pi}\alpha_{qi} + \alpha_{pi}\dot{\alpha}_{qi} = 0$$

This has the precise form Eq. (2.49), proving the objectivity of the deformation rate tensor d_{ij}. Now subtracting the expressions of $v'_{p,q}$ and $v'_{q,p}$, we get

$$v'_{p,q} - v'_{q,p} = \alpha_{qj}\alpha_{pi}(v_{j,i} - v_{i,j}) + \dot{\alpha}_{pi}\alpha_{qi} - \dot{\alpha}_{qi}\alpha_{pi}$$
$$\Rightarrow w'_{ij} = \alpha_{qj}\alpha_{pi}w_{ij} + \dot{\alpha}_{pi}\alpha_{qi}; \text{as } \dot{\alpha}_{pi}\alpha_{qi} + \alpha_{pi}\dot{\alpha}_{qi} = 0$$

This is not of the form of Eq.(2.49). Thus, the spin tensor w_{ij} is not objective. The term $\Omega_{pq} = \dot{\alpha}_{pi}\alpha_{qi}$ responsible for the non-objective character of the spin tensor w_{ij} is the *relative angular velocity of the two frames*.

Exercise

Theoretical

1. Calculate the first two material derivatives of
   (a) The acceleration vector $a_k(x,t)$;
   (b) The second-order displacement gradient $\dfrac{\partial^2 x_i}{\partial x_j \partial x_k}$.

2. Show that the acceleration vector, $\vec{a}$, may be expressed in vectorial form:
$$\vec{a} = \frac{\partial \vec{v}}{\partial t} + \left(\vec{v}\cdot\vec{\nabla}\right)\vec{v} = \frac{\partial \vec{v}}{\partial t} + \vec{w}\times\vec{v} + \vec{\nabla}\left(\frac{v^2}{2}\right)$$

   Show that the above form is valid for any arbitrary curvilinear coordinates.

3. Prove that path line is identical with the stream line in steady motion.
4. Find the necessary and sufficient condition for a surface to be material.
5. Prove that the normal components of the deformation rate tensor are the stretchings, and the mixed components are half the shearings in rectangular co-ordinates.
6. Prove that the spin is the angular velocity of the principal axes of the deformation-rate tensor.
7. Prove that at each point of a body there exists at least one material direction that is stationary instantaneously.
8. Calculate the first two material derivatives of the Lagrangian and Eulerian strain rate tensors and the spin tensor.
9. Find the equations of continuity and vorticity for the following special types of motions:
   (a) Isochoric;
   (b) Irrotational.

   Show that when the motion is isochoric and irrotational, then the velocity field is derivable from an analytic function.

2.11 Objective Tensor

10. For a simply connected region, find the integrability conditions for
$$d_{ij} = \frac{1}{2}\left[\frac{\partial v_i}{\partial x_j} + \frac{\partial v_j}{\partial x_i}\right]_t.$$

11. Investigate the objectivity of the following quantities:
    (a) Eulerian strain rate.
    (b) The material derivative of the deformation rate tensor.
    (c) the acceleration.
    (d) The material derivative of the spin tensor.
    (e) The material derivative of the area vector.

Numerical

1. Let the motion equations be given in component form by the Lagrangian description
$$x_1 = X_1 e^t + X_3\left(e^t - 1\right), x_2 = X_2 + X_3\left(e^t - e^{-t}\right); x_3 = X_3.$$

    For the motion determine the velocity and acceleration fields, and express these in both Lagrangian and Eulerian forms.
    Answer: Material form
$$v_1 = (X_1 + X_3)e^t, v_2 = X_3\left(e^t + e^{-t}\right);$$
$$v_3 = 0; a_1 = (X_1 + X_3)e^t, a_2 = X_3\left(e^t - e^{-t}\right); a_3 = 0.$$

    Spatial form
$$v_1 = (x_1 + x_3)e^t, v_2 = x_3\left(e^t + e^{-t}\right); v_3 = 0;$$
$$a_1 = (x_1 + x_3)e^t, a_2 = x_3\left(e^t - e^{-t}\right); a_3 = 0.$$

2. Find k such that $v_1 = kx_3(x_2 - 2)^2, v_2 = -x_1 x_2$, and $v_3 = kx_1$ may be velocity components of an incompressible continuum.
    [**Answer:** $k = 1$]

3. The motion of a continuous medium is specified by the component equations
$$x_1 = \frac{1}{2}(X_1 + X_2)e^t + \frac{1}{2}(X_1 - X_2)e^{-t};$$
$$x_2 = \frac{1}{2}(X_1 + X_2)e^t - \frac{1}{2}(X_1 - X_2)e^{-t}; x_3 = X_3.$$

(a) Show that the Jacobian determinant J does not vanish, and solve for the inverse equations $X = X(x,t)$
(b) Calculate the velocity and acceleration components in terms of the material coordinates.
(c) Using the inverse equations developed in part (a), express the velocity and acceleration components in terms of spatial coordinates.

Answer: (a) $J = 1$

$$X_1 = \frac{1}{2}(x_1 + x_2)e^{-t} + \frac{1}{2}(x_1 - x_2)e^{t}; X_2 = \frac{1}{2}(x_1 + x_2)e^{-t} - \frac{1}{2}(x_1 - x_2)e^{t}; X_3 = x_3.$$

(b) $v_1 = \frac{1}{2}(X_1 + X_2)e^{t} - \frac{1}{2}(X_1 - X_2)e^{-t};$

$v_2 = \frac{1}{2}(X_1 + X_2)e^{t} + \frac{1}{2}(X_1 - X_2)e^{-t}; v_3 = 0.$

$a_1 = \frac{1}{2}(X_1 + X_2)e^{t} + \frac{1}{2}(X_1 - X_2)e^{-t};$

$a_2 = \frac{1}{2}(X_1 + X_2)e^{t} - \frac{1}{2}(X_1 - X_2)e^{-t}; a_3 = 0.$

(c) $v_1 = x_2; v_2 = x_1; v_3 = 0; a_1 = x_1; a_2 = x_2; a_3 = 0$

4. A continuum body has a motion defined by the equations

$$x_1 = X_1 + 2X_2 t^2; x_2 = X_2 + 2X_1 t^2; x_3 = X_3.$$

(a) Determine the velocity components at $t = 1.5$ s of the particle which occupied the point (2, 3, 4) when $t = 1.0$ s.
(b) Determine the equation of the path along which the particle designated in part (a) moves.
(c) Calculate the acceleration components of the same particle at time $t = 2$ s.

Answer: (a) $v_1 = 2; v_2 = 8; v_3 = 0$ (b) $4x_1 - x_2 = 5$ in the plane $x_3 = 4$,
(c) $a_1 = \frac{4}{3}; a_2 = \frac{16}{3}; a_3 = 0$

5. If the motion $x = x(X, t)$ is given in component form by the equations

$$x_1 = X_1(1+t); x_2 = X_2(1+t)^2; x_3 = X_3(1+t^2)$$

Determine expressions for the velocity and acceleration components in terms of both Lagrangian and Eulerian coordinates.

2.11 Objective Tensor

Answer:

$$v_1 = X_1 = \frac{x_1}{1+t}; v_2 = 2X_2(1+t) = \frac{2x_2}{1+t}; v_3 = 2X_3t = \frac{2x_3t}{1+t^2};$$

$$a_1 = 0; a_2 = 2X_2 = \frac{2x_2}{(1+t)^2}; a_3 = 2X_3 = \frac{2x_3}{1+t^2}$$

6. A velocity field is given in Lagrangian form by

$$v_1 = 2t + X_1, v_2 = X_2 e^t, v_3 = X_3 - t$$

Integrate these equations to obtain $x = x(X,t)$ with $x = X$ at $t = 0$, and using that result compute the velocity and acceleration components in the Eulerian (spatial) form.

Answer:

$$v_1 = \frac{x_1 + 2t + t^2}{1+t}; v_2 = x_2; v_3 = \frac{2x_3 - 2t - t^2}{2(1+t)}; a_1 = 2; a_2 = x_2; a_3 = -1.$$

7. The temperature field in a continuum is given by the expression

$$\theta = e^{-3t/x^2} \quad \text{where} \quad x^2 = x_1^2 + x_2^2 + x_3^2$$

The velocity field of the medium has components

$$v_1 = x_2 + 2x_3, v_2 = x_3 - x_1, v_3 = x_1 + 3x_2$$

Determine the material derivative of the temperature field.

Answer:

$$\frac{d\theta}{dt} = -e^{-3t}\left(3x^2 + 6x_1 x_3 + 8x_2 x_3\right)/x^4$$

8. In a certain region of a fluid the flow velocity has components

$$v_1 = A\left(x_1^3 + x_1 x_2^2\right)e^{-kt}, v_2 = A\left(x_2^3 + x_2 x_1^2\right)e^{-kt}, v_3 = 0$$

where A and B are constants. Use the (spatial) material derivative operator to determine the acceleration components at the point (1, 1, 0) when $t = 0$.

Answer: $a_1 = -2A(k-5A), a_2 = -A(k-5A), a_3 = 0$

9. The displacement field is given in terms of the spatial variables and time by the equations $u_1 = x_2 t^2, u_2 = x_3 t, u_3 = x_1 t.$ Using the (spatial) material derivative operator, determine the velocity components.

Answer: $v_1 = (2x_2 t + x_3 t^2 + x_1 t^3)/(1-t^4), v_2 = (x_3 + x_1 t + 2x_2 t^3)/(1-t^4)$
$v_3 = (x_1 + 2x_2 t^2 + x_3 t^3)/(1-t^4)$

10. Show that for the flow $v_i = \dfrac{x_i}{1+t}$ the streamlines and path lines coincide. Find the streak line of a continuum particle for the velocity field.

11. Show that $\dfrac{x^2}{a^2}\tan^2\tan^2 t + \dfrac{y^2}{b^2}\cot^2\cot^2 t = 1$ is a possible form for the bounding surface of a liquid.

12. Prove that $\dfrac{x^2}{a^2}f_1(t) + \dfrac{y^2}{b^2}f_2(t) + \dfrac{z^2}{c^2}f_3(t) = 1$, is a possible form of a bounding surface of the liquid provided

$$f_1(t)f_2(t)f_3(t) = \text{constant}.$$

13. Show that the variable ellipsoid

$$F(x,y,z,t) = \dfrac{x^2}{a^2 e^{-t}\cos(t+\pi/4)} + \dfrac{y^2}{b^2 e^{t}\sin(t+\pi/4)} + \dfrac{z^2}{c^2 \sec 2t} - 1 = 0$$

is a possible form for the bounding surface of a liquid at any time and determine the velocity component of any particle on this boundary.

14. Show that the variable ellipsoid $\dfrac{x^2}{a^2 k^2 t^4} + kt^2\left\{\dfrac{y^2}{b^2} + \dfrac{z^2}{c^2}\right\} = 1$ is a possible form for the bounding surface of a liquid at any time t.

15. Show that a surface of the form $ax^4 + by^4 + cz^4 - \mu(t) = 0$ is a possible form of a boundary surface of a homogeneous liquid at any time t, the velocity potential of the liquid motion being

$$\phi = (\beta - \gamma)x^2 + (\gamma - \alpha)y^2 + (\alpha - \beta)z^2,$$

where, $\mu, \alpha, \beta, \gamma$ are given functions of time and a, b, c are suitable functions of time.

16. Show that $\phi = x f(r)$ is a possible form for the velocity potential of an incompressible liquid motion.

17. For the motion $x_1 = X_1, x_2 = X_2 + X_1(e^{-2t} - 1), x_3 = X_3 + X_1(e^{-3t} - 1)$. Compute strain tensor e_{ij}, rotation tensor r_{ij} spin tensor w_{ij} and strain-rate tensor d_{ij}.

18. A velocity field is given by $v_1 = 0, v_2 = A(x_1 x_2 - x_3^2)e^{-Bt}$, $v_3 = A(x_2^2 - x_1 x_3)e^{-Bt}$. Compute spin tensor w_{ij} and strain-rate tensor d_{ij} for the point $(1,0,3)$ when $t = 0$.

Answer: The spin tensors are given by

2.11 Objective Tensor

$$(w_{ij}) = \begin{pmatrix} w_{11} & w_{12} & w_{13} \\ w_{21} & w_{22} & w_{23} \\ w_{31} & w_{32} & w_{33} \end{pmatrix} = \begin{pmatrix} 0 & 0 & \dfrac{3A}{2} \\ 0 & 0 & -3A \\ -\dfrac{3A}{2} & 3A & 0 \end{pmatrix}$$

The spin rate tensors are given by

$$(d_{ij}) = \begin{pmatrix} d_{11} & d_{12} & d_{13} \\ d_{21} & d_{22} & d_{23} \\ d_{31} & d_{32} & d_{33} \end{pmatrix} = \begin{pmatrix} 0 & 0 & -\dfrac{3A}{2} \\ 0 & A & -3A \\ -\dfrac{3A}{2} & -3A & -A \end{pmatrix}$$

19. A velocity field is given by $v_1 = Ax_3 - Bx_2, v_2 = Bx_1 - Cx_3, v_3 = Cx_2 - Ax_1$ the vortex lines are straight lines. Further show that above velocity field represents a rigid body rotational motion.

    Answer: The spin tensors are given by

$$(w_{ij}) = \begin{pmatrix} w_{11} & w_{12} & w_{13} \\ w_{21} & w_{22} & w_{23} \\ w_{31} & w_{32} & w_{33} \end{pmatrix} = \begin{pmatrix} 0 & -B & A \\ B & 0 & -C \\ -A & C & 0 \end{pmatrix} \neq 0$$

The spin rate tensors are given by

$$(d_{ij}) = \begin{pmatrix} d_{11} & d_{12} & d_{13} \\ d_{21} & d_{22} & d_{23} \\ d_{31} & d_{32} & d_{33} \end{pmatrix} = \begin{pmatrix} 0 & 0 & 0 \\ 0 & 0 & 0 \\ 0 & 0 & 0 \end{pmatrix}$$

20. In a plane motion of a continuum, the velocity field is given by

    $v_1 = -U\dfrac{a^2(x_1^2 - x_2^2)}{(x_1^2 + x_2^2)^2}$, $v_2 = 2U\dfrac{a^2 x_1 x_2}{(x_1^2 + x_2^2)^2}$, $v_3 = -U$. Find spin tensor w_{ij} and strain-rate tensor d_{ij} and acceleration at (x_1, x_2, x_3).

    Answer: The spin tensors are given by

$$(w_{ij}) = \begin{pmatrix} w_{11} & w_{12} & w_{13} \\ w_{21} & w_{22} & w_{23} \\ w_{31} & w_{32} & w_{33} \end{pmatrix} = \begin{pmatrix} 0 & 0 & 0 \\ 0 & 0 & 0 \\ 0 & 0 & 0 \end{pmatrix}$$

The spin rate tensors are given by

$$(d_{ij}) = \begin{pmatrix} d_{11} & d_{12} & d_{13} \\ d_{21} & d_{22} & d_{23} \\ d_{31} & d_{32} & d_{33} \end{pmatrix} = \frac{2Ua^2}{(x_1^2+x_2^2)^2} \begin{pmatrix} -x_1(x_1^2-3x_2^2) & -x_2(3x_1^2-x_2^2) & 0 \\ -x_2(3x_1^2-x_2^2) & x_1(x_1^2-3x_2^2) & 0 \\ 0 & 0 & 0 \end{pmatrix}$$

The acceleration at (x_1, x_2, x_3) are given by

$$f_1 = -\frac{2U^2 a^4 x_1}{(x_1^2+x_2^2)^{-3}},\ f_2 = -\frac{2U^2 a^4 x_2}{(x_1^2+x_2^2)^{-3}},\ f_3 = 0$$

21. Determine the streamlines, for the velocity field, in a plane motion of a continuum, given by

$$v_1 = U\frac{a^2(x_1^2-x_2^2)}{(x_1^2+x_2^2)^2},\ v_2 = 2U\frac{a^2 x_1 x_2}{(x_1^2+x_2^2)^2},\ v_3 = 0,$$

where U and a are constants. Determine:
   (a) Components of the deformation rate tensor.
   (b) Components of the spin tensor and vorticity vector,
   (c) Invariants of the deformation rate tensor.

22. Determine the streamlines, for the velocity field, in a plane motion of a continuum, given by

$$v_1 = -U\left[1 - \frac{a^2(x_1^2-x_2^2)}{(x_1^2+x_2^2)^2}\right],\ v_2 = 2U\frac{a^2 x_1 x_2}{(x_1^2+x_2^2)^2},\ v_3 = 0,$$

where U and are constants and (x_1, x_2, x_3) are the spatial rectangular co-ordinates. Determine:
   (i) Components of the deformation rate tensor.
   (ii) Components of the spin tensor and vorticity vector,
   (iii) Invariants of the deformation rate tensor.

23. Determine the streamlines and vortex lines, for the velocity field, in a plane motion of a continuum, given by

$$v_1 = -\frac{U}{4\pi}\frac{x_1^2}{x_2(x_1^2+x_2^2)},\ v_2 = \frac{U}{4\pi}\frac{x_2^2}{x_1(x_1^2+x_2^2)},\ v_3 = 0,$$

where U and a are constants and (x_1, x_2, x_3) are the spatial rectangular co-ordinates. Determine:
   (i) Components of the deformation rate tensor.
   (ii) Components of the spin tensor and vorticity vector,
   (iii) Invariants of the deformation rate tensor.

2.11 Objective Tensor

24. For the steady velocity field given by $v_1 = x_1^2 x_2 + x_2^3$, $v_2 = -(x_1^3 + x_1 x_2^2)$, $v_3 = 0$ determine expressions for the principal values of the rate of deformation tensor at an arbitrary point $P(x_1, x_2, x_3)$. Prove also that the streamlines are circular.

    Answer: $x_1^2 + x_2^2, 0, -(x_1^2 + x_2^2)$

25. For velocity field given by $v_1 = 2x_3, v_2 = 2x_3, v_3 = 0$, determine the principal strain rates and principal directions. Determine maximum shear rate.

    Answer: $\sqrt{2}, \left(-\dfrac{1}{2}, -\dfrac{1}{2}, \dfrac{1}{\sqrt{2}}\right); \left(\dfrac{1}{\sqrt{2}}, -\dfrac{1}{\sqrt{2}}, 0\right); \left(\dfrac{1}{2}, \dfrac{1}{2}, \dfrac{1}{\sqrt{2}}\right).$

26. The velocity field of a continuum is prescribed by

    $\dot{x} = f(y) - y\, g(r);\ \dot{y} = x\, g(r);\ \dot{z} = 0$ where, $r = (x^2 + y^2)^{1/2}$.

    Determine:
    (a) The deformation rate tensor and its invariants;
    (b) Vorticity vector.
    Give a discussion of the type of motion that the body is undergoing.

27. If $\Phi = A(x^2 + y^2 + z^2)^{-\frac{3}{2}} z \tan^{-1} \tan^{-1} \dfrac{y}{x}$ be the velocity potential of the irrotational motion of a continuum and then show that lines of flow will lie on series of surfaces $(x^2 + y^2 + z^2)^3 = c(x^2 + y^2)^2$.

■ ■ ■

3

Theory of Stress

Continuum mechanics deals with deformable bodies, as opposed to rigid bodies. A solid is a deformable body that possesses shear strength, a solid can support shear forces (forces parallel to the material surface on which they act). Fluids, on the other hand, do not sustain shear forces. For the study of the mechanical behavior of solids and fluids these are assumed to be continuous bodies, which mean that the matter fills the entire region of space it occupies, despite the fact that matter is made of atoms, has voids, and is discrete. Therefore, when continuum mechanics refers to a point or particle in a continuous body it does not describe a point in the inter-atomic space or an atomic particle, rather an idealized part of the body occupying that point.

Stress is a measure of *force intensity,* either within or on the bounding surface of a body subjected to loads. It should be noted that in continuum mechanics a body is considered stress free if the only forces present are those interatomic forces required to hold the body together. And so it follows that the stresses that concern us here are those which result from the application of forces by an external agent.

3.1 Forces of Continuum: Body and Surface Forces, Mass Density

Following the classical dynamics of Newton and Euler, the motion of a material body is produced by the action of externally applied forces which are assumed to be of two kinds:

(*i*) Body force F_B
(*ii*) Surface force F_C

Body Force: *Body forces* are forces originating from sources outside of the body that act on all volume elements (or mass) of the body and distributed throughout the body. Saying that body forces are due to outside sources implies that the interaction between different parts of the body (internal forces) is manifested through the contact forces alone. These forces arise from the presence of the body in force fields, *e.g.* gravitational field(gravitational forces) or electromagnetic field (electromagnetic forces), or from inertial forces when bodies are in motion. As the mass of a continuous body is assumed to be continuously distributed, any force originating from the mass is also continuously distributed. Thus, body forces are specified by vector fields which are assumed to be continuous over the entire volume of the body, *i.e.* acting on every point in it.

Body forces are represented by a body force density $b(x,t)$ (force per unit of mass), which is a frame-indifferent vector field or by the symbol $p(x,t)$ (force per unit volume).

In the case of gravitational forces, the intensity of the force depends on, or is proportional to, the mass density $\rho(x,t)$ of the material, and it is specified in terms of force per unit mass (b_i) or per unit volume (p_i). These two specifications are related through the material density by the equation

$$\rho b_i = p_i \qquad (3.1)$$

Similarly, the intensity of electromagnetic forces depends upon the strength (electric charge) of the electromagnetic field.

The total body force applied to a continuous body is expressed as

$$F_B = \iiint_V b\,dm = \iiint_V \rho b\,dV. \qquad (3.2)$$

Surface force: Those forces which act upon and are distributed in some fashion over a surface element of the body, regardless of whether that element is part of the bounding surface, or an arbitrary element of surface within the body, are called *surface forces* or *contact forces*. *Surface forces*, expressed as force per unit area, can act either on the bounding surface of the body, as a result of mechanical contact with other bodies, or on imaginary internal surfaces that bound portions of the body, as a result of the mechanical interaction between the parts of the body to either side of the surface.

The total contact force on the particular internal surface is then expressed as the sum (surface integral) of the contact forces on all differential surfaces dS:

$$F_C = \iint_S T^{(n)}\,dS. \qquad (3.3)$$

In continuum mechanics a body is considered stress-free if the only forces present are those inter-atomic forces (ionic, metallic, and van der Waals forces) required to hold the body together and to keep its shape in the absence of all external influences, including gravitational attraction. Stresses generated during manufacture of the body to a specific configuration are also excluded when considering stresses in a body. Therefore, the stresses considered in continuum mechanics are only those produced by deformation of the body, only relative changes in stress are considered, not the absolute values of stress.

(a) Thus, the total force F applied to a body or to a portion of the body can be expressed as:

$$F = F_B + F_C = \iint_S T^{(n)} dS + \iiint_V \rho b dV. \qquad (3.4)$$

By stress principle, this is the form of the first law of motion.

(b) Body forces and contact forces acting on the body lead to corresponding moments of force (torques) relative to a given point. Thus, the total applied torque M (with respect to the origin of the coordinate system) in the body can be given by

$$M = M_B + M_C = \iint_S r \times T^{(n)} dS + \iiint_V r \times \rho b dV. \qquad (3.5)$$

In certain situations, not commonly considered in the analysis of the mechanical behavior or materials, it becomes necessary to include two other types of forces:

(i) *Body moments* and

(ii) *Couple stresses* (surface couples, contact torques).

Body moments, or body couples, are moments per unit volume or per unit mass applied to the volume of the body. Couple stresses are moments per unit area applied on a surface. Both are important in the analysis of stress for a polarized dielectric solid under the action of an electric field, materials where the molecular structure is taken into consideration (*e.g.* bones), solids under the action of an external magnetic field, and the dislocation theory of metals.

3.2 Euler-Cauchy Stress Principle

The Euler–Cauchy stress principle states that *upon any surface (real or imaginary) that divides the body, the action of one part of the body on the other is equivalent (equipollent) to the system of distributed forces and couples on the surface dividing the body*, and it is represented by a field $\vec{T}^{(n)}$, called the stress vector, defined on the surface S and assumed to depend continuously on the surface's unit vector $\vec{n}$.

We consider a homogeneous, isotropic material body $\mathcal{B}$ having a bounding surface S, and a volume V. To formulate the Euler–Cauchy stress principle, consider an imaginary surface passing through an internal material point P dividing the continuous body into two segments, (one may use either the cutting plane diagram or the diagram with the arbitrary volume inside the continuum enclosed by the surface S). Only surface forces will be discussed in this article as they are relevant to the Cauchy stress tensor.

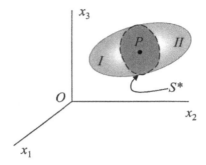

Fig. 3.1: Typical continuum volume showing cutting plane S^* passing through point P

The idea of stress arises from the consideration of the internal forces which the particles of one part of the deformed body exert on the particles of the adjacent part through the separating boundary surface in the form of restoring forces or contact forces $\vec{F}$. Consider a deformed body, every part of which is held in equilibrium under the action of external surface forces. We consider a configuration occupied by a body $\mathcal{B}$ at time t. We imagine a closed surface in the interior of $\mathcal{B}$. Let P be any point on $\mathcal{B}$ and (x_1, x_2, x_3) be its rectangular Cartesian coordinates. Let δS be small element of surface of arbitrary size and shape surrounding point with outward drawn normal $\vec{n}$ to S (outward with respect to S). We distinguish between two sided of δS according to the direction of $\vec{n}$ —the positive side and the negative side of δS.

When the body is subjected to external surface forces or *contact forces* $\vec{F}$, following Euler's equations of motion, internal contact forces and moments are transmitted from point to point in the body, and from one segment to the other through the dividing surface S, due to the mechanical contact of one portion of the continuum onto the other. Thus the surface forces distributed over the surface element δS of S containing P with normal vector $\vec{n}$ can be reduced to a single restoring contact force $\delta \vec{F}^{(n)}$ acting at P along a definite direction together with a single surface moment $\delta \vec{G}^{(n)}$. The forces $\delta \vec{F}^{(n)}$ is a function of the element δS and $\vec{n}$.

Cauchy's stress principle asserts that as becomes very small and tends to zero, in any manner keeping the point P always within it, the ratio $\dfrac{\delta \vec{F}^{(n)}}{\delta S}$ becomes $\dfrac{d\vec{F}^{(n)}}{dS}$, i.e.,

$$\lim_{\delta S \to 0} \frac{\delta \vec{F}^{(n)}}{\delta S} = \frac{d\vec{F}^{(n)}}{dS} = \vec{T}^{(n)}(x_1, x_2, x_3), \qquad (3.6)$$

3.2 Euler-Cauchy Stress Principle

and the couple stress vector $\vec{M}^{(n)}$ given by

$$\lim_{\delta S \to 0} \frac{\delta \vec{G}^{(n)}}{\delta S} = \frac{d\vec{G}^{(n)}}{dS} = \vec{M}^{(n)}. \tag{3.7}$$

In specific fields of continuum mechanics the couple stress is assumed not to vanish; however, classical branches of continuum mechanics address non-polar materials which do not consider couple stresses and body moments. The vector $\vec{T}^{(n)}$ is called the *surface traction*, also called *stress vector*, *traction*, or *traction vector* at $P(x_1, x_2, x_3)$ which represents a force per unit area acting on the material S by the material lying outside S.

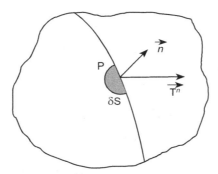

Fig. 3.2: Traction vector $\vec{T}^{(n)}$ acting at point P of plane element δS, whose normal is $\vec{n}$

The frame-indifferent stress vector $\vec{T}^{(n)}(x_1, x_2, x_3)$ is used to emphasize the fact that the stress vector depends not only on the position (x_1, x_2, x_3) of the given point in the continuum but also depends upon the orientation of the particular surface element throughout whose exterior unit normal is $\vec{n}$. Also $\vec{T}^{(n)}(x_1, x_2, x_3)$ is the external surface force per unit area acting at a point (x_1, x_2, x_3) on the surface element on the material inside the surface element.

The stress vector $\vec{T}^{(n)}(x_1, x_2, x_3)$ is considered positive, when it is directed on the same side of the surface element as in the positive normal $\vec{n}$ and it tends to restore the material.

Depending on the orientation of the plane under consideration, the stress vector $\vec{T}^{(n)}(x_1, x_2, x_3)$ may not necessarily be perpendicular to that plane, *i.e.* parallel to $\vec{n}$, and can be resolved into two components:

(i) The normal component directed along the normal $\vec{n}$, called normal stress and is denoted by $\vec{N}^{(n)}$. The normal stress is considered positive if its sense coincides with the sense of outward normal to the surface element at a given point, it tends to pull the material.

(ii) The tangential component directed along the tangent to the surface element (parallel to this plane), called tangential stress and is denoted by $\vec{S}^{(n)}$. The tangential stress is sometimes called shearing stress, since they set up in material by a shearing motion or displacement in which the parallel layers of matter slide relative to each other. The shear stress can be further decomposed into two mutually perpendicular vectors.

3.2.1 Cauchy's Postulate

According to the *Cauchy Postulate*, the stress vector $\vec{T}^{(n)}$ remains unchanged for all surfaces passing through the point P and having the same normal vector $\vec{n}$ at , i.e., having a common tangent at P. This means that the stress vector is a function of the normal vector $\vec{n}$ only, and is not influenced by the curvature of the internal surfaces.

3.2.2 Cauchy's Fundamental Lemma

A consequence of Cauchy's postulate is *Cauchy's Fundamental Lemma* also called the *Cauchy reciprocal theorem*, which states that the stress vectors acting on opposite sides of the same surface are equal in magnitude and opposite in sign. Let

$\vec{T}^{(n)}(x_1, x_2, x_3)$ = the action of the part II on part I transmitted through the surface element with normal $\vec{n}$ at (x_1, x_2, x_3)

$\vec{T}^{(-n)}(x_1, x_2, x_3)$ = the reaction of the part I on part II transmitted through the same surface element at (x_1, x_2, x_3).

Cauchy's fundamental lemma is equivalent to Newton's third law of motion of action and reaction, and is expressed as

$$\vec{T}^{(-n)}(x_1, x_2, x_3) = \vec{T}^{(n)}(x_1, x_2, x_3).$$

Theorem 3.1: (Cauchy's Fundamental Theorem for Stress)
The stress vector at a point on any arbitrary plane surface is a linear function of three stress vectors acting on any three mutually perpendicular planes through that point.

Proof: Consider the deformed continuous media in a state of motion. Let $P(x_1, x_2, x_3)$ be any point in the medium. To determine the dependence of the stress and couple stress vectors on the exterior normal, we next apply the principles of balance of momenta to a small tetrahedron *PABC* of volume δV with its vertex $P(x_1, x_2, x_3)$ in the interior off and having three of its

3.2 Euler-Cauchy Stress Principle

orthogonal faces faces *PBC, PAC, PAB* on the coordinate surfaces and the fourth oblique plane face *ABC* (Fig. 3.2) with unit normal $\vec{n} = (n_1, n_2, n_3)$ at a small distance h from P. This tetrahedron is sometimes called the *Cauchy tetrahedron*.

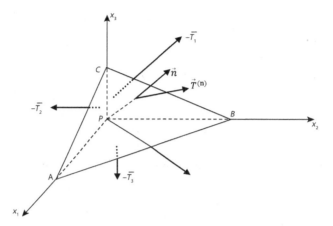

Fig. 3.3: Free body diagram of tetrahedron element having its vertex at *P*.

Consider the motion of the specific portion of the continuum which occupies the tetrahedron at time t. Let us denote $\delta S_1, \delta S_2, \delta S_3$ and δS respectively, the area of faces *PBC, PAC, PAB* and *ABC*. Now

(*i*) The volume of the tetrahedron $\delta V = \dfrac{1}{3} h . \delta S$. (3.8)

(*ii*) The area of the faces of the tetrahedron perpendicular to the axes can be found by projecting δS into each face

δS_1 = area of *PBC* parallel to $x_2 x_3$ plane

= projection of δS on the plane parallel to $x_2 x_3$ plane = $n_1 \delta S$.

Similarly $\delta S_2 = n_2 . \delta S, \delta S_3 = n_3 . \delta S$. Now tetrahedral portion of *PAC* of the continuum moves under the action of body forces and surface forces in the shape of stress vectors transmitted across each of the four boundary faces by the material outside the tetrahedron.

Let $\vec{T}^{(n)}$ denote the stress vector at P transmitted across a plane normal to $\vec{n}$ and parallel to face *ABC*. Since exterior normals to three orthogonal faces *PBC, PAC, PAB* of the tetrahedron are directed oppositely to the positive direction of x_i axes the stress vector $\vec{T}_i$ across these faces are taken with negative signs. Accordingly, let $-\vec{T}_1, -\vec{T}_2, -\vec{T}_3$ denote stress vectors at P transmitted across the faces *PBC, PAC, PAB* normal to x_1, x_2 and x_3 axis, respectively. On account of the assumed continuity of all these physical quantities, the average stress vector exerted across the oblique face ABC of the tetrahedron normal to $\vec{n}$ by the outside material on the inside material can be taken as

$$\vec{T}^{(n)*} = \vec{T}^{(n)} + \mu, \text{ where, } \mu \to 0 \text{ as } h \to 0 \tag{3.9}$$

so that $\vec{T}^{(n)*} \to \vec{T}^{(n)}$ as $h \to 0$. The total thrust exerted across face ABC of the tetrahedron by outside material on the inside material is $\vec{T}^{(n)*} \delta S$. The corresponding total thrusts exerted across the three orthogonal faces PBC, PAC, PAB by the material outside tetrahedron on the material inside it are $-\vec{T}_1^* \delta S_1, -\vec{T}_2^* \delta S_2, -\vec{T}_3^* \delta S_3$ where $-\vec{T}_i^* \to -\vec{T}_i$ as $h \to 0$. Resultant surface force acting on the material inside tetrahedron across all its four faces is

$$\vec{T}^{(n)*} \delta S - \vec{T}_1^* \delta S_1 - \vec{T}_2^* \delta S_2 - \vec{T}_3^* \delta S_3.$$

If $\vec{F}$ denote the body force acting on inside material then the resultant body force acting on the material is $\rho \delta V \vec{F}$, where ρ is the density of the portion of the material.

If $\vec{a}$ be the acceleration of the material within tetrahedron per unit mass, then the equilibrium of forces, *i.e.* Euler's first law of motion (Newton's second law of motion), gives:

$$\vec{T}^{(n)*} \delta S - \vec{T}_1^* \delta S_1 - \vec{T}_2^* \delta S_2 - \vec{T}_3^* \delta S_3 + \rho \delta V \vec{F} = \rho \delta V \vec{a}$$

or, $\quad \vec{T}^{(n)*} \delta S - \vec{T}_1^* n_1 \delta S - \vec{T}_2^* n_2 \delta S - \vec{T}_3^* n_3 \delta S + \rho \dfrac{1}{3} h. \delta S \vec{F} = \rho \dfrac{1}{3} h. \delta S \vec{a}$

or, $\quad \vec{T}^{(n)*} - \vec{T}_1^* n_1 - \vec{T}_2^* n_2 - \vec{T}_3^* n_3 + \rho \dfrac{1}{3} h \vec{F} = \rho \dfrac{1}{3} h \vec{a}$

or, $\quad \vec{T}^{(n)} - \vec{T}_1 n_1 - \vec{T}_2 n_2 - \vec{T}_3 n_3 = 0, \text{ as } h \to 0$

or, $\quad \vec{T}^{(n)} = \vec{T}_1 n_1 + \vec{T}_2 n_2 + \vec{T}_3 n_3 = \vec{T}_i n_i. \tag{3.10}$

Therefore, the stress vector at a point on any arbitrary plane surface is a linear function of three stress vectors acting on any three mutually perpendicular planes through that point.

3.3 Cauchy's Stress Formula: Stress Tensor

The state of stress at a point in the body is defined by all the stress vectors $\vec{T}^{(n)}$ associated with all planes (infinite in number) that pass through that point. However, according to Cauchy's fundamental theorem, also called

3.3 Cauchy's Stress Formula: Stress Tensor

Cauchy's stress theorem, merely by knowing the stress vectors on three mutually perpendicular planes, the stress vector on any other plane passing through that point can be found through coordinate transformation equations.

We have seen that the stress vector $\vec{T}^{(n)}$ at a point $P(x_1, x_2, x_3)$ across a plane with a unit normal $\vec{n} = (n_1, n_2, n_3)$ can be written as

$$\vec{T}^{(n)} = \vec{T}_1 n_1 + \vec{T}_2 n_2 + \vec{T}_3 n_3 = \vec{T}_i n_i, \qquad (3.11)$$

where, $\vec{T}_1$ denotes the stress vector exerted across the plane perpendicular to the x_1 axis. Similarly, for $\vec{T}_2$ and $\vec{T}_3$. Since $\vec{T}_1$ can be resolved into three mutually perpendicular direction $(\vec{e}_1, \vec{e}_2, \vec{e}_3)$ parallel to the coordinate axis and hence we can write

$$\vec{T}_1 = T_{11}\vec{e}_1 + T_{12}\vec{e}_2 + T_{13}\vec{e}_3$$

where, T_{11}, T_{12}, T_{13} are certain scalar functions of (x_1, x_2, x_3, t). Therefore, in general

$$\vec{T}_i = T_{i1}\vec{e}_1 + T_{i2}\vec{e}_2 + T_{i3}\vec{e}_3 = T_{ij}\vec{e}_j. \qquad (3.12)$$

Therefore, T_{ij} is the jth component of the stress vector $\vec{T}_i$ at the point P across a plane normal to x_i axis. Therefore, using equation (3.12) we can write equation (3.11) as

$$\vec{T}^{(n)} = \vec{T}_i n_i = \left(T_{ij}\vec{e}_j\right)n_i = \left(T_{ij}n_i\right)\vec{e}_j = \left(T_{ji}n_j\right)\vec{e}_i = T_i^{(n)}\vec{e}_i \qquad (3.13)$$

where, $T_i^{(n)} = T_{ji}n_j$ is the component of $\vec{T}^{(n)}$ in the direction of x_i axis. Then the stress vector at a point P across any arbitrary plane with normal $\vec{n} = (n_1, n_2, n_3)$ is a linear combination of the line stress component T_{ij} acting across three mutually perpendicular plane at the point parallel to the coordinate axes. The formula given by the equation (3.13) is known as Cauchy's stress formula.

(i) In matrix notation, the Cauchy's stress formula can be written as

$$\begin{pmatrix} T_1^{(n)} \\ T_2^{(n)} \\ T_3^{(n)} \end{pmatrix} = \begin{pmatrix} T_{11} & T_{12} & T_{13} \\ T_{21} & T_{22} & T_{23} \\ T_{31} & T_{32} & T_{33} \end{pmatrix} \begin{pmatrix} n_1 \\ n_2 \\ n_3 \end{pmatrix} \qquad (3.14)$$

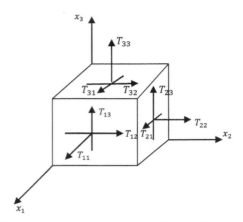

Fig. 3.4: Cartesian stress components shown in their positive sense.

(ii) The components T_{11}, T_{22}, T_{33} are the normal stresses pulling the material lying on the negative side of the respective plane elements parallel to coordinate planes. The remaining components T_{12}, T_{23}, etc. are shearing stresses.

(iii) The normal component of the stress on any plane with normal $\bar{n} = (n_1, n_2, n_3)$ is given by

$$N^{(n)} = T_i^{(n)} n_i = \left(T_{ji} n_j\right) n_i = T_{ij} n_i n_j$$

(iv) If $S^{(n)}$ be the magnitude of the shearing stress, then

$$(N^{(n)})^2 + (S^{(n)})^2 = (T_1^{(n)})^2 + (T_2^{(n)})^2 + (T_3^{(n)})^2$$

or,

$$(S^{(n)})^2 = (T_1^{(n)})^2 + (T_2^{(n)})^2 + (T_3^{(n)})^2 - (N^{(n)})^2$$

$$= (T_{j1} n_j)^2 + (T_{j2} n_j)^2 + (T_{j3} n_j)^2 - (T_{ij} n_i n_j)^2 \qquad (3.15)$$

(v) The Cauchy stress tensor obeys the tensor transformation law under a change in the system of coordinates. A graphical representation of this transformation law is the Mohr's circle for stress.

(vi) The Cauchy stress tensor is used for stress analysis of material bodies experiencing small deformations: It is a central concept in the linear theory of elasticity. For large deformations, also called finite deformations, other measures of stress are required, such as Piola-Kirchhoff stress tensor, the Biot stress tensor, and the Kirchhoff stress tensor.

(vii) According to the principle of conservation of linear momentum, if the continuum body is in static equilibrium can be demonstrated that

3.3 Cauchy's Stress Formula: Stress Tensor

the components of the Cauchy stress tensor in every material point in the body satisfy the equilibrium equations (Cauchy equation's of motion for zero acceleration). At the same time, according to the principle of conservation of angular momentum, equilibrium requires that the summation of moments with respect to an arbitrary point is zero, which leads to the conclusion that the stress tensor is symmetric, thus having only six independent stress components, instead of the original nine.

Example 3.1: Let the components of the stress tensor at a point P be given in matrix form by

$$(T_{ij}) = \begin{pmatrix} 7 & 0 & -2 \\ 0 & 5 & 0 \\ -2 & 0 & 4 \end{pmatrix}$$

in units of mega-Pascals. Determine stress vector on the plane at the point P whose unit normal is $\hat{n} = \left(\dfrac{2}{3}, -\dfrac{2}{3}, \dfrac{1}{3}\right)$.

Solution: From the Cauchy's stress formula (2.14) we have

$$\begin{pmatrix} T_1^{(n)} \\ T_2^{(n)} \\ T_3^{(n)} \end{pmatrix} = \begin{pmatrix} T_{11} & T_{12} & T_{13} \\ T_{21} & T_{22} & T_{23} \\ T_{31} & T_{32} & T_{33} \end{pmatrix} \begin{pmatrix} n_1 \\ n_2 \\ n_3 \end{pmatrix} = \begin{pmatrix} 7 & 0 & -2 \\ 0 & 5 & 0 \\ -2 & 0 & 4 \end{pmatrix} \begin{pmatrix} \dfrac{2}{3} \\ -\dfrac{2}{3} \\ \dfrac{1}{3} \end{pmatrix} = \begin{pmatrix} 4 \\ -\dfrac{10}{3} \\ 0 \end{pmatrix},$$

which is the stress vector on the plane at the point P whose unit normal is $\hat{n}$.

Example 3.2: Let the components of the stress tensor at P be given in matrix form by

$$(T_{ij}) = \begin{pmatrix} 0 & 1 & 2 \\ 1 & b & 1 \\ 2 & 1 & 0 \end{pmatrix}.$$

in units of mega-Pascals, where b is a constant. Determine b so that stress vector on some plane at the point will be zero. Determine the direction cosines of the normal to the plane.

Solution: From the Cauchy's stress formula (3.14) we have

$$\begin{pmatrix} T_1^{(n)} \\ T_2^{(n)} \\ T_3^{(n)} \end{pmatrix} = \begin{pmatrix} T_{11} & T_{12} & T_{13} \\ T_{21} & T_{22} & T_{23} \\ T_{31} & T_{32} & T_{33} \end{pmatrix} \begin{pmatrix} n_1 \\ n_2 \\ n_3 \end{pmatrix} \Rightarrow \begin{pmatrix} 0 & 1 & 2 \\ 1 & b & 1 \\ 2 & 1 & 0 \end{pmatrix} \begin{pmatrix} n_1 \\ n_2 \\ n_3 \end{pmatrix} = \begin{pmatrix} 0 \\ 0 \\ 0 \end{pmatrix}$$

For non-trivial solution, for the system of linear equations

$$\begin{vmatrix} 0 & 1 & 2 \\ 1 & b & 1 \\ 2 & 1 & 0 \end{vmatrix} = 0 \Rightarrow -1(-2) + 2(1-2b) = 0 \Rightarrow b = 1.$$

For $b=1$, the system of linear equations reduces to

$$-2n_1 = n_2 = -2n_3 \Rightarrow \frac{n_1}{1} = \frac{n_2}{-2} = \frac{n_3}{1} = \frac{\sqrt{n_1^2 + n_2^2 + n_3^2}}{\sqrt{1^2 + (-2)^2 + 1^2}} = \frac{1}{\sqrt{6}}.$$

Therefore, the direction cosines of the normal to the plane is $\vec{n} = (n_1, n_2, n_3) = \frac{1}{\sqrt{6}}(1, -2, 1)$.

Example 3.3: Let the components of the stress tensor at P be given in matrix form by

$$(T_{ij}) = \begin{pmatrix} T & aT & bT \\ aT & T & cT \\ bT & cT & T \end{pmatrix}$$

where, a, b, c are constants and T is some stress value. Determine the constants a, b, c so that stress vector on the octahedral plane normal to $\left(\frac{1}{\sqrt{3}}, \frac{1}{\sqrt{3}}, \frac{1}{\sqrt{3}}\right)$ vanishes.

Solution: From the Cauchy's stress formula (2.14) we have

$$\begin{pmatrix} T_1^{(n)} \\ T_2^{(n)} \\ T_3^{(n)} \end{pmatrix} = \begin{pmatrix} T_{11} & T_{12} & T_{13} \\ T_{21} & T_{22} & T_{23} \\ T_{31} & T_{32} & T_{33} \end{pmatrix} \begin{pmatrix} n_1 \\ n_2 \\ n_3 \end{pmatrix} \Rightarrow \begin{pmatrix} T & aT & bT \\ aT & T & cT \\ bT & cT & T \end{pmatrix} \frac{1}{\sqrt{3}} \begin{pmatrix} 1 \\ 1 \\ 1 \end{pmatrix} = \begin{pmatrix} 0 \\ 0 \\ 0 \end{pmatrix}$$

$$\Rightarrow a + b + 1 = 0, c + a + 1 = 0, b + c + 1 = 0.$$

Solving, these equations, we get $a = b = c = -\frac{1}{2}$.

3.3 Cauchy's Stress Formula: Stress Tensor

Example 3.4: The state of stress throughout a continuum is given with respect to Cartesian axes OX_1, OX_2, OX_3 by

$$(T_{ij}) = \begin{pmatrix} 3x_1x_2 & 5x_2^2 & 0 \\ 5x_2^3 & 0 & 2x_3 \\ 0 & 2x_3 & 0 \end{pmatrix}.$$

Determine the stress vector acting at a point $P(2, 1, \sqrt{3})$ on the plane tangent to the cylindrical surface $x_2^2 + x_3^2 = 4$ at the point $P(2, 1, \sqrt{3})$.

Solution: The given equation of the cylindrical surface (Fig. 3.5) can be written as

$$\Phi(x_1, x_2, x_3) = x_2^2 + x_3^2 - 4 = 0$$

$$\Rightarrow \nabla\Phi(x_1, x_2, x_3) = (0, 2x_2, 2x_3) = (0, 2, 2\sqrt{3}) \text{ at } P(2, 1, \sqrt{3})$$

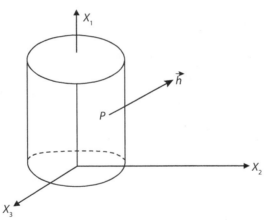

Fig. 3.5: Cylindrical surface $x_2^2 + x_3^2 - 4 = 0$

The unit normal to the surface at P is given by

$$\vec{n} = (n_1, n_2, n_3) = \frac{1}{|\nabla\Phi(2,1,\sqrt{3})|} \nabla\Phi(2,1,\sqrt{3}) = \frac{1}{2}(0, 2, \sqrt{3})$$

The state of stress throughout a continuum is given by

$$(T_{ij}) = \begin{pmatrix} 3x_1x_2 & 5x_2^2 & 0 \\ 5x_2^3 & 0 & 2x_3 \\ 0 & 2x_3 & 0 \end{pmatrix} = \begin{pmatrix} 6 & 5 & 0 \\ 5 & 0 & 2\sqrt{3} \\ 0 & 2\sqrt{3} & 0 \end{pmatrix} \text{ at } P(2,1,\sqrt{3})$$

Finally, by the Cauchy's stress formula (3.14) gives the stress vector at P on the plane perpendicular to $\vec{n} = (n_1, n_2, n_3)$ is given by

$$\begin{pmatrix} T_1^{(n)} \\ T_2^{(n)} \\ T_3^{(n)} \end{pmatrix} = \begin{pmatrix} T_{11} & T_{12} & T_{13} \\ T_{21} & T_{22} & T_{23} \\ T_{31} & T_{32} & T_{33} \end{pmatrix} \begin{pmatrix} n_1 \\ n_2 \\ 3 \end{pmatrix} = \begin{pmatrix} 6 & 5 & 0 \\ 5 & 0 & 2\sqrt{3} \\ 0 & 2\sqrt{3} & 0 \end{pmatrix} \frac{1}{2}\begin{pmatrix} 0 \\ 2 \\ \sqrt{3} \end{pmatrix} = \frac{1}{2}\begin{pmatrix} 5 \\ 6 \\ 2\sqrt{3} \end{pmatrix}.$$

Example 3.5: Let the components of the stress tensor at P be given in matrix form by

$$(T_{ij}) = \begin{pmatrix} -a & 0 & d \\ 0 & b & c \\ d & e & c \end{pmatrix}.$$

Determine the unit normal of a plane parallel to x_3 axis on which resultant stress vector is tangential to the plane.

Solution: From the Cauchy's stress formula (3.14) we have

$$\begin{pmatrix} T_1^{(n)} \\ T_2^{(n)} \\ T_3^{(n)} \end{pmatrix} = \begin{pmatrix} T_{11} & T_{12} & T_{13} \\ T_{21} & T_{22} & T_{23} \\ T_{31} & T_{32} & T_{33} \end{pmatrix} \begin{pmatrix} n_1 \\ n_2 \\ 0 \end{pmatrix} \Rightarrow \begin{pmatrix} T_1^{(n)} \\ T_2^{(n)} \\ T_3^{(n)} \end{pmatrix} = \begin{pmatrix} -a & 0 & d \\ 0 & b & c \\ d & e & c \end{pmatrix} \begin{pmatrix} n_1 \\ n_2 \\ 0 \end{pmatrix}$$

$$\Rightarrow T_1^{(n)} = -an_1, T_2^{(n)} = bn_2, T_3^{(n)} = (dn_1 + cn_2).$$

The normal component of the stress on any plane with normal $\bar{n} = (n_1, n_2, n_3)$, satisfying $n_1^2 + n_2^2 + n_3^2 = 1$ is given by

$$N^{(n)} = T_i^{(n)} n_i = (-an_1)n_1 + (bn_2)n_2 + (dn_1 + cn_2).0 = 0$$

$$\Rightarrow -an_1^2 + bn_2^2 = 0; \text{and} n_2^2 + n_3^2 = 1$$

$$\Rightarrow n_1^2 = \frac{b}{a+b}, n_2^2 = \frac{a}{a+b}, n_3 = 0$$

$$\Rightarrow n_1 = \sqrt{\frac{b}{a+b}}, n_2 = \sqrt{\frac{a}{a+b}}, n_3 = 0.$$

3.4 Cauchy's Stress Quadric

The state of stress in the neighbourhood of any point can be understood more clearly by a geometrical treatment due to Cauchy.

3.4 Cauchy's Stress Quadric

Consider a point $P(x_i)$ in the deformed state of a continuum body. Let T_{ij} be the stress tensor at $P_0(X_i)$ with respect to a system of axes OX_1, OX_2, OX_3 fixed in space. We introduce a local system of axes $P\zeta_1, P\zeta_2, P\zeta_3$ with the origin at $P(x_i)$ and parallel to the axes OX_1, OX_2, OX_3 respectively. For a given set of stress tensor T_{ij}, we can construct a quadric surface with its centre at $P(x_i)$ given by

$$T_{ij}\zeta_i\zeta_j = 1. \tag{3.16}$$

The quadric surface given by equation (3.16) is known as Cauchy's stress quadric at $P(x_i)$

Properties of Stress Quadric

Property 1. The normal stress across any plane through the centre of stress quadric is equal to the inverse of the square of the central radius vector of the quadric normal to the plane.

Consider a point in the deformed state of a continuum body. Let T_{ij} be the stress tensor at P with respect to a fixed system of axes OX_1, OX_2, OX_3 fixed in space.

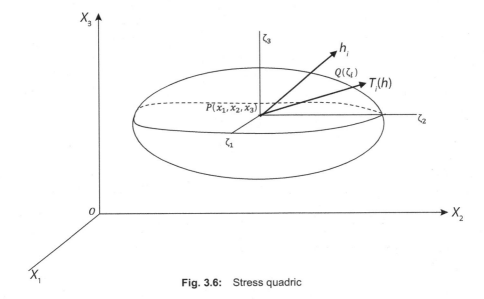

Fig. 3.6: Stress quadric

We introduce a local system of axes $P\zeta_1, P\zeta_2, P\zeta_3$ with the origin at P and parallel to the axes OX_1, OX_2, OX_3 respectively. Let the equation of the stress quadric with its centre at $P(x_1, x_2, x_3)$ be

$$T_{ij}\zeta_i\zeta_j = 1.$$

Let $Q(\zeta_1,\zeta_2,\zeta_3)$ be a point on its surface. Consider a plane element through normal to PQ. Let r denote the length PQ, i.e., $|PQ|=r$. The direction cosines of the line PQ are $\left(\dfrac{\zeta_1}{r},\dfrac{\zeta_2}{r},\dfrac{\zeta_3}{r}\right)=(n_1,n_2,n_3)$. The normal stresses at $P(x_1,x_2,x_3)$ across the plane whose normal (n_1,n_2,n_3), then

$$N^{(n)} = T_{ij}n_in_j = T_{ij}\frac{\zeta_i}{r}\frac{\zeta_j}{r} = \frac{1}{r^2}T_{ij}\zeta_i\zeta_j = \frac{1}{r^2} \tag{3.17}$$

as $Q(\zeta_1,\zeta_2,\zeta_3)$ lies on the stress $T_{ij}\zeta_i\zeta_j = 1$. This shows that, the normal stress across any plane through the centre of stress quadric is equal to the inverse of the square of the central radius vector of the quadric normal to the plane.

Property 2: The stress vector across any plane through the centre of stress quadric is directed parallel to the normal to the surface of the quadric at a point where the central radius vector of the quadric normal to the plane meets the quadric.

Consider a point $P(x_1,x_2,x_3)$ in the deformed state of a continuum body. Let T_{ij} be the stress tensor at P with respect to a fixed system of axes OX_1, OX_2, OX_3, fixed in space.

We introduce a local system of axes $P\zeta_1, P\zeta_2, P\zeta_3$ with the origin at P and parallel to the axes OX_1, OX_2, OX_3 respectively. Let the equation of the stress quadric with its centre at $P(x_1,x_2,x_3)$ be

$$T_{ij}\zeta_i\zeta_j = 1.$$

Let $Q(\zeta_1,\zeta_2,\zeta_3)$ be any point on this surface. Consider a plane element through P normal to PQ. Let r denote the length PQ, i.e., $|PQ|=r$. The direction cosines of the line PQ are $\left(\dfrac{\zeta_1}{r},\dfrac{\zeta_2}{r},\dfrac{\zeta_3}{r}\right)=(n_1,n_2,n_3)$. Let $T_i^{(n)}$ denote the stress vector across this plane, then by the Cauchy stress formula (3.13), we get

$$T_i^{(n)} = T_{ij}n_j = T_{ij}\frac{\zeta_j}{r} \Rightarrow T_{ij}\zeta_j = rT_i^{(n)} \tag{3.18}$$

Let us consider a quadric function

$$2F(\zeta_1,\zeta_2,\zeta_3) = T_{ij}\zeta_i\zeta_j$$

3.4 Cauchy's Stress Quadric

So the equation of the stress quadric becomes $2F(\zeta_1,\zeta_2,\zeta_3)=1$. It follows from the above result that

$$\frac{\partial}{\partial \zeta_i}\left[2F(\zeta_1,\zeta_2,\zeta_3)\right]=\frac{\partial}{\partial \zeta_i}\left[T_{kl}\zeta_k\zeta_l\right]=T_{kl}\left[\frac{\partial \zeta_k}{\partial \zeta_i}\zeta_l+\zeta_k\frac{\partial \zeta_l}{\partial \zeta_i}\right]$$

$$=T_{kl}\left[\delta_{ki}\zeta_l+\zeta_k\delta_{li}\right]=T_{kl}\delta_{ki}\zeta_l+T_{kl}\zeta_k\delta_{li}$$

$$=T_{il}\zeta_l+T_{ki}\zeta_k=T_{ij}\zeta_j+T_{ji}\zeta_j=T_{ij}\zeta_j+T_{ij}\zeta_j=2T_{ij}\zeta_j$$

$$\frac{\partial}{\partial \zeta_i}\left[F(\zeta_1,\zeta_2,\zeta_3)\right]=T_{ij}\zeta_j=rT_i^{(n)} \tag{3.19}$$

But $\frac{\partial}{\partial \zeta_i}\left[F(\zeta_1,\zeta_2,\zeta_3)\right]$ are direction ratios of the normal to the quadric surface $2F(\zeta_1,\zeta_2,\zeta_3)=1$ at the point. It follows that relative displacement is directed along the normal to the quadric surface at $Q(\zeta_1,\zeta_2,\zeta_3)$.

Example 3.6: Determine the Cauchy's stress quadric at P for a state of stress

$$(T_{ij})=\begin{pmatrix} a & 0 & 0 \\ 0 & b & 0 \\ 0 & 0 & c \end{pmatrix}.$$

where a, b, c are all of the same sign.

Solution: Consider a point $P(x_i)$ in the deformed state of a continuum body. The stress tensor T_{ij} at $P(x_i)$ with respect to a system of axes OX_1, OX_2, OX_3, fixed in space are given by

$$(T_{ij})=\begin{pmatrix} a & 0 & 0 \\ 0 & b & 0 \\ 0 & 0 & c \end{pmatrix}.$$

Since all the shearing stresses $T_{ij}, i \neq j$ vanish, the coordinate axes are the principal axes of stresses. We introduce a local system of axes $P\zeta_1, P\zeta_2, P\zeta_3$ with the origin at $P(x_i)$ and parallel to the axes X_1, OX_2, OX_3 respectively. For the given set of stress tensor T_{ij}, a quadric surface with its centre at $P(x_i)$ with respect to the principal axes is given by

$$T_{ij}\zeta_i\zeta_j=1 \quad \Rightarrow \quad T_{11}\zeta_1^2+T_{22}\zeta_2^2+T_{33}\zeta_3^2=1.$$

or,
$$a\zeta_1^2 + b\zeta_2^2 + c\zeta_3^2 = 1$$

which is the required Cauchy's stress quadric at $P(x_i)$ and it represents an ellipsoid.

3.5 Transformation of Stress Components

Let T_{ij} be the stress tensor at a point with respect to a system of axes OX_1, OX_2, OX_3. Let this system of axes be rotated about O to obtain a new system of axes OX_1', OX_2', OX_3'. Let T_{ij}' be the stress tensor with respect to the new system of axes OX_1', OX_2', OX_3'. If (l_1, l_2, l_3) be the direction cosines of OX_1', (m_1, m_2, m_3), be the direction cosines of OX_2', and (n_1, n_2, n_3) be direction cosines of OX_3' then we can prove by analogy with theory of strain

$$T_{11}' = T_{ij} l_i l_j; \quad T_{22}' = T_{ij} m_i m_j; \quad T_{33}' = T_{ij} n_i n_j$$

$$T_{23}' = T_{ij} m_i n_j; \quad T_{31}' = T_{ij} n_i l_j; \quad T_{12}' = T_{ij} l_i m_j.$$

At the point P let the rectangular Cartesian coordinate systems $P X_1 X_2 X_3$ and $P X_1' X_2' X_3'$ be related to one another by the transformation dyadic $A = a_{ij} \hat{e}_i \hat{e}_j$.

In matrix form, the stress vector transformation is written

$$\begin{pmatrix} T_1'^{(n)} \\ T_2'^{(n)} \\ T_3'^{(n)} \end{pmatrix} = \begin{pmatrix} a_{11} & a_{12} & a_{13} \\ a_{21} & a_{22} & a_{23} \\ a_{31} & a_{32} & a_{33} \end{pmatrix} \begin{pmatrix} T_1^{(n)} \\ T_2^{(n)} \\ T_3^{(n)} \end{pmatrix}$$

Explicitly, the matrix multiplications are given respectively by

$$\begin{pmatrix} T_{11}' & T_{12}' & T_{13}' \\ T_{21}' & T_{22}' & T_{23}' \\ T_{31}' & T_{32}' & T_{33}' \end{pmatrix} = \begin{pmatrix} a_{11} & a_{12} & a_{13} \\ a_{21} & a_{22} & a_{23} \\ a_{31} & a_{32} & a_{33} \end{pmatrix} \begin{pmatrix} T_{11} & T_{12} & T_{13} \\ T_{21} & T_{22} & T_{23} \\ T_{31} & T_{32} & T_{33} \end{pmatrix} \begin{pmatrix} a_{11} & a_{21} & a_{31} \\ a_{12} & a_{22} & a_{32} \\ a_{13} & a_{23} & a_{33} \end{pmatrix}$$

Example 3.7: The state of stress at a point with respect to Cartesian axes OX_1, OX_2, OX_3 is given by

$$(T_{ij}) = \begin{pmatrix} 15 & -10 & 0 \\ -10 & 5 & 0 \\ 0 & 0 & 20 \end{pmatrix}$$

3.6 Principal Stress

Determine the stress tensor T'_{ij} for related axes $OX'_1 OX'_2 OX'_3$ for which transformation matrix is

$$(a_{ij}) = \begin{pmatrix} \frac{3}{5} & 0 & -\frac{4}{5} \\ 0 & 1 & 0 \\ \frac{4}{5} & 0 & \frac{3}{5} \end{pmatrix}$$

Solution: By the stress transformation law, we get

$$\begin{pmatrix} T_{11}' & T_{12}' & T_{13}' \\ T_{21}' & T_{22}' & T_{23}' \\ T_{31}' & T_{32}' & T_{33}' \end{pmatrix} = \begin{pmatrix} a_{11} & a_{12} & a_{13} \\ a_{21} & a_{22} & a_{23} \\ a_{31} & a_{32} & a_{33} \end{pmatrix} \begin{pmatrix} T_{11} & T_{12} & T_{13} \\ T_{21} & T_{22} & T_{23} \\ T_{31} & T_{32} & T_{33} \end{pmatrix} \begin{pmatrix} a_{11} & a_{21} & a_{31} \\ a_{12} & a_{22} & a_{32} \\ a_{13} & a_{23} & a_{33} \end{pmatrix}$$

$$= \begin{pmatrix} \frac{3}{5} & 0 & -\frac{4}{5} \\ 0 & 1 & 0 \\ \frac{4}{5} & 0 & \frac{3}{5} \end{pmatrix} \begin{pmatrix} 15 & -10 & 0 \\ -10 & 5 & 0 \\ 0 & 0 & 20 \end{pmatrix} \begin{pmatrix} \frac{3}{5} & 0 & \frac{4}{5} \\ 0 & 1 & 0 \\ -\frac{4}{5} & 0 & \frac{3}{5} \end{pmatrix} = \begin{pmatrix} \frac{91}{5} & -6 & -\frac{12}{5} \\ -6 & 5 & -8 \\ -\frac{12}{5} & -8 & \frac{84}{5} \end{pmatrix}.$$

3.6 Principal Stress

There are certain invariants associated with the stress tensor, whose values do not depend upon the coordinate system chosen, or the area element upon which the stress tensor operates. These are the three eigenvalues of the stress tensor, which are called the principal stresses. The element of plane area on which principal stress is acting is called principal plane and the direction of principal stress is called principal direction of stress of principal axis of stress.

Determination of Principal Strain and Principal Direction of Strain

Consider a point $P(x_1, x_2, x_3)$ inside a continuum body. Consider an element of plane drawn through the point P with unit normal $\hat{n} = (n_1, n_2, n_3)$. Let $T_i^{(n)}$ denote the stress vector acting across this area whose magnitude is T. Let T_{ij} be the stress tensor at P. In order that $T_i^{(n)}$ is a principal stress, it must act along the normal n_i, so that

$$T_i^{(n)} = T n_i, \text{ where, } n_i n_j = 1. \quad (3.20)$$

Now the stress vector is related to stress tensor by the Cauchy stress formula (3.13) as

$$T_{ij}n_j = T_i^{(n)} = Tn_i = Tn_j\delta_{ij}, \text{ from Eq.}(2.20). \qquad (3.21)$$

or, $\qquad (T_{ij} - T\delta_{ij})n_j = 0; \quad j = 1,2,3$

Expanding we get

$$(T_{11} - T)n_1 + T_{12}n_2 + T_{13}n_3 = 0$$

$$T_{21}n_1 + (T_{22} - T)n_2 + T_{23}n_3 = 0 \qquad (3.22)$$

$$T_{31}n_1 + T_{32}n_2 + (T_{33} - T)n_3 = 0$$

This is set of three homogeneous linear equation for n_1, n_2, n_3 which has to satisfy the condition $n_i n_i = 1$, i.e.,

$$n_1^2 + n_2^2 + n_3^2 = 1. \qquad (3.23)$$

The condition for the existence for a non-trivial solution of the equation (3.22)

$$\begin{vmatrix} T_{11} - T & T_{12} & T_{13} \\ T_{21} & T_{22} - T & T_{23} \\ T_{31} & T_{32} & T_{33} - T \end{vmatrix} = 0 \qquad (3.24)$$

The cubic equation (3.24) is called the characteristic equation has three roots T_1, T_2, T_3 are called the principle stresses. Corresponding to each T_i we can solve the system of equation (3.22) subject to (3.23) as

$$\begin{pmatrix} T_{11} - T & T_{12} & T_{13} \\ T_{21} & T_{22} - T & T_{23} \\ T_{31} & T_{32} & T_{33} - T \end{pmatrix} \begin{pmatrix} n_1 \\ n_2 \\ n_3 \end{pmatrix} = \begin{pmatrix} 0 \\ 0 \\ 0 \end{pmatrix}$$

and find (n_1, n_2, n_3) which gives the corresponding principal direction.

(i) Since T_{ij} is symmetric second order tensor the roots of the characteristic equation (3.24) are real and hence all principal stress and direction are real.

(ii) Principal direction of stress corresponding to distinct principal stress are orthogonal to each other.

(iii) When two roots say T_1, T_2 are equal we calculate the principle axis corresponding to $T = T_3$. The other two axes are given by any two mutually perpendicular line which are also perpendicular to the third axis.

(iv) When $T_1 = T_2 = T_3$ any three mutually perpendicular lines to the point P may be taken as the principal axis.

3.6 Principal Stress

Example 3.8:

Let the components of the stress tensor at P be given in matrix form by

$$(T_{ij}) = \begin{pmatrix} 1 & 2 & 3 \\ 2 & 4 & 6 \\ 3 & 6 & 1 \end{pmatrix}.$$

in units of mega-Pascals. Determine

(i) Stress vector $\vec{T}^{(n)}$ at a point P normal to x_1 axis
(ii) Stress vector $\vec{T}^{(n)}$ at a point P on the plane whose normal has direction ratios 1:–1:2
(iii) Stress vector $\vec{T}^{(n)}$ at a point P on the plane through P parallel to the plane $2x_1 - 2x_2 - x_3 = 0$
(iv) Normal components $N^{(n)}$ of stress vector on the plane $2x_1 - 2x_2 - x_3 = 0$
(v) Principal stresses at the point P
(vi) Principal directions of stress at P.

Solution: In the given problem, the stress tensor T_{ij} at the given point P are given by

$$(T_{ij}) = \begin{pmatrix} 1 & 2 & 3 \\ 2 & 4 & 6 \\ 3 & 6 & 1 \end{pmatrix}.$$

(i) The direction cosines of the x_1 axis is $\vec{n} = (n_1, n_2, n_3) = (1, 0, 0)$, so that $n_1 = 1, n_2 = 0$ and $n_3 = 0$. Therefore, the stress vector $\vec{T}^{(n)}$ at a point P is given by

$$T_1^{(n)} = T_{1j} n_j = T_{11} n_1 + T_{12} n_2 + T_{13} n_3 = 1.1 + 2.0 + 3.0 = 1$$

$$T_2^{(n)} = T_{2j} n_j = T_{21} n_1 + T_{22} n_2 + T_{23} n_3 = 2.1 + 4.0 + 6.0 = 2$$

$$T_3^{(n)} = T_{3j} n_j = T_{31} n_1 + T_{32} n_2 + T_{33} n_3 = 3.1 + 6.0 + 1.0 = 3$$

Therefore, the stress vector $\vec{T}^{(n)}$ at a point P normal to x_1 axis is $\vec{T}^{(n)} = (1, 2, 3)$. This vector represents the components of the force per unit area (traction) on the plane defined by $\vec{n} = (n_1, n_2, n_3) = (1, 0, 0)$.

(ii) The unit outward normal to this plane having direction ratios 1:-1:2 is $\bar{n} = (n_1, n_2, n_3) = \frac{1}{\sqrt{6}}(1,-1,2)$. Therefore, the stress vector $\bar{T}^{(n)}$ at a point P is given by

$$T_1^{(n)} = T_{1j}n_j = T_{11}n_1 + T_{12}n_2 + T_{13}n_3 = 1 \cdot \frac{1}{\sqrt{6}} + 2 \cdot (-\frac{1}{\sqrt{6}}) + 3 \cdot \frac{2}{\sqrt{6}} = \frac{5}{\sqrt{6}}$$

$$T_2^{(n)} = T_{2j}n_j = T_{21}n_1 + T_{22}n_2 + T_{23}n_3 = 2 \cdot \frac{1}{\sqrt{6}} + 4 \cdot (-\frac{1}{\sqrt{6}}) + 6 \cdot \frac{2}{\sqrt{6}} = \frac{10}{\sqrt{6}}$$

$$T_3^{(n)} = T_{3j}n_j = T_{31}n_1 + T_{32}n_2 + T_{33}n_3 = 3 \cdot \frac{1}{\sqrt{6}} + 6 \cdot \left(-\frac{1}{\sqrt{6}}\right) + 1 \cdot \frac{2}{\sqrt{6}} = -\frac{1}{\sqrt{6}}.$$

Therefore, the stress vector $\bar{T}^{(n)}$ at a point P on the plane whose normal has direction ratios 1:-1:2 is $\bar{T}^{(n)} = (\frac{5}{\sqrt{6}}, \frac{10}{\sqrt{6}}, -\frac{1}{\sqrt{6}})$. This vector represents the components of the force per unit area (traction) on the plane defined by $\bar{n} = (n_1, n_2, n_3) = \frac{1}{3}(2,-2,-1)$.

(iii) The equation of the plane is $2x_1 - 2x_2 - x_3 = 0$, and the unit outward normal to this plane is $\bar{n} = (n_1, n_2, n_3) = \frac{1}{\sqrt{6}}(1,-1,2)$ Therefore, the stress vector $\bar{T}^{(n)}$ at a point P is given by

$$\begin{pmatrix} T_1^{(n)} \\ T_2^{(n)} \\ T_3^{(n)} \end{pmatrix} = \begin{pmatrix} T_{11} & T_{12} & T_{13} \\ T_{21} & T_{22} & T_{23} \\ T_{31} & T_{32} & T_{33} \end{pmatrix} \begin{pmatrix} n_1 \\ n_2 \\ n_3 \end{pmatrix} = \begin{pmatrix} 1 & 2 & 3 \\ 2 & 4 & 6 \\ 3 & 6 & 1 \end{pmatrix} \frac{1}{3}\begin{pmatrix} 2 \\ -2 \\ -1 \end{pmatrix} = -\frac{1}{3}\begin{pmatrix} 5 \\ 10 \\ 7 \end{pmatrix}.$$

Therefore, the stress vector $\bar{T}^{(n)}$ at a point P on the plane through P parallel to the plane $2x_1 - 2x_2 - x_3 = 0$ is $\bar{T}^{(n)} = (-\frac{5}{3}, -\frac{10}{3}, -\frac{7}{3})$. This vector represents the components of the force per unit area (traction) on the plane defined by $\bar{n} = (n_1, n_2, n_3) = \frac{1}{3}(2,-2,-1)$.

In this example, we clearly see the dependency of the cutting plane and the stress vector. Here, we have considered two different cutting planes at the same point and found that three distinct traction vectors arose from the given stress tensor components.

3.6 Principal Stress

(iv) The Normal components $N^{(n)}$ of stress vector on the plane $2x_1 - 2x_2 - x_3 = 0$, whose unit outward normal $\vec{n} = (n_1, n_2, n_3) = \frac{1}{3}(2,-2,-1)$ is given by

$$N^{(n)} = T_{ij} n_i n_j = \frac{1}{3}(2,-2,-1) \begin{pmatrix} 1 & 2 & 3 \\ 2 & 4 & 6 \\ 3 & 6 & 1 \end{pmatrix} \frac{1}{3} \begin{pmatrix} 2 \\ -2 \\ -1 \end{pmatrix} = \frac{17}{9}.$$

(v) The Principal stresses T_1, T_2, T_3 at the point P are the roots of the characteristic equation

$$\begin{vmatrix} T_{11}-T & T_{12} & T_{13} \\ T_{21} & T_{22}-T & T_{23} \\ T_{31} & T_{32} & T_{33}-T \end{vmatrix} = \begin{vmatrix} 1-T & 2 & 3 \\ 2 & 4-T & 6 \\ 3 & 6 & 1-T \end{vmatrix} = 0$$

or, $\qquad T^3 - 6T^2 - 40T = 0 \Rightarrow T = -4, 0, 10.$

(vi) The principal directions of stress at P are given by the equation (3.22) as

$$(T_{11}-T)n_1 + T_{12}n_2 + T_{13}n_3 = 0$$

$$T_{21}n_1 + (T_{22}-T)n_2 + T_{23}n_3 = 0$$

$$T_{31}n_1 + T_{32}n_2 + (T_{33}-T)n_3 = 0$$

For $T = T_1 = -4$, the above system of equations become

$$\begin{pmatrix} 1+4 & 2 & 3 \\ 2 & 4+4 & 6 \\ 3 & 6 & 1+4 \end{pmatrix} \begin{pmatrix} n_1^{(1)} \\ n_2^{(1)} \\ n_3^{(1)} \end{pmatrix} = \begin{pmatrix} 0 \\ 0 \\ 0 \end{pmatrix} \Rightarrow \begin{pmatrix} n_1^{(1)} \\ n_2^{(1)} \\ n_3^{(1)} \end{pmatrix} = \begin{pmatrix} 1 \\ 2 \\ -3 \end{pmatrix}$$

For $T = T_2 = 0$, the above system of equations become

$$\begin{pmatrix} 1-0 & 2 & 3 \\ 2 & 4-0 & 6 \\ 3 & 6 & 1-0 \end{pmatrix} \begin{pmatrix} n_1^{(2)} \\ n_2^{(2)} \\ n_3^{(2)} \end{pmatrix} = \begin{pmatrix} 0 \\ 0 \\ 0 \end{pmatrix} \Rightarrow \begin{pmatrix} n_1^{(2)} \\ n_2^{(2)} \\ n_3^{(2)} \end{pmatrix} = \begin{pmatrix} 1 \\ 0 \\ 1 \end{pmatrix}$$

For $T = T_3 = 10$, the above system of equations become

$$\begin{pmatrix} 1-10 & 2 & 3 \\ 2 & 4-10 & 6 \\ 3 & 6 & 1-10 \end{pmatrix} \begin{pmatrix} n_1^{(3)} \\ n_2^{(3)} \\ n_3^{(3)} \end{pmatrix} = \begin{pmatrix} 0 \\ 0 \\ 0 \end{pmatrix} \Rightarrow \begin{pmatrix} n_1^{(3)} \\ n_2^{(3)} \\ n_3^{(3)} \end{pmatrix} = \begin{pmatrix} 3 \\ 6 \\ 5 \end{pmatrix}$$

3.7 Stress Invariants

The principal stresses T_1, T_2, T_3 are the roots of the characteristic equation

$$\begin{vmatrix} T_{11}-T & T_{12} & T_{13} \\ T_{21} & T_{22}-T & T_{23} \\ T_{31} & T_{32} & T_{33}-T \end{vmatrix} = 0. \tag{3.25}$$

Expanding, we get

$$T^3 - \Theta_1 T^2 + \Theta_2 T - \Theta_3 = 0, \tag{3.26}$$

where,

$$\Theta_1 = T_{11} + T_{22} + T_{33} \tag{3.27}$$

$$\Theta_2 = \begin{vmatrix} T_{11} & T_{12} \\ T_{21} & T_{22} \end{vmatrix} + \begin{vmatrix} T_{22} & T_{23} \\ T_{32} & T_{33} \end{vmatrix} + \begin{vmatrix} T_{33} & T_{31} \\ T_{13} & T_{11} \end{vmatrix} \tag{3.28}$$

$$\Theta_3 = \begin{vmatrix} T_{11} & T_{12} & T_{13} \\ T_{21} & T_{22} & T_{23} \\ T_{31} & T_{32} & T_{33} \end{vmatrix}. \tag{3.29}$$

Relations between the roots and the coefficients of the cubic equation (3.26) are

$$\Theta_1 = T_1 + T_2 + T_3; \Theta_2 = T_1 T_2 + T_2 T_3 + T_3 T_1; \Theta_3 = T_1 T_2 T_3.$$

Since the principal stresses T_1, T_2, T_3 at a point do not depend on the choice of the coordinate axes, $\Theta_1, \Theta_2, \Theta_3$ given by equations (3.26),-(3.28) are invariant with respect to an orthogonal transformation of coordinates $\Theta_1, \Theta_2, \Theta_3$ are called first, second, third stress invariants respectively.

Example 3.9: Let the components of the stress tensor at P be given in matrix form by

$$(T_{ij}) = \begin{pmatrix} 6 & -3 & 2 \\ -3 & 6 & 0 \\ 0 & 0 & 8 \end{pmatrix}$$

in units of mega-Pascals. Evaluate directly stress invariants from stress tensor. Determine principal stresses and show that stress invariants calculated from principal stresses are the same.

Solution: Using equation (3.27), we get

3.7 Stress Invariants

$$\Theta_1 = T_{11} + T_{22} + T_{33} = 6 + 6 + 8 = 20.$$

$$\Theta_2 = \begin{vmatrix} T_{11} & T_{12} \\ T_{21} & T_{22} \end{vmatrix} + \begin{vmatrix} T_{22} & T_{23} \\ T_{32} & T_{33} \end{vmatrix} + \begin{vmatrix} T_{33} & T_{31} \\ T_{13} & T_{11} \end{vmatrix}$$

$$= \begin{vmatrix} 6 & -3 \\ -3 & 6 \end{vmatrix} + \begin{vmatrix} 6 & 0 \\ 0 & 8 \end{vmatrix} + \begin{vmatrix} 8 & 0 \\ 2 & 6 \end{vmatrix} = 27 + 48 + 48 = 123$$

$$\Theta_3 = \begin{vmatrix} T_{11} & T_{12} & T_{13} \\ T_{21} & T_{22} & T_{23} \\ T_{31} & T_{32} & T_{33} \end{vmatrix} = \begin{vmatrix} 6 & -3 & 2 \\ -3 & 6 & 0 \\ 0 & 0 & 8 \end{vmatrix} = 216.$$

The principal stresses T_1, T_2, T_3 at the point P are the roots of the characteristic equation

$$\begin{vmatrix} T_{11} - T & T_{12} & T_{13} \\ T_{21} & T_{22} - T & T_{23} \\ T_{31} & T_{32} & T_{33} - T \end{vmatrix} = \begin{vmatrix} 6 - T & -3 & 2 \\ -3 & 6 - T & 0 \\ 0 & 0 & 8 - T \end{vmatrix} = 0$$

or,
$$T^3 - 20T^2 + 123T - 216 = 0 \Rightarrow T = 3, 8, 9.$$

Therefore, the stress invariants calculated from principal stresses are given by

$$\Theta_1 = T_1 + T_2 + T_3 = 20; \Theta_2 = T_1 T_2 + T_2 T_3 + T_3 T_1 = 123; \Theta_3 = T_1 T_2 T_3 = 216.$$

Example 3.10: The principal stresses at a point P be given as $T_1 = 9, T_2 = 8, T_3 = 3$. If the stress tensor at that point is given by

$$(T_{ij}) = \begin{pmatrix} T_{11} & -3 & 0 \\ -3 & 6 & 0 \\ 0 & 0 & T_{33} \end{pmatrix}$$

in units of mega-Pascals. Find the values of T_{11} and T_{33}.

Solution: Here, we use the stress invariants. Using equation (3.27), we get

$$\Theta_1 = T_{11} + T_{22} + T_{33} = T_1 + T_2 + T_3$$

$$\Rightarrow T_{11} + T_{33} = 9 + 8 + 3 - 6 = 14.$$

$$\Rightarrow T_{11} + 6 + T_{33} = 9 + 8 + 3$$

Using equation (3.28), we get

$$\Theta_2 = \begin{vmatrix} T_{11} & T_{12} \\ T_{21} & T_{22} \end{vmatrix} + \begin{vmatrix} T_{22} & T_{23} \\ T_{32} & T_{33} \end{vmatrix} + \begin{vmatrix} T_{33} & T_{31} \\ T_{13} & T_{11} \end{vmatrix} = T_1 T_2 + T_1 T_3 + T_2 T_3$$

$$\Rightarrow \begin{vmatrix} T_{11} & -3 \\ -3 & 6 \end{vmatrix} + \begin{vmatrix} 6 & 0 \\ 0 & T_{33} \end{vmatrix} + \begin{vmatrix} T_{33} & 0 \\ 0 & T_{11} \end{vmatrix} = 9.8 + 9.3 + 8.3$$

$$\Rightarrow 6T_{11} + 6T_{33} + T_{11} T_{33} = 132$$

$$\Rightarrow 6T_{11} + 6(14 - T_{11}) + T_{11}(14 - T_{11}) = 132$$

$$\Rightarrow T_{11}^2 - 14T_{11} + 48 = 0 \Rightarrow T_{11} = 6, 8.$$

Therefore, $T_{33} = 8, 6$ Using equation (2.29), we get

$$\Theta_3 = \begin{vmatrix} T_{11} & T_{12} & T_{13} \\ T_{21} & T_{22} & T_{23} \\ T_{31} & T_{32} & T_{33} \end{vmatrix} = T_1 T_2 T_3$$

$$\Rightarrow \begin{vmatrix} T_{11} & -3 & 0 \\ -3 & 6 & 0 \\ 0 & 0 & T_{33} \end{vmatrix} = 9.8.3 = 216$$

We see that $T_{11} = 6$ and $T_{33} = 8$ satisfy the above equation. Hence the values of T_{11} and T_{33} are $T_{11} = 6$ and $T_{33} = 8$.

3.8 Extremum of Stress Values

The stress vector on an arbitrary plane at P may be resolved into a component normal to the plane having a magnitude σ_N, along with a shear component which acts in the plane and has a magnitude σ_S, as shown in Fig. 3.7. (Here, σ_N and σ_S are *not* vectors, but scalar magnitudes of vector components. The subscripts N and S are to be taken as part of the component symbols.)

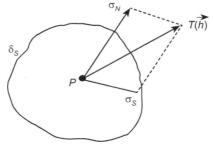

Fig. 3.7: Traction vector components normal and in-plane (shear) at P on the plane whose normal is ni

3.8 Extremum of Stress Values

Theorem 3.2: (**Extremum normal stress**)
The extremum values of normal stress at a point of a continuum are principal stresses.

Proof: Consider a point $P(x_i)$ in the deformed state of a continuum body. Consider an elementary plane area through the point $P(x_i)$ with unit normal $\bar{n} = (n_1, n_2, n_3)$. Let us take the rectangular Cartesian coordinate axes along the principal directions of stress at the point P. Let T_1, T_2, T_3 be the principal stresses at the point and ordered so that $T_1 > T_2 > T_3$. If T_{ij} be the stress tensors at a point then

$$T_{11} = T_1, T_{22} = T_2, T_{33} = T_3, T_{12} = T_{31} = T_{23} = 0. \tag{3.30}$$

If $N^{(n)}$ be the normal stress at the point P acting in the direction $\bar{n} = (n_1, n_2, n_3)$, then

$$N^{(n)} = T_{ij} n_i n_j = T_{11} n_1^2 + T_{22} n_2^2 + T_{33} n_3^2 = T_1 n_1^2 + T_2 n_2^2 + T_3 n_3^2. \tag{3.31}$$

The direction cosines $\bar{n} = (n_1, n_2, n_3)$ satisfy the equation

$$n_i n_j = n_1^2 + n_2^2 + n_3^2 = 1. \tag{3.32}$$

We require the extremum values of the normal stress $N^{(n)}$ for variations of n_1, n_2, n_3 subject to the constraint (2.32). For this, let us construct a Lagrangian function

$$F(n_1, n_2, n_3) = N^{(n)} - \lambda(n_1^2 + n_2^2 + n_3^2 - 1)$$

$$= T_1 n_1^2 + T_2 n_2^2 + T_3 n_3^2 - \lambda\left(n_1^2 + n_2^2 + n_3^2 - 1\right), \tag{3.33}$$

where, l is the Lagrangian parameter. For extreme value of F, we have

$$\frac{\partial F}{\partial n_1} = 0 \Rightarrow T_1 n_1 - \lambda n_1 = 0; \quad \frac{\partial F}{\partial n_2} = 0 \Rightarrow T_2 n_2 - \lambda n_2 = 0; \quad \frac{\partial F}{\partial n_3} = 0 \Rightarrow T_3 n_3 - \lambda n_3 = 0.$$

$$\Rightarrow n_1 (T_1 n_1 - \lambda n_1) + n_2 (T_2 n_2 - \lambda n_2) + n_3 (T_3 n_3 - \lambda n_3) = 0$$

$$\Rightarrow \lambda \left(n_1^2 + n_2^2 + n_3^2 \right) = T_1 n_1^2 + T_2 n_2^2 + T_3 n_3^2 \Rightarrow \lambda = N^{(n)}. \tag{3.34}$$

Therefore,
$$T_1 n_1 - \lambda n_1 = 0 \Rightarrow \left(T_1 - N^{(n)} \right) n_1 = 0 \tag{3.35}$$

$$T_2 n_2 - \lambda n_2 = 0 \Rightarrow \left(T_2 - N^{(n)} \right) n_2 = 0 \tag{3.36}$$

$$T_3 n_3 - \lambda n_3 = 0 \Rightarrow \left(T_3 - N^{(n)}\right) n_3 = 0 \qquad (3.37)$$

Equations (3.35)-(3.37) determine three unknowns n_1, n_2, n_3 for which is extremum. The trivial zero solution of the above system of equations is $n_1 = 0, n_2 = 0, n_3 = 0$ which is not compatible with the condition (3.32), and should be rejected. One type of non-trivial solution is

$$n_1 = 0, n_2 = 0, n_3 \neq 0,$$

so that, from equation (3.32), $n_3^2 = 1 \Rightarrow n_3 = \pm 1$. Equations (3.35) and (3.36) are automatically satisfied. Equation (3.37) gives $N^{(n)} = T_3$. Similarly, the solution $n_1 = 0, n_2 \neq 0, n_3 = 0$, of equations (3.35)-(3.37) gives $N^{(n)} = T_2$ and the solution $n_1 \neq 0, n_2 = 0, n_3 = 0$, of equations (3.35)-(3.37) gives $N^{(n)} = T_1$. Since $T_1 > T_2 > T_3$, maximum value of the normal stress $N^{(n)} = T_1$ and minimum value of $N^{(n)} = T_3$. Therefore, extremum values of the normal stress at a point are always principal stresses acting across planes for which shearing stress components vanish identically.

Theorem 3.2: (Extremum Shearing Stress):

The maximum shearing stress acts on the plane that bisects the angle between the greatest and the smallest principal stress planes. Its value is one half the difference between the greatest and the smallest principal stresses.

Proof: Consider a point $P(x_i)$ in the deformed state of a continuum body. Consider an elementary plane area through the point $P(x_i)$ with unit normal $\bar{n} = (n_1, n_2, n_3)$. Let us take the rectangular Cartesian coordinate axes along the principal directions of stress at the point P. Let T_1, T_2, T_3 be the principal stresses at the point P and ordered so that $T_1 > T_2 > T_3$. If T_{ij} be the stress tensors at a point P then

$$T_{11} = T_1, T_{22} = T_2, T_{33} = T_3, T_{12} = T_{31} = T_{23} = 0. \qquad (3.38)$$

If $N^{(n)}$ be the normal stress at the point P acting in the direction $\bar{n} = (n_1, n_2, n_3)$, then

$$N^{(n)} = T_{ij} n_i n_j = T_{11} n_1^2 + T_{22} n_2^2 + T_{33} n_3^2 = T_1 n_1^2 + T_2 n_2^2 + T_3 n_3^2. \qquad (3.39)$$

The direction cosines $\bar{n} = (n_1, n_2, n_3)$ satisfy the equation

$$n_i n_j = n_1^2 + n_2^2 + n_3^2 = 1. \qquad (3.40)$$

3.8 Extremum of Stress Values

If $T_i^{(n)}$ denote the stress at the point P across a plane element through P with unit normal n_i, then it is given by

$$T_i^{(n)} = T_{ji}n_j;\ T_1^{(n)} = T_{1j}n_j = T_1n_1;\ T_2^{(n)} = T_{2j}n_j = T_2n_2;\ T_\#^{(n)} = T_{3j}n_j = T_3n_3. \quad (3.41)$$

Let T be magnitude of the stress vector $T_i^{(n)}$, then

$$T^2 = (T_1^{(n)})^2 + (T_2^{(n)})^2 + (T_3^{(n)})^2 = (T_1n_1)^2 + (T_2n_2)^2 + (T_3n_3)^2. \quad (3.42)$$

If $S^{(n)}$ be the magnitude of the shearing stress at the point P, then

$$T^2 = (N^{(n)})^2 + (S^{(n)})^2 = (T_1^{(n)})^2 + (T_2^{(n)})^2 + (T_3^{(n)})^2$$

or, $\quad (S^{(n)})^2 = (T_1^{(n)})^2 + (T_2^{(n)})^2 + (T_3^{(n)})^2 - (N^{(n)})^2$

$$= (T_1n_1)^2 + (T_2n_2)^2 + (T_3n_3)^2 - (T_1n_1^2 + T_2n_2^2 + T_3n_3^2)^2. \quad (3.43)$$

We require the extremum values of the normal stress $S^{(n)}$ for variations of n_1, n_2, n_3 subject to the constraint (3.40). For this, let us construct a Lagrangian function

$$F(n_1,n_2,n_3) = (S^{(n)})^2 - \lambda\left(n_1^2 + n_2^2 + n_3^2 - 1\right) = (T_1n_1)^2 + (T_2n_2)^2 + (T_3n_3)^2$$

$$-(T_1n_1^2 + T_2n_2^2 + T_3n_3^2)^2 - \lambda\left(n_1^2 + n_2^2 + n_3^2 - 1\right), \quad (3.44)$$

where, λ is the Lagrangian parameter. For extreme value of F, we have

$$\frac{\partial F}{\partial n_1} = 0 \Rightarrow T_1^2 n_1 - 2T_1 n_1(T_1n_1^2 + T_2n_2^2 + T_3n_3^2) - \lambda n_1 = 0$$

$$\frac{\partial F}{\partial n_2} = 0 \Rightarrow T_2^2 n_2 - 2T_2 n_2(T_1n_1^2 + T_2n_2^2 + T_3n_3^2) - \lambda n_2 = 0$$

$$\frac{\partial F}{\partial n_3} = 0 \Rightarrow T_3^2 n_3 - 2T_3 n_3(T_1n_1^2 + T_2n_2^2 + T_3n_3^2) - \lambda n_3 = 0$$

$$\Rightarrow n_1\left[T_1^2 n_1 - 2T_1 n_1(T_1n_1^2 + T_2n_2^2 + T_3n_3^2) - \lambda n_1\right]$$

$$+ n_2[T_2^2 n_2 - 2T_2 n_2(T_1n_1^2 + T_2n_2^2 + T_3n_3^2) - \lambda n_2]$$

$$+ n_3[T_3^2 n_3 - 2T_3 n_3(T_1n_1^2 + T_2n_2^2 + T_3n_3^2) - \lambda n_3] = 0$$

$$\Rightarrow \lambda\left(n_1^2 + n_2^2 + n_3^2\right) = \left(T_1^2 n_1^2 + T_2^2 n_2^2 + T_3^2 n_3^2\right) - 2(T_1n_1^2 + T_2n_2^2 + T_3n_3^2)^2$$

$$\Rightarrow \lambda = (S^{(n)})^2 - (T_1n_1^2 + T_2n_2^2 + T_3n_3^2)^2 = (S^{(n)})^2 - (N^{(n)})^2. \quad (3.45)$$

Therefore,
$$T_1^2 n_1 - 2T_1 n_1 (T_1 n_1^2 + T_2 n_2^2 + T_3 n_3^2) - \lambda n_1 = 0$$
$$\Rightarrow T_1^2 n_1 - 2T_1 n_1 N^{(n)} - n_1 (S^{(n)})^2 - (N^{(n)})^2 = 0$$
$$\Rightarrow [T_1^2 - 2T_1 N^{(n)} - (S^{(n)})^2 + (N^{(n)})^2] n_1 = 0. \qquad (3.46)$$

Similarly, we get
$$[T_2^2 - 2T_2 N^{(n)} - (S^{(n)})^2 + (N^{(n)})^2] n_2 = 0 \qquad (3.47)$$

and
$$[T_3^2 - 2T_3 N^{(n)} - (S^{(n)})^2 + (N^{(n)})^2] n_3 = 0. \qquad (3.48)$$

Equations (3.46)-(3.48) determine three unknowns n_1, n_2, n_3 for which $S^{(n)}$ is extremum. The trivial zero solution of the above system of equations is $n_1 = 0, n_2 = 0, n_3 = 0$ which is not compatible with the condition (3.40), and should be rejected. One type of non-trivial solution is

$$n_1 = 0, n_2 = 0, n_3 \neq 0,$$

so that, from equation (3.40), $n_3^2 = 1 \Rightarrow n_3 = \pm 1$. For this values of n_i, equation (3.43) becomes

$$(S^{(n)})^2 = T_1^2 n_1^2 + T_2^2 n_2^2 + T_3^2 n_3^2 - (T_1 n_1^2 + T_2 n_2^2 + T_3 n_3^2)^2$$
$$= 0 \Rightarrow (S^{(n)})^2 = 0 \Rightarrow S^{(n)} = 0.$$

Therefore, from equation (3.39), we get $N^{(n)} = |T_3|$. Also the equation (3.48) becomes

$$T_3^2 - 2T_3 N^{(n)} - (S^{(n)})^2 + (N^{(n)})^2 = 0$$

which is automatically satisfied by the values of $S^{(n)}$ and $N^{(n)}$. Thus solution of equations (3.46)–(3.48) is therefore

$$n_1 = 0, n_2 = 0, n_3 = \pm 1 \text{ for which } S^{(n)} = 0. \qquad (3.49)$$

By cyclic permutation we obtain two more solutions of this type

$$n_1 = \pm 1, n_2 = 0, n_3 = 0 \text{ for which } S^{(n)} = 0.$$

$$n_1 = 0, n_2 = \pm 1, n_3 = 0 \text{ for which } S^{(n)} = 0.$$

This merely varies the known fact that planar elements normal to the principal direction of stress are free from shear stress. Thus minimum values of S are zero and they are associated with the principal directions. To determine the direction associated with the maximum values of $S^{(n)}$ we consider second type of solution by assuming

3.8 Extremum of Stress Values

$$n_1 = 0, n_2 \neq 0, n_3 \neq 0,$$

so that the equation (3.46) is automatically satisfied. Also,

$$n_2^2 + n_3^2 = 1; N^{(n)} = T_2 n_2^2 + T_3 n_3^2.$$

Equations (3.47) and (3.48) become

$$T_2^2 - 2T_2 N^{(n)} - (S^{(n)})^2 + (N^{(n)})^2 = 0; \quad T_3^2 - 2T_3 N^{(n)} - (S^{(n)})^2 + (N^{(n)})^2 = 0$$

$$\Rightarrow T_2^2 - T_3^2 - 2T_2 N^{(n)} + 2T_3 N^{(n)} = 0$$

$$\Rightarrow T_2 + T_3 = 2N^{(n)} = 2\left[T_2 n_2^2 + T_3 n_3^2\right]$$

$$\Rightarrow \frac{T_2 + T_3}{2} = T_2 n_2^2 + T_3(1 - n_2^2) \Rightarrow n_2^2 = \frac{\frac{T_2 + T_3}{2} - T_3}{T_2 - T_3} = \frac{1}{2}$$

$$\Rightarrow n_2 = \pm\frac{1}{\sqrt{2}} \Rightarrow n_3 = \pm\sqrt{1 - n_2^2} = \pm\frac{1}{\sqrt{2}}.$$

Equation (3.43) gives

$$(S^{(n)})^2 = \frac{T_2^2 + T_3^2}{2} - \left(\frac{T_2 + T_3}{2}\right)^2 = \frac{1}{4}\left[2T_2^2 + 2T_3^2 - T_2^2 - T_3^2 - 2T_2 T_3\right]$$

$$\Rightarrow (S^{(n)})^2 = \frac{1}{4}(T_2 - T_3)^2 \Rightarrow |S^{(n)}| = \frac{1}{2}|T_2 - T_3|.$$

Thus, the solution of equations (3.46)–(3.48) is therefore

$$n_1 = 0, n_2 = \pm\frac{1}{\sqrt{2}}, n_3 = \pm\frac{1}{\sqrt{2}} \text{ for which } S^{(n)} | = \frac{1}{2}(T_2 - T_3) \quad (3.49)$$

By cyclic permutation we obtain two more solutions of this type

$$n_1 = \pm\frac{1}{\sqrt{2}}, n_2 = 0, n_3 = \pm\frac{1}{\sqrt{2}} \text{ for which } |S^{(n)}| = \frac{1}{2}(T_1 - T_3).$$

$$n_1 = \pm\frac{1}{\sqrt{2}}, n_2 = \pm\frac{1}{\sqrt{2}}, n_3 = 0 \text{ for which } |S^{(n)}| = \frac{1}{2}(T_1 - T_2).$$

Since $T_1 > T_2 > T_3$, T_1 s the largest and T_3 is smallest. Thus the maximum value of the shearing stress is given by

$$\left|S^{(n)}\right| = \frac{1}{2}(T_1 - T_3). \quad (3.50)$$

and it acts on the plane element containing x_2 principal axis and bisecting the angle between x_1 and x_3 axes. Therefore, the maximum shearing stress at any point of the continuum is equal to one-half the difference between algebraically, the largest and smallest principal stresses and acts on the plane that bisects the angle between the directions corresponding to largest and smallest principal stresses.

Example 3.11:

Let the components of the stress tensor at P be given in matrix form by

$$\left(T_{ij}\right) = \begin{pmatrix} 1 & 0 & 2 \\ 0 & 1 & 0 \\ 2 & 0 & -2 \end{pmatrix}$$

in units of mega-Pascals. Find principal stresses and show that principal directions which correspond to largest and smallest principal stresses are both perpendicular to x_2 axis.

Solution: The Principal stresses T_1, T_2, T_3 at the point P are the roots of the characteristic equation

$$\begin{vmatrix} T_{11}-T & T_{12} & T_{13} \\ T_{21} & T_{22}-T & T_{23} \\ T_{31} & T_{32} & T_{33}-T \end{vmatrix} = \begin{vmatrix} 1-T & 0 & 2 \\ 0 & 1-T & 0 \\ 2 & 0 & -2-T \end{vmatrix} = 0$$

or, $\quad T^3 - 3T^2 - 6T + 8 = 0 \Rightarrow T = -2, 1, 4.$

Example 3.12:
Let the components of the stress tensor at be given in matrix form by

$$\left(T_{ij}\right) = \begin{pmatrix} 3 & 1 & 2 \\ 1 & 0 & 2 \\ 1 & 2 & 0 \end{pmatrix}.$$

in units of mega-Pascals. Determine principal stresses and principal direction. Find maximum value of the shearing stresses.

Solution: The Principal stresses T_1, T_2, T_3 at the point P are the roots of the characteristic equation

$$\begin{vmatrix} T_{11}-T & T_{12} & T_{13} \\ T_{21} & T_{22}-T & T_{23} \\ T_{31} & T_{32} & T_{33}-T \end{vmatrix} = \begin{vmatrix} 3-T & 1 & 2 \\ 1 & 0-T & 2 \\ 1 & 2 & 0-T \end{vmatrix} = 0$$

or, $\quad T^3 - 3T^2 - 6T + 8 = 0 \Rightarrow T = -2, 1, 4.$

The principal directions of stress at P are given by the equation (3.22) as

$$(T_{11} - T)n_1 + T_{12}n_2 + T_{13}n_3 = 0$$

$$T_{21}n_1 + (T_{22} - T)n_2 + T_{23}n_3 = 0$$

$$T_{31}n_1 + T_{32}n_2 + (T_{33} - T)n_3 = 0$$

For $T = T_1 = -2$, the above system of equations become

$$\begin{pmatrix} 3+2 & 1 & 2 \\ 1 & 0+2 & 2 \\ 1 & 2 & 0+2 \end{pmatrix} \begin{pmatrix} n_1^{(1)} \\ n_2^{(1)} \\ n_3^{(1)} \end{pmatrix} = \begin{pmatrix} 0 \\ 0 \\ 0 \end{pmatrix} \Rightarrow \begin{pmatrix} n_1^{(1)} \\ n_2^{(1)} \\ n_3^{(1)} \end{pmatrix} = \begin{pmatrix} 0 \\ 1 \\ -1 \end{pmatrix}$$

For $T = T_2 = 1$, the above system of equations become

$$\begin{pmatrix} 3-1 & 1 & 2 \\ 1 & 0-1 & 2 \\ 1 & 2 & 0-1 \end{pmatrix} \begin{pmatrix} n_1^{(2)} \\ n_2^{(2)} \\ n_3^{(2)} \end{pmatrix} = \begin{pmatrix} 0 \\ 0 \\ 0 \end{pmatrix} \Rightarrow \begin{pmatrix} n_1^{(2)} \\ n_2^{(2)} \\ n_3^{(2)} \end{pmatrix} = \begin{pmatrix} 1 \\ -1 \\ -1 \end{pmatrix}$$

For $T = T_3 = 4$, the above system of equations become

$$\begin{pmatrix} 3-4 & 1 & 2 \\ 1 & 0-4 & 2 \\ 1 & 2 & 0-4 \end{pmatrix} \begin{pmatrix} n_1^{(3)} \\ n_2^{(3)} \\ n_3^{(3)} \end{pmatrix} = \begin{pmatrix} 0 \\ 0 \\ 0 \end{pmatrix} \Rightarrow \begin{pmatrix} n_1^{(3)} \\ n_2^{(3)} \\ n_3^{(3)} \end{pmatrix} = \begin{pmatrix} 2 \\ 1 \\ 1 \end{pmatrix}$$

Therefore, the principal directions are given by

$$\frac{1}{\sqrt{2}}(0,1,-1); \frac{1}{\sqrt{3}}(1,-1,-1); \frac{1}{\sqrt{6}}(2,1,1).$$

Since $4 > 1 > -2$, $T_1 = 4$ is the largest and $T_3 = -2$ is smallest. Thus the maximum value of the shearing stress is given by

$$\left| S^{(n)} \right| = \frac{1}{2}(T_1 - T_3) = \frac{1}{2}[4 - (-2)] = 3.$$

3.9 Mohr's Circles for Stress

Mohr's circle, named after Christian Otto Mohr, is a two-dimensional graphical representation of the transformation law for the Cauchy stress tensor. Mohr's circle also tells us the principal angles (orientations) of the principal stresses without your having to plug an angle into stress transformation equations.

The abscissa, $N^{(n)}$, and ordinate, $S^{(n)}$, of each point on the circle, are the magnitudes of the normal stress and shear stress components, respectively, acting on the rotated coordinate system. In other words, the circle is the locus of points that represent the state of stress on individual planes at all their orientations, where the axes represent the principal axes of the stress element.

The Mohr circle can be applied to any symmetric 2×2 tensor matrix, including the strain and moment of inertia tensors.

3.9.1 Motivation for the Mohr Circle

In engineering, the stress distribution within an object, for instance stresses in a rock mass around a tunnel, airplane wings, or building columns, is determined through a stress analysis. Calculating the stress distribution implies the determination of stresses at every point (material particle) in the object. According to Cauchy, the *stress at* distribution implies the determination of stresses at every point (material particle) in the object. According to Cauchy, the *stress at any point* in an object (Fig. 3.8), assumed as a continuum, is completely defined by the nine stress components T_{ij} of a second order tensor of type (2,0) known as the Cauchy stress tensor T:

$$T = \begin{pmatrix} T_{11} & T_{12} & T_{13} \\ T_{21} & T_{22} & T_{23} \\ T_{31} & T_{32} & T_{33} \end{pmatrix} = \begin{pmatrix} T_{xx} & T_{xy} & T_{xz} \\ T_{yx} & T_{yy} & T_{yz} \\ T_{zx} & T_{zy} & T_{zz} \end{pmatrix} = \begin{pmatrix} T_x & T_{xy} & T_{xz} \\ T_{yx} & T_y & T_{yz} \\ T_{zx} & T_{zy} & T_z \end{pmatrix}$$

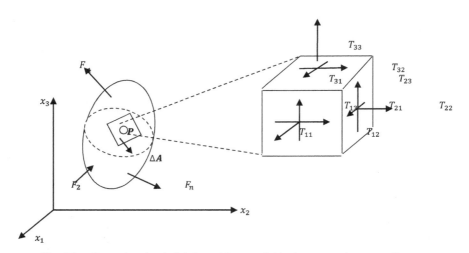

Fig. 3.8: Stress in a loaded deformable material body assumed as a continuum.

3.9 Mohr's Circles for Stress

After the stress distribution within the object has been determined with respect to a coordinate system (x, y), it may be necessary to calculate the components of the stress tensor at a particular material point P with respect to a rotated coordinate system (x', y'), i.e., the stresses acting on a plane with a different orientation passing through that point of interest—forming an angle with the coordinate system (x, y), (Fig. 3.9).

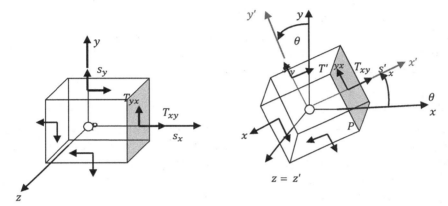

Fig. 3.9: Stress transformation at a point in a continuum under plane stress conditions.

For example, it is of interest to find the maximum normal stress and maximum shear stress, as well as the orientation of the planes where they act upon. To achieve this, it is necessary to perform a tensor transformation under a rotation of the coordinate system. From the definition of tensor, the Cauchy stress tensor obeys the tensor transformation law. A graphical representation of this transformation law for the Cauchy stress tensor is the Mohr circle for stress.

3.9.2 Sign Conventions

There are two separate sets of sign conventions that need to be considered when using the Mohr Circle: One sign convention for stress components in the "physical space", and another for stress components in the "Mohr-Circle-space". There is no standard sign convention, and the choice of a particular sign convention is influenced by convenience for calculation and interpretation for the particular problem in hand. A more detailed explanation of these sign conventions is presented below.

From the convention of the Cauchy stress tensor (Fig. 3.8 and Fig. 3.9), the first subscript in the stress components denotes the face on which the stress component acts, and the second subscript indicates the direction of the stress component. Thus T_{xy} is the shear stress acting on the face with normal vector in the positive direction of the x-axis, and in the positive direction of the y-axis.

In the Mohr-circle-space sign convention, normal stresses have the same sign as normal stresses in the physical-space sign convention: positive normal stresses act outward to the plane of action, and negative normal stresses act inward to the plane of action.

Shear stresses, however, have a different convention in the Mohr-circle space compared to the convention in the physical space. In the Mohr-circle-space sign convention, positive shear stresses rotate the material element in the counterclockwise direction, and negative shear stresses rotate the material in the clockwise direction. This way, the shear stress component T_{xy} is positive in the Mohr-circle space, and the shear stress component T_{yx} is negative in the Mohr-circle space.

3.9.3 Mohr's Circle for Two-dimensional State of Stress

In two dimensions, the stress tensor at a given material point P, with respect to any two perpendicular directions is completely defined by only three stress components. For the particular coordinate system (x, y), these stress components are: the normal stresses $N^{(x)}$ and $N^{(y)}$, and the shear stress T_{xy}. To derive the equation of the Mohr circle for the two-dimensional cases of plane stress and plane strain, first consider a two-dimensional infinitesimal material element around a material point P (Fig. 3.10), with a unit area in the direction parallel to the y-z plane, *i.e.*, perpendicular to the page.

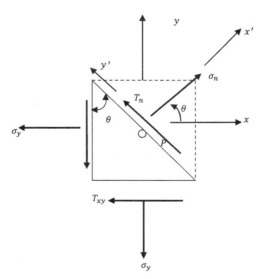

Fig. 3.10: Stress components at a plane passing through a point in a continuum under plane stress conditions.

3.9 Mohr's Circles for Stress

From the balance of angular momentum, the symmetry of the Cauchy stress tensor can be demonstrated. This symmetry implies that $T_{xy} = T_{yx}$. Thus, the Cauchy stress tensor can be written as:

$$N = \begin{pmatrix} T_{11} & T_{12} & 0 \\ T_{21} & T_{22} & 0 \\ 0 & 0 & 0 \end{pmatrix} = \begin{pmatrix} T_x & T_{xy} \\ T_{yx} & T_y \end{pmatrix}$$

The objective is to use the Mohr circle to find the stress components $S^{(n)}$ and $N^{(n)}$, on a rotated coordinate system (x', y'), i.e., on a differently oriented plane passing through P and perpendicular to the x-y plane (Fig. 3.10). The rotated coordinate system (x', y') makes an angle θ with the original coordinate system (x, y). From equilibrium of forces in the direction of $S^{(n)}$ (x'-axis), we have

$$\sum F_{x'} = S^{(n)} dA - S^{(x)} dA \cos^2 \theta - S^{(y)} dA \sin^2 \theta - T_{xy} dA \sin \theta \cos \theta$$

$$S^{(n)} = S^{(x)} \cos^2 \theta + S^{(y)} \sin^2 \theta + 2T_{xy} \sin \theta \cos \theta$$

$$= \frac{S^{(x)} + S^{(y)}}{2} - \frac{S^{(x)} - S^{(y)}}{2} \cos 2\theta + T_{12} \sin 2\theta. \tag{3.51}$$

Now, from equilibrium of forces in the direction of $N^{(n)}$ (y-axis) (Figure 3.10), and knowing that the area of the plane where $N^{(n)}$ acts is dA, we have:

$$\sum F_{y'} = N^{(n)} dA + S^{(x)} dA \sin \theta \cos \theta - S^{(y)} dA \sin \theta \cos \theta$$
$$- T_{xy} dA \cos^2 \theta + T_{xy} dA \sin^2 \theta = 0 \tag{3.52}$$

$$N^{(n)} = -\left(S^{(x)} - S^{(y)}\right) \sin \theta \cos \theta + T_{xy} \left(\cos^2 \theta - \sin^2 \theta\right)$$
$$= -\frac{1}{2}\left(S^{(x)} - S^{(y)}\right) \sin 2\theta + T_{xy} \cos 2\theta. \tag{3.53}$$

From equilibrium of forces on the infinitesimal element, the magnitudes of the normal stress $S^{(n)}$ and the shear stress $N^{(n)}$ are given by:

$$S^{(n)} = \frac{S^{(x)} + S^{(y)}}{2} - \frac{S^{(x)} - S^{(y)}}{2} \cos 2\theta + T_{xy} \sin 2\theta. \tag{3.54}$$

$$N^{(n)} = -\frac{1}{2}\left(S^{(x)} - S^{(y)}\right) \sin 2\theta + T_{xy} \cos 2\theta. \tag{3.55}$$

Both equations can also be obtained by applying the tensor transformation law on the known Cauchy stress tensor, which is equivalent to performing the static equilibrium of forces in the direction of $S^{(n)}$ and $N^{(n)}$. These two equations are the parametric equations of the Mohr circle. In these equations,

2θ is the parameter, and $S^{(n)}$ and $N^{(n)}$ are the coordinates. This means that by choosing a coordinate system with abscissa and ordinate, giving values to the parameter θ will place the points obtained lying on a circle.

Eliminating the parameter 2θ from these parametric equations will yield the non-parametric equation of the Mohr circle. This can be achieved by rearranging the equations for $S^{(n)}$ and $N^{(n)}$, first transposing the first term in the first equation and squaring both sides of each of the equations then adding them. Thus, we have

$$\left[S^{(n)} - \frac{T_{11}+T_{22}}{2}\right]^2 + N^{(n)2} = \left[\frac{T_{11}-T_{22}}{2}\right]^2 + T_{12}^{\,2} \tag{3.56}$$

$$\left[S^{(n)} - S_{avg}\right]^2 + N^{(n)2} = r^2$$

where, $S_{avg} = \frac{T_{11}+T_{22}}{2}$, $r = \sqrt{\left(\frac{T_{11}-T_{22}}{2}\right)^2 + T_{12}^{\,2}}$. This is the equation of a circle (the Mohr circle) of the form

$$(x-a)^2 + (y-b)^2 = r^2$$

with radius r centered at a point with coordinates $(a,b) = (S_{avg}, 0)$ in the $\left(S^{(n)}, N^{(n)}\right)$ coordinate.

Solving equation (3.24) by hand requires finding the roots of a cubic equation, so we consider the easier 2-D case, which yields a quadric equation

$$\begin{vmatrix} T_{11} - T & T_{12} \\ T_{21} & T_{22} - T \end{vmatrix} = 0 \Rightarrow T^2 - (T_{11}+T_{22})T + \{T_{11}T_{22} - T_{12}^{\,2}\} = 0$$

$$\Rightarrow T = \frac{(T_{11}+T_{22}) \pm \sqrt{(T_{11}+T_{22})^2 - 4\{T_{11}T_{22} - T_{12}^{\,2}\}}}{2}$$

$$\Rightarrow T = \frac{T_{11}+T_{22}}{2} \pm \frac{\sqrt{(T_{11}+T_{22})^2 - 4T_{11}T_{22} + 4T_{12}^{\,2}}}{2}$$

$$\Rightarrow T = \left(\frac{T_{11}+T_{22}}{2}\right) \pm \sqrt{\left(\frac{T_{11}-T_{22}}{2}\right)^2 + T_{12}^{\,2}}. \tag{3.57}$$

An inspection of the diagram below shows that the first term in brackets in equation (3.57) is the mean normal stress

3.9 Mohr's Circles for Stress

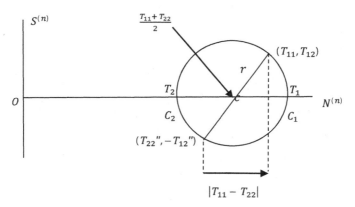

Fig. 3.11: Mohr's circle for the dimensional state of stress

(*i.e.*, the center of the Mohr's circle) and the second term in brackets is the maximum possible shear stress (*i.e.*, the radius of the Mohr's circle). So the principal stresses lie at the end of a horizontal diameter through the Mohr's circle. The terms $c = \dfrac{T_{11}+T_{22}}{2}$, $r = \sqrt{\left(\dfrac{T_{11}-T_{22}}{2}\right)^2 + T_{12}^{\;2}}$, $I_1 = T_{11} + T_{22}$, and $I_2 = T_{11}T_{22} - T_{12}^{\;2}$ are called invariants and are independent of the frame of reference.

Finding principal normal and shear stresses: The magnitude of the principal stresses are the abscissas of the points C_1 and C_2 (Figure 3.11) where the circle intersects the $N^{(n)}$-axis. The magnitude of the major principal stress S_1 is always the greatest absolute value of the abscissa of any of these two points. Likewise, the magnitude of the minor principal stress S_2 is always the lowest absolute value of the abscissa of these two points. As expected, the ordinates of these two points are zero, corresponding to the magnitude of the shear stress components on the principal planes. Alternatively, the values of the principal stresses can be found by

$$S_1 = S_{max} = S_{avg} + r;\quad S_2 = S_{min} = S_{avg} - r$$

where the magnitude of the average normal stress S_{avg} is the abscissa of the centre C, given by $S_{avg} = \dfrac{T_{11}+T_{22}}{2}$ and the length of the radius r of the circle (based on the equation of a circle passing through two points), is $r = \sqrt{\left(\dfrac{T_{11}-T_{22}}{2}\right)^2 + T_{12}^{\;2}}$.

The maximum and minimum shear stresses correspond to the abscissa of the highest and lowest points on the circle, respectively. These points are located at the intersection of the circle with the vertical line passing through the center of the circle C. Thus, the magnitude of the maximum and minimum shear stresses are equal to the value of the circle's radius

$$S_{max,min} = \pm r.$$

3.9.4 Mohr's Circle for Three-dimensional State of Stress

To construct the Mohr circle for a general three-dimensional case of stresses at a point, the values of the principal stresses (T_1, T_2, T_3) and (n_1, n_2, n_3) their principal directions must be first evaluated.

Consider again the state of stress at P referenced to principal axes (Figure 3.10) and let the principal stresses be ordered according to $T_1 > T_2 > T_3$. As before, we may express the normal and shear components $N^{(n)}$ and $S^{(n)}$ of the stress vector on any plane at P in terms of the components of the normal to that plane by the following equations

$$N^{(n)} = T_1 n_1^2 + T_2 n_2^2 + T_3 n_3^2$$

$$(N^{(n)})^2 + (S^{(n)})^2 = (T_1 n_1)^2 + (T_2 n_2)^2 + (T_3 n_3)^2$$

which, along with the condition $n_1^2 + n_2^2 + n_3^2 = 1$, provide us with three equations for the three direction cosines. Solving these equations, (using Gauss elimination method) we obtain

$$n_1^2 = \frac{\left(N^{(n)} - T_2\right)\left(N^{(n)} - T_3\right) + (S^{(n)})^2}{(T_1 - T_2)(T_1 - T_3)}. \tag{3.58}$$

$$n_2^2 = \frac{\left(N^{(n)} - T_3\right)\left(N^{(n)} - T_1\right) + (S^{(n)})^2}{(T_2 - T_3)(T_2 - T_1)} \tag{3.59}$$

$$n_3^2 = \frac{\left(N^{(n)} - T_1\right)\left(N^{(n)} - T_2\right) + (S^{(n)})^2}{(T_3 - T_1)(T_3 - T_2)}. \tag{3.60}$$

In these equations, T_1, T_2, T_3 are known; $N^{(n)}$ and $S^{(n)}$ are functions of the direction cosines n_i. Our intention here is to interpret these equations graphically by representing conjugate pairs of $N^{(n)}$, $S^{(n)}$ values, which satisfy equation (3.58)-(3.60), as a point in the *stress plane* having N as absicca and $S^{(n)}$ as ordinate (see Fig. 3.12).

3.9 Mohr's Circles for Stress

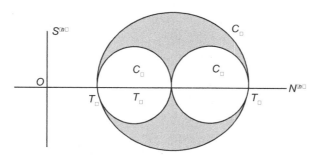

Fig 3.12: Mohr's circle for three-dimensional state of stress.

To develop this graphical interpretation of the three-dimensional stress state in terms of $N^{(n)}$ and $S^{(n)}$, we note that the denominator for the expression of n_1^2 is positive since both $T_1 - T_2 > 0$ and $T_1 - T_3 > 0$, and also that $n_1^2 > 0$, all of which tells us that

$$\left(N^{(n)} - T_2\right)\left(N^{(n)} - T_3\right) + (S^{(n)})^2 \geq 0. \tag{3.61}$$

For the case where the equality sign holds, this equation (3.61) may be rewritten as

$$\left(N^{(n)} - T_2\right)\left(N^{(n)} - T_3\right) + (S^{(n)})^2 = 0$$

$$\Rightarrow (N^{(n)})^2 - [T_2 + T_3]N^{(n)} + \left(S^{(n)}\right)^2 = 0$$

$$\Rightarrow \left[N^{(n)} - \frac{1}{2}(T_2 + T_3)\right]^2 + \left(S^{(n)}\right)^2 = \left[\frac{1}{2}(T_2 - T_3)\right]^2$$

which is the equation of the circle in the $N^{(n)}$, $S^{(n)}$ plane, with its center at the point $\frac{1}{2}(T_2 + T_3)$ on the $N^{(n)}$ axis, and having radius $\frac{1}{2}(T_2 - T_3)$. We label this circle C_1 and display it in Figure 3.12. For the case in which the inequality sign holds for Eq. (3.61), we observe that conjugate pairs of values of $N^{(n)}$ and $S^{(n)}$ which satisfy this relationship result in stress points having coordinates exterior to circle C_1. Thus, combinations of $N^{(n)}$ and $S^{(n)}$ which satisfy Eq. (3.58) lie on, or exterior to, circle C_1 in Figure 3.12.

Examining Eq. (3.59), we note that the denominator is negative since $T_2 - T_3 > 0$ and $T_2 - T_1 < 0$. The direction cosines are real numbers, so that $n_2^2 \geq 0$ and we have

$$\left(N^{(n)} - T_3\right)\left(N^{(n)} - T_1\right) + (S^{(n)})^2 \leq 0. \tag{3.62}$$

For the case where the equality sign holds, this equation (3.62) may be rewritten as

$$\left(N^{(n)} - T_3\right)\left(N^{(n)} - T_1\right) + (S^{(n)})^2 = 0$$

$$\Rightarrow (N^{(n)})^2 - [T_3 + T_1]N^{(n)} + \left(S^{(n)}\right)^2 = 0$$

$$\Rightarrow \left[N^{(n)} - \frac{1}{2}(T_1 + T_3)\right]^2 + \left(S^{(n)}\right)^2 = \left[\frac{1}{2}(T_1 - T_3)\right]^2$$

which is the equation of the circle in the $N^{(n)}$, $S^{(n)}$ plane, with its center at the point $\frac{1}{2}(T_1 + T_3)$ on the $N^{(n)}$ axis, and having radius $\frac{1}{2}(T_1 - T_3)$. We label this circle C_2 and display it in Figure 3.12 and the stress points which satisfy the inequality of Eq. (3.62) lie *interior* to it. Following the same general procedure, we rearrange Eq. (3.60) into an expression from which we extract the equation of the third circle, C_3 in Figure 3.12, namely

$$\left[N^{(n)} - \frac{1}{2}(T_1 + T_2)\right]^2 + \left(S^{(n)}\right)^2 = \left[\frac{1}{2}(T_1 - T_2)\right]^2$$

Admissible stress points in the $N^{(n)}$, $S^{(n)}$ plane lie *on* or *exterior* to this circle. The three circles defined above, and shown in Figure 3.12, are called *Mohr's circles for stress*. All possible pairs of values of $N^{(n)}$ and $S^{(n)}$ at P which satisfy Eq. (3.58)-(3.60) lie on these circles or within the shaded areas enclosed by them. In addition, it is clear from the Mohr's circles diagram that the maximum shear stress value at P is the radius of circle C_2, which confirms the result presented in Eq. (3.50).

These equations for the Mohr circles show that all admissible stress points $(S^{(n)}, N^{(n)})$ lie on these circles or within the shaded area enclosed by them (see Figure 3.12). Stress points $(S^{(n)}, N^{(n)})$ satisfying the equation for circle C_1 lie on, or outside circle C_1. Stress points $(S^{(n)}, N^{(n)})$ satisfying the equation for circle C_2 lie on, or inside circle C_2. And finally, stress points $(S^{(n)}, N^{(n)})$ satisfying the equation for circle C_3 lie on, or outside circle C_3.

3.10 Plane Stress

After performing a stress analysis on a material body assumed as a continuum, the components of the Cauchy stress tensor at a particular material point are

3.10 Plane Stress

known with respect to a co-ordinate system. The Mohr circle is then used to determine graphically the stress components acting on a rotated coordinate system, *i.e.*, acting on a differently oriented plane passing through that point. This construction involves three circles, and it is a generalization of Mohr's circle for plane stress.

The transformation equations for plane stress can be represented in a graphical form known as Mohr's circle. This graphical representation is very useful in depending the relationships between normal and shear stresses acting on any inclined plane at a point in a stresses body. When one—and only one—principal stress is zero, we have a state of *plane stress* for which the plane of the two nonzero principal stresses is the *designated plane*. This is an important state of stress because it represents the physical situation occurring at an unloaded point on the bounding surface of a body under stress. The zero principal stress may be any one of the three principal stresses as indicated by the corresponding Mohr's circles of Figure 3.13

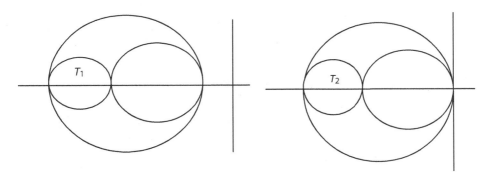

Fig. 3.13(a): Mohr's circle for plane stress $T_1 = 0$

Fig. 3.13(b): Mohr's circle for plane stress $T_2 = 0$

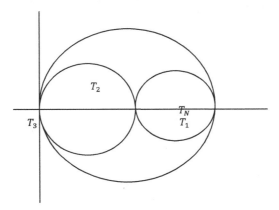

Fig. 3.13(c): Mohr's circle for plane stress $T_3 = 0$

If the principal stresses are not ordered and the direction of the zero principal stress is arbitrarily chosen as x_3, we have plane stress parallel to the $x_1 x_2$ plane and the stress matrix takes the form

$$(T_{ij}) = \begin{pmatrix} T_{11} & T_{12} & 0 \\ T_{21} & T_{22} & 0 \\ 0 & 0 & 0 \end{pmatrix}$$

or, with respect to principal axes, the form

$$(T_{ij}^*) = \begin{pmatrix} T_1 & 0 & 0 \\ 0 & T_2 & 0 \\ 0 & 0 & 0 \end{pmatrix}$$

The pictorial description of this plane stress situation is portrayed by the block element of a continuum body shown in Figure 3.14(a),

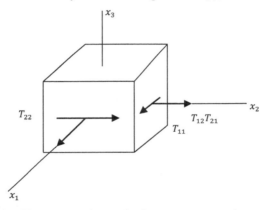

Fig. 3.14.(a): Plane stress element having nonzero x_1 and x_2 components.

and is sometimes represented by a single Mohr's circle (Fig. 3.14(b)), the locus of which identifies stress points (having coordinates $N^{(n)}$ and $S^{(n)}$) for unit normals lying in the $x_1 x_2$-plane only.

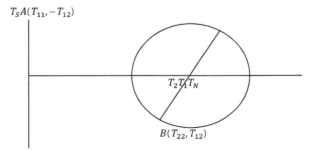

Fig. 3.14.(b): Mohr's circle for the stress components.

3.10 Plane Stress

The equation of the circle in Figure 3.14(b) is

$$\left[N^{(n)} - \frac{1}{2}(T_{11} + T_{22})\right]^2 + \left(S^{(n)}\right)^2 = \left[\frac{1}{2}(T_{11} - T_{22})\right]^2 + T_{12}^2$$

from which the center of the circle is noted to be at $N^{(n)} = \frac{1}{2}(T_{11} + T_{22})$, and the maximum shear stress in the $x_1 x_2$-plane to be the radius of the circle. Points A and B on the circle represent the stress states for area elements having unit normal and, respectively. For an element of area having a unit normal in an arbitrary direction at point P, we must include the two dashed circles shown in Figure 3.14(c) to completely specify the stress state.

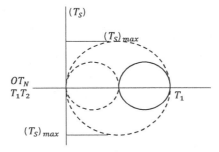

Fig. 3.14.(c): General Mohr's circles for the plane stress element.

With respect to axes $OX_1'X_2'X_3'$ rotated by the angle θ about the X_3 axis relative to $OX_1X_2X_3$ as shown in Figure 3.15, the transformation equations for plane stress in the X_1X_2 plane are given by the general tensor transformation formula. Using the table of direction cosines for this situation as listed in Table 3.1, we may express the primed stress components in terms of the rotation angle θ and the unprimed components by

$$T_{11}' = \frac{T_{11} + T_{22}}{2} + \frac{T_{11} - T_{22}}{2} \cos 2\theta + T_{12} \sin 2\theta$$

$$T_{22}' = \frac{T_{11} + T_{22}}{2} - \frac{T_{11} - T_{22}}{2} \cos 2\theta - T_{12} \sin 2\theta$$

$$T_{12}' = -\frac{T_{11} - T_{22}}{2} \sin 2\theta + T_{12} \cos 2\theta$$

Table 3.1: Transformation table for general plane stress.

	x_1	x_2	x_3
x_1'	$\sin\theta$	$\cos\theta$	0
x_2'	$-\sin\theta$	$\cos\theta$	0
x_3'	0	0	1

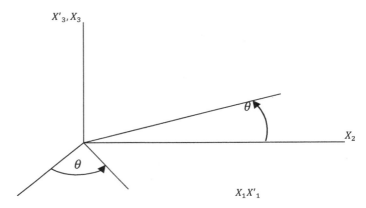

Fig. 3.15: Representative rotation of axes for plane stress.

In addition, if the principal axes of stress are chosen for the primed directions, it is easily shown that the two nonzero principal stress values are given by

$$T_{(1)}, T_{(2)} = \frac{T_{11}+T_{22}}{2} \pm \sqrt{\left[\frac{1}{2}(T_{11}-T_{22})\right]^2 + T_{12}^2}.$$

Example 3.13: A specimen is loaded with equal tensile and shear stresses. This case of plane stress may be represented by the matrix

$$(T_{ij}) = \begin{pmatrix} T_0 & T_0 & 0 \\ T_0 & T_0 & 0 \\ 0 & 0 & 0 \end{pmatrix}$$

where T_0 is a constant stress. Determine the principal stress values and plot the Mohr's circles.

Solution: For this stress state, the determinant is given by

$$\begin{vmatrix} T_0 - T & T_0 & 0 \\ T_0 & T_0 - T & 0 \\ 0 & 0 & -T \end{vmatrix} = 0 \Rightarrow T_{(1)} = 2T_0, T_{(2)} = 0$$

so that, in principal axes form, the stress matrix is

$$(T_{ij}) = \begin{pmatrix} 2T_0 & 0 & 0 \\ 0 & 0 & 0 \\ 0 & 0 & 0 \end{pmatrix}.$$

3.10 Plane Stress

The Mohr's circle diagram is shown in Figure 3.16.

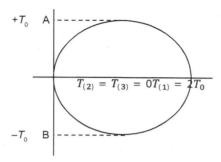

Fig 3.16: Mohr's circle for principal stresses, $T_{(1)} = 2T_0, T_{(2)} = T_{(3)} = 0$

Here, because of the double-zero root, one of the three Mohr's circles degenerates into a point (the origin) and the other two circles coincide. Also, we note that physically this is simply a one-dimensional tension in the direction and that the maximum shear stress values (shown by points A and B) occur on the x_1 and x_1 coordinate planes which make 45° with the principal direction. As the stress element is rotated away from the principal (or maximum shear) directions, the normal and shear stress components will always lie on Mohr's Circle.

Exercise

Theoretical

1. Explain the concept of stress vector at a point of a continuum medium.
2. State, with reference to contact forces, Euler's first law of motion for a continuous medium. Obtain stress equation of motion due to Cauchy from this law.
3. What is non-polar material? Show that for a non-polar material, Cauchy's second law of motion gives the symmetric stress tensor.
4. Derive Cauchy's first law of motion for a continuous medium from Euler's first law of motion.
5. State and prove fundamental theorem of stress.
6. Define Principal stress. Write down three Principal invariants of stress tensor.
7. Write down Caychy's stress formula and explain it.
8. Define stress quadric of Cauchy at a point of a continuous medium and prove the following as its properties:

(i) The normal stress across any plane through the centre of stress quadric is equal to the inverse of the square of the central radius vector of the quadric normal to the plane.

(ii) The stress vector across any plane through the centre of stress quadric is directed parallel to the normal to the surface of the quadric at a point where the central radius vector of the quadric normal to the plane meets the quadric.

9. Prove that, the stress vector at a point P of a surface with an exterior unit normal $\bar{n}$ is a linear function of the stress vectors acting on the coordinate surfaces through the same point P, the coefficients being the cosine directors of $\bar{n}$.

10. Prove that the stress vector acting on any plane through a point is fully characterized as a linear function of the stress tensor at the point.

11. Prove that, the extremum values of normal stress at a point of a continuum are principal stresses.

12. Prove that, the maximum shearing stress at any point of the continuum is equal to one-half the difference between algebraically, the largest and smallest principal stresses and acts on the plane that bisects the angle between the directions corresponding to largest and smallest principal stresses.

Numerical

1. Let the components of the stress tensor at P be given in matrix form by

$$(T_{ij}) = \begin{pmatrix} 3 & 2 & 2 \\ 2 & 4 & 0 \\ 2 & 0 & 2 \end{pmatrix}.$$

in units of mega-Pascals. Determine

(i) Stress vector $\bar{T}^{(n)}$ at a point P normal to x_1 axis

(ii) Stress vector $\bar{T}^{(n)}$ at a point P on the plane whose normal has direction ratios $1:-3:2$

(iii) Stress vector $\bar{T}^{(n)}$ at a point P on the plane through parallel to the plane $x_1 + 2x_2 + 3x_3 = 1$

(iv) Normal components of stress vector on the plane $x_1 + 2x_2 + 3x_3 = 1$

3.10 Plane Stress

    (v) Principal stresses at the point P

    (vi) Principal directions of stress at P.

2. The principal stresses at a point P be given as $T_1 = 1, T_2 = -1, T_3 = 3$. If the stress tensor at that point is given by

$$\left(T_{ij}\right) = \begin{pmatrix} T_{11} & 0 & 0 \\ 0 & 1 & 2 \\ 0 & 2 & T_{33} \end{pmatrix}$$

in units of mega-Pascals. Find the values of T_{11} and T_{33}.

    [**Ans:** Either 1,1 or -3,5.]

3. The principal stresses at a point P be given as $T_1 = 1, T_2 = -1, T_3 = 3$. If the stress tensor at that point is given by

$$\left(T_{ij}\right) = \begin{pmatrix} T_{11} & 0 & 0 \\ 0 & 1 & 2 \\ 0 & 2 & T_{33} \end{pmatrix}$$

in units of mega-Pascals. Find the values of T_{11} and T_{33}.

    Ans: $T_{11=1}$ and $T_{33=1}$

4. Let the components of the stress tensor at P be given in matrix form by

$$\left(T_{ij}\right) = \begin{pmatrix} 3 & 1 & 2 \\ 1 & 0 & 2 \\ 1 & 2 & 0 \end{pmatrix}.$$

in units of mega-Pascals. Determine principal stresses and principal direction. Find maximum value of the shearing stresses.

    [**Ans:** Principal stresses are 10, 5, -15, $|S^{(n)}| = 12.5$.]

5. Let the components of the stress tensor at P be given in matrix form by

$$\left(T_{ij}\right) = \begin{pmatrix} 3 & 1 & 1 \\ 1 & 0 & 2 \\ 1 & 2 & 0 \end{pmatrix}.$$

in units of mega-Pascals. Determine principal stresses and principal direction. Find maximum value of the shearing stresses.

    [**Ans:** Principal stresses are 1, 4, -2, $|S^{(n)}| = 3$.]

6. The components of the stress tensor at P be given in matrix form by

$$(a)(T_{ij}) = \begin{pmatrix} 3 & 1 & 1 \\ 1 & 0 & 2 \\ 1 & 2 & 0 \end{pmatrix}. (b)(T_{ij}) = \begin{pmatrix} 18 & 0 & 24 \\ 0 & -50 & 0 \\ 24 & 0 & 32 \end{pmatrix}. (c)(T_{ij}) = \begin{pmatrix} 0 & 5 & 5 \\ 5 & 0 & -5 \\ 5 & -5 & -10 \end{pmatrix}.$$

Determine principal stresses and corresponding principal directions. Find maximum value of the shearing stresses.

(a) [**Ans:** -2,1,4] (b) [**Ans:** 50,0,-50](c) [**Ans:** -15,0,5]

7. The state of stress at a point in a continuous medium is given by the stress matrix

$$(T_{ij}) = \begin{pmatrix} T & aT & bT \\ aT & T & cT \\ bT & cT & T \end{pmatrix}$$

where a, b, c are constants and T is some stress value. Determine the constants a, b, c so that the stress vector on a plane normal to $\left(\dfrac{1}{\sqrt{3}}; \dfrac{1}{\sqrt{3}}; \dfrac{1}{\sqrt{3}} \right)$ vanishes.

8. The state of stress at a point with respect to Cartesian axes $O X_1 X_2 X_3$ is given by

$$(T_{ij}) = \begin{pmatrix} 15 & -10 & 0 \\ -10 & 5 & 0 \\ 0 & 0 & 20 \end{pmatrix}$$

Determine the stress tensor T'_{ij} for related axes $O X_1' X_2' X_3'$ for which transformation matrix is

$$(a_{ij}) = \begin{pmatrix} \dfrac{3}{5} & 0 & -\dfrac{4}{5} \\ 0 & 1 & 0 \\ \dfrac{4}{5} & 0 & \dfrac{3}{5} \end{pmatrix}.$$

■ ■ ■

4

Fundamental Principles of Continuum Mechanics

In this Chapter we are concerned with the fundamental principles which states that some physical quantity is either conserved or balanced. Basic equations of continuum mechanics are derived from three broad principles namely

(*i*) Principle of conservation of mass
(*ii*) Balance of momentum
(*iii*) Conservation of energy.

4.1 Principle of Conservation of Mass

Every material body, as well as every portion of such a body is endowed with a non-negative, scalar measure, called the *mass* of the body or of the portion under consideration. Physically, the mass is associated with the inertia property of the body, that is, its tendency to resist a change in motion. The measure of mass may be a function of the space variables and time. If Δm is the mass of a small volume ΔV in the current configuration, and if we assume that Δm is absolutely continuous, the limit

$$\rho = \lim_{\Delta V \to 0} \frac{\Delta m}{\Delta V}$$

defines the scalar field $\rho = \rho(x,t)$ called the *mass density* of the body for that configuration at time t. Therefore, the mass m of the entire body is given by

$$m = \iiint_V \rho(x,t) dV.$$

In the same way, we define the mass of the body in the referential (initial) configuration in terms of the density field $\rho_0 = \rho_0(X,t)$ by the integral

$$m = \iiint_{V_0} \rho_0(X,t)dV_0$$

The law of *conservation of mass* asserts that the total mass of any portion of a continuum medium remains unchanged during the motion, that is, remains constant in every configuration. When this is valid for an arbitrarily small neighbourhood of each material point, we say that the mass is conserved locally. This leads to equation of continuity.

4.1.1 Equation of Continuity in Lagrangian Method

In this method the principle of conservation of mass in the form that the mass of a specific portion of the moving continuum enclosed in a volume does not change as it moves.

Consider a specific portion of the continuum occupying at the initial instant $t = 1$ an arbitrary volume V_0 in the undeformed state. Let $P_0(X_1, X_2, X_3)$ be any point in it and $\rho_0 = \rho(X_1, X_2, X_3)$ be the density at P_0. Let dV_0 be the element of the volume at P_0 and mass of this element is $\rho_0 dV_0$. The total mass of the continuum which fills the volume at V_0 is $\iiint_{V_0} \rho_0 dV$.

At subsequent time $t > 0$, different particles of the continuum forming the volume V_0 move in such a manner that they form some other volume V in the deformed state. Let the particle at the initial position P_0 occupy the subsequent position $P(x_1, x_2, x_3)$ and let ρ be the density of the medium at P. In the Lagrangian description, the equation of motion of a continuum body is expressed by

$$x_i = x_i(X_1, X_2, X_3, t), \text{ and } \rho = \rho(X_1, X_2, X_3, t) \quad (4.1)$$

Obviously, ρdV is the mass of an element of volume dV at P in V. The total matter which fills the volume v at time t is $\iiint_V \rho_0 dV$. Also, we know that

$$dV = JdV_0 = \frac{\partial(x_1, x_2, x_3)}{\partial(X_1, X_2, X_3)} dV_0 \quad (4.2)$$

By principle of conservation of mass, the matter which fills the volume V is the same as the matter which fills volume V_0, therefore,

$$\iiint_{V_0} \rho_0 dV_0 = \iiint_V \rho dV = \iiint_V \rho \frac{\partial(x_1, x_2, x_3)}{\partial(X_1, X_2, X_3)} dV_0$$

or,

$$\iiint_{V_0} \left[\rho_0 - \rho \frac{\partial(x_1, x_2, x_3)}{\partial(X_1, X_2, X_3)} \right] dV_0 = 0$$

4.1 Principle of conservation of Mass

Since the volume V_0 is arbitrary, it follows that integrand must vanish at every point of the continuum, i.e.,

$$\rho_0 - \rho \frac{\partial(x_1, x_2, x_3)}{\partial(X_1, X_2, X_3)} = 0, i.e., \rho_0 = \rho \frac{\partial(x_1, x_2, x_3)}{\partial(X_1, X_2, X_3)} = \rho J \qquad (4.3)$$

which is the required equation of continuity in material method.

4.1.2 Equation of Continuity in Eulerian Method

In this method the principle of conservation of mass is expressed in the form that the rate at which the mass of the continuum within any fixed closed surface increases is equal to the rate at which the net mass of the continuum flows in across the boundary surface.

Consider any fixed closed arbitrary surface S enclosing a volume V lying entirely in a region through which continuum moves. Let $P(x_1, x_2, x_3)$ be any point in it and let ρ be the density of the material at time t so that

$$\rho = \rho(x_1, x_2, x_3, t) \qquad (4.4)$$

Obviously, ρdV is the mass of an element of volume dv at P in V. The total matter which fills the volume at time t is $\iiint_V \rho dV$. The net local rate of the increase of this mass in volume V is

$$\frac{\partial}{\partial t} \iiint_V \rho dV = \iiint_V \frac{\partial \rho}{\partial t} dV \qquad (4.5)$$

as volume V is a fixed region of space and space coordinate (x_1, x_2, x_3) are independent of t. Let Q be any point on the surface S. Let dS be a surface element at Q at time t. The normal component of velocity $\vec{v}$ along the direction of outward normal $\vec{n}$ is $\vec{n} \cdot \vec{v}$.

(i) Mass of the continuum leaving volume V by flowing out across dS per unit time is $\rho \vec{n} \cdot \vec{v} \, dS$

(ii) Therefore mass of the continuum entering into volume V by flowing in across dS per unit time is $-\rho \vec{n} \cdot \vec{v} dS$.

The rate at which the net mass of the continuum flows in across total boundary surface S is

$$= -\iint_S \rho \vec{n} \cdot \vec{v} dS.$$

From the principle of conservation of mass

$$\iiint_V \frac{\partial \rho}{\partial t} dV = -\iint_S \rho \vec{n} \cdot \vec{v} dS = -\iint_S \vec{n} \cdot (\rho \vec{v}) dS = -\iiint_V \vec{\nabla} \cdot (\rho \vec{v}) dV$$

(by applying Gauss divergence theorem)

or,

$$\iiint_V \left[\frac{\partial \rho}{\partial t} + \vec{\nabla} \cdot (\rho \vec{v})\right] dV = 0.$$

Since the volume V is arbitrary, the integrand must vanish at every point of the medium. Hence

$$\frac{\partial \rho}{\partial t} + \vec{\nabla} \cdot (\rho \vec{v}) = 0$$

or,

$$\frac{\partial \rho}{\partial t} + (\vec{v} \cdot \vec{\nabla} \rho) + \rho \vec{\nabla} \cdot \vec{v} = 0$$

or,

$$\left[\frac{\partial}{\partial t} + (\vec{v} \cdot \vec{\nabla})\right]\rho + \rho \vec{\nabla} \cdot \vec{v} = 0$$

or,

$$\frac{d\rho}{dt} + \rho \vec{\nabla} \cdot \vec{v} = 0; \quad \frac{d}{dt} \equiv \frac{\partial}{\partial t} + (\vec{v} \cdot \vec{\nabla}) \tag{4.6}$$

where, $\dfrac{d}{dt}$ = differentiation following the motion of the continuum. This equation (4.6) is the Euler's equation of motion. Equation (4.6) is the equation of *conservation of mass*, also known as the *equation of continuity*.

In Cartesian coordinates, Eq. (4.6) can be written as

$$\frac{\partial \rho}{\partial t} + v_1 \frac{\partial \rho}{\partial x_1} + v_2 \frac{\partial \rho}{\partial x_2} + v_3 \frac{\partial \rho}{\partial x_3} + \rho \left(\frac{\partial v_1}{\partial x_1} + \frac{\partial v_2}{\partial x_2} + \frac{\partial v_3}{\partial x_3}\right) = 0.$$

In cylindrical coordinates, it becomes

$$\frac{\partial \rho}{\partial t} + v_r \frac{\partial \rho}{\partial r} + \frac{v_\theta}{r} \frac{\partial \rho}{\partial \theta} + v_z \frac{\partial \rho}{\partial z} + \rho \left(\frac{\partial v_r}{\partial r} + \frac{1}{r}\frac{\partial v_\theta}{\partial \theta} + \frac{v_r}{r} + \frac{\partial v_z}{\partial z}\right) = 0.$$

In spherical coordinates, it becomes

$$\frac{\partial \rho}{\partial t} + v_r \frac{\partial \rho}{\partial r} + \frac{v_\theta}{r} \frac{\partial \rho}{\partial \theta} + \frac{v_\varphi}{r \sin \theta} \frac{\partial \rho}{\partial \varphi} +$$

$$\rho \left(\frac{\partial v_r}{\partial r} + \frac{1}{r}\frac{\partial v_\theta}{\partial \theta} + \frac{2v_r}{r} + \frac{1}{r \sin \theta}\frac{\partial v_\varphi}{\partial \varphi} + \frac{v_\theta \cot \theta}{r}\right) = 0.$$

4.1 Principle of conservation of Mass

For an incompressible material, the material derivative of the density is zero and the mass conservation equation reduces to simply $\vec{\nabla} \cdot \vec{v} = 0$.

In rectangular Cartesian coordinates:

$$\frac{\partial v_1}{\partial x_1} + \frac{\partial v_2}{\partial x_2} + \frac{\partial v_3}{\partial x_3} = 0$$

In cylindrical coordinates:

$$\frac{\partial v_r}{\partial r} + \frac{1}{r}\frac{\partial v_\theta}{\partial \theta} + \frac{v_r}{r} + \frac{\partial v_z}{\partial z} = 0.$$

In spherical coordinates:

$$\frac{\partial v_r}{\partial r} + \frac{1}{r}\frac{\partial v_\theta}{\partial \theta} + \frac{2v_r}{r} + \frac{1}{r\sin\theta}\frac{\partial v_\varphi}{\partial \varphi} + \frac{v_\theta \cot\theta}{r} = 0.$$

Example 4.1: Prove that liquid motion is possible when velocities at a point (x_1, x_2, x_3) are given by $v_1 = \dfrac{3x_1^2 - r^2}{r^5}$, $v_2 = \dfrac{3x_1 x_2}{r^5}$, $v_3 = \dfrac{3x_1 x_3}{r^5}$ where, $r^2 = x_1^2 + x_2^2 + x_3^2$.

Solution: For the given velocity field $\vec{v} = (v_1, v_2, v_3)$, we have

$$v_{1,1} = \frac{\partial v_1}{\partial x_1} = \frac{\partial}{\partial x_1}\left(\frac{3x_1^2 - r^2}{r^5}\right) = \frac{3x_1}{r^7}\left(3r^2 - 5x_1^2\right)$$

$$v_{2,2} = \frac{\partial v_2}{\partial x_2} = \frac{\partial}{\partial x_2}\left(\frac{3x_1 x_2}{r^5}\right) = \frac{3x_1}{r^7}\left(r^2 - 5x_2^2\right)$$

$$v_{3,3} = \frac{\partial v_3}{\partial x_3} = \frac{\partial}{\partial x_3}\left(\frac{3x_1 x_3}{r^5}\right) = \frac{3x_1}{r^7}\left(r^2 - 5x_3^2\right)$$

For incompressible liquid velocity component must satisfy equation of continuity $v_{i,i} = \dfrac{\partial v_1}{\partial x_1} + \dfrac{\partial v_2}{\partial x_2} + \dfrac{\partial v_3}{\partial x_3} = 0$. Now

$$\frac{\partial v_1}{\partial x_1} + \frac{\partial v_2}{\partial x_2} + \frac{\partial v_3}{\partial x_3} = \frac{3x_1}{r^7}\left(5r^2 - 5r^2\right) = 0.$$

Equation of continuity is satisfied. Hence, liquid motion is possible.

4.2 Balance of Linear Momentum

The principle of balance of momentum states that the time rate of change of linear momentum of any portion $V(t)$ of a continuum in motion is equal to the total applied force acting on that portion.

Let Ox_1, Ox_2, Ox_3 be a fixed rectangular coordinate system. Let the specific portion of the deformed continuum occupied the volume $V(t)$ bounded by the close surface $S(t)$ at time t. Let $\rho(x_1, x_2, x_3, t)$ denote the density and $\vec{v}(x_1, x_2, x_3, t)$ denote the velocity at a point $P(x_1, x_2, x_3)$ in $V(t)$ at time t. Then the total linear momentum of the continuum in $V(t)$ is given by

$$\vec{P}(t) = \int_V \rho \vec{v}(x_1, x_2, x_3, t) d\tau \tag{4.7}$$

where $d\tau$ is an elementary of volume in $V(t)$. Let $\vec{F}(x_1, x_2, x_3, t)$ be the field of body force per unit mass in volume $V(t)$ and $\overline{T^{(n)}}(x_1, x_2, x_3, t)$ be the surface force per unit area of the surface $S(t)$. This surface force is the stress vector acting across the surface element with unique outward normal $\bar{n}$ exerted by the surrounding material on the portion $V(t)$. Therefore the total force acting on the material occupying the region $V(t)$ at time t is given by

$$\vec{R}(t) = \int_V \rho \vec{F}(x_1, x_2, x_3, t) d\tau + \int_S \overline{T^{(n)}}(x_1, x_2, x_3, t) dS \tag{4.8}$$

By the principle of balance of linear momentum we have $\dfrac{d\vec{P}(t)}{dt} = \vec{R}(t)$, i.e.,

$$\frac{d}{dt} \int_V \rho \vec{v}(x_1, x_2, x_3, t) d\tau = \int_V \rho \vec{F}(x_1, x_2, x_3, t) d\tau + \int_S \overline{T^{(n)}}(x_1, x_2, x_3, t) dS$$

or, $$\int_V \rho \frac{d\vec{v}}{dt} d\tau = \int_V \rho \vec{F}(x_1, x_2, x_3, t) d\tau + \int_S \overline{T^{(n)}}(x_1, x_2, x_3, t) dS \tag{4.9}$$

The equation (4.9) can be written as

$$\int_V \rho \frac{dv_i}{dt} d\tau = \int_V \rho F_i d\tau + \int_S T_i^{(n)}(x_1, x_2, x_3, t) dS$$

$$= \int_V \rho F_i d\tau + \int_S T_{ij}(x_1, x_2, x_3, t) n_j dS; \, i, j = 1, 2, 3 \tag{4.10}$$

where, $T_{ij}(x_1, x_2, x_3, t)$ being the stress tensor. We get it from Cauchy stress tensor. Using Gauss integral formula to the second term on the right hand side equation (4.10) gives us

$$\int_V \left[\rho \frac{dv_i}{dt} - \rho F_i - \frac{\partial T_{ij}}{\partial x_j} \right] d\tau = 0 \tag{4.11}$$

Since the equation (4.11) holds for arbitrary volume V, therefore at a point of continuity of the integrand

$$\rho\frac{dv_i}{dt} - \rho F_i - \frac{\partial T_{ij}}{\partial x_j} = 0 \qquad (4.12)$$

Equation (4.12) is known as the Cauchy's equation of motion for a continuum.

4.3 Balance of Angular Momentum

The principle of balance of angular momentum states that the time rate of change of total angular momentum of any portion of a nonpolar continuum about any fixed point is equal to the resultant moment about the same point of the external forces acting on this portion.

Take the fixed point at the origin. Let ρ be the density $\vec{v}$ and be the velocity of a material point whose position vector is $\vec{r}$, then the angular momentum of an arbitrary volume $V(t)$ of the continuum about the origin at the time is given by

$$\vec{H}(t) = \int_V (\vec{r} \times \vec{v}) \rho d\tau \qquad (4.13)$$

If $\vec{F}$ be the external body force per unit mass of the continuum assumed to be non-polar and $\overline{T^{(n)}}$ be the stress vector across the surface S of V. The resultant moment $\overline{M}(t)$ of the external force acting on the portion of the continuum with respect to the origtin at time t is given by

$$\overline{M}(t) = \int_V \rho \vec{r} \times \vec{F}(x_1, x_2, x_3, t) d\tau + \int_S \vec{r} \times \overline{T^{(n)}}(x_1, x_2, x_3, t) dS \qquad (4.14)$$

The principle of balance of angular momentum gives us $\dfrac{d\overline{H}(t)}{dt} = \overline{M}(t)$, i.e.,

$$\frac{d}{dt}\overline{H}(t) = \frac{d}{dt}\int_V (\vec{r} \times \vec{v}) \rho d\tau = \int_V \rho \vec{r} \times \vec{F} d\tau + \int_S \vec{r} \times \overline{T^{(n)}} dS$$

or, $\quad \int_V \left(\dfrac{d}{dt}\vec{r} \times \vec{v}\right) \rho d\tau + \int_V \left(\vec{r} \times \dfrac{d}{dt}\vec{v}\right) \rho d\tau = \int_V \rho \vec{r} \times \overline{F} d\tau + \int_S \vec{r} \times \overline{T^{(n)}} dS$

or, $\quad \int_V (\vec{r} \times \vec{f}) \rho d\tau = \int_V \rho \vec{r} \times \overline{F} d\tau + \int_S \vec{r} \times \overline{T^{(n)}} dS \qquad (4.15)$

where, $\vec{v} = \dfrac{d\vec{r}}{dt}$ and $\vec{f} = \dfrac{d\vec{v}}{dt}$. The ith component of the equation (4.15) is

$$\int_V \rho \varepsilon_{ijk} x_j f_k d\tau = \int_V \rho \varepsilon_{ijk} x_j F_k d\tau + \int_S \varepsilon_{ijk} x_j T_k^{(n)} dS$$

$$\int_V \rho \varepsilon_{ijk} x_j f_k d\tau = \int_V \rho \varepsilon_{ijk} x_j F_k d\tau + \int_S \varepsilon_{ijk} x_j T_{lk} n_l dS \quad (4.16)$$

by using Cauchy's formula. By using Gauss divergence theorem we get

$$\int_S \varepsilon_{ijk} x_j T_{lk} n_l dS = \int_V \frac{\partial}{\partial x_l}(\varepsilon_{ijk} x_j T_{lk}) d\tau$$

$$= \int_V \varepsilon_{ijk} \left[\frac{\partial x_j}{\partial x_l} T_{lk} + x_j \frac{\partial T_{lk}}{\partial x_l} \right] d\tau$$

$$= \int_V \varepsilon_{ijk} \left[T_{jk} + x_j \frac{\partial T_{lk}}{\partial x_l} \right] d\tau$$

Therefore, from equation (4.16), we get

$$\int_V \varepsilon_{ijk} x_j \left[\rho f_k - \rho F_k - \frac{\partial T_{lk}}{\partial x_l} \right] d\tau = \int_V \varepsilon_{ijk} T_{jk} d\tau$$

or, $$\int_V \varepsilon_{ijk} T_{jk} d\tau = 0$$

since the L.H.S. vanishes by Cauchy's equation of motion. Therefore at a point of continuity at the integrand we must have

$$\varepsilon_{ijk} T_{jk} = 0. \quad (4.17)$$

For $i = 1,$ $T_{23} + (-1)T_{32} = 0; T_{23} = T_{32}$

For $i = 2,$ $T_{31} + (-1)T_{13} = 0; T_{13} = T_{31}$

For $i = 3,$ $T_{12} + (-1)T_{21} = 0; T_{12} = T_{21}$

The stress tensor T_{ij} is symmetric, i.e.,

$$T_{ij} = T_{ji} \quad (4.18)$$

The equation (4.18) is known as Cauchy's second equation of motion. Thus we see that for non-polar fluid the stress tensor is symmetric and hence the stress at a point is completely specified by 6 stress components.

Deduction 4.1 (Stress equation of equilibrium): If the continuum is in equilibrium under a system of forces $\vec{F}$ per unit mass then the velocity component v_i vanish and Cauchy's equation of motion reduces to

$$\frac{\partial T_{ij}}{\partial x_j} + \rho F_i = 0 \quad (4.19)$$

i.e., $$\frac{\partial T_{11}}{\partial x_1} + \frac{\partial T_{12}}{\partial x_2} + \frac{\partial T_{13}}{\partial x_3} + \rho F_1 = 0$$

$$\frac{\partial T_{21}}{\partial x_1} + \frac{\partial T_{22}}{\partial x_2} + \frac{\partial T_{23}}{\partial x_3} + \rho F_2 = 0$$

$$\frac{\partial T_{31}}{\partial x_1} + \frac{\partial T_{32}}{\partial x_2} + \frac{\partial T_{33}}{\partial x_3} + \rho F_3 = 0$$

4.4 Thermodynamics of Continuous Media

This section is concerned with the thermodynamics of continuous media. Thermodynamics is a study primarily centered around the principles of energy and entropy and their various applications. In the usual courses that are offered in physics or in engineering curricula, seldom is a genuinely dynamical phenomenon treated. A proper name for such studies is thermo-statics.

Thermodynamics of continuous media starts with the principle of conservation of energy ,we discuss the idea of potential energy and strain energy. Thermodynamic non-dissipative stress and pressure are derived and discussed, and transition to global thermostatics is made.

4.4.1 Principle of Conservation of Energy

The principle of conservation of energy, also known as the first law of thermodynamics, states that energy can neither be created nor destroyed but can only be changed it in form in the nonrealistic mechanics. Thus if an amount of it(energy) is supplied to a system, part of the energy may be utilised in doing work by the system and the remainder goes to increase the total energy (kinetic energy, internal energy) of the system. We may state the principle of conservation of energy as the time rate of change of kinetic energy and the internal energy of a given portion of a thermomechanical continuum, as it moves, is equal to the sum of the rate of work done by the given external body forces and surface forces acting on the considered portion and the rate of heat energy added to it.

Referred to a fixed rectangular axis, let $\rho(x_1, x_2, x_3, t)$ be the density, $\vec{v}(x_1, x_2, x_3, t)$ be the velocity at a point that belonging to a specified portion say V, of the continuum and let S be the boundary of V. Then the K.E. of the matter in

$$K(t) = \int_V \frac{1}{2} \rho(x_1, x_2, x_3, t) v_k v_k d\tau. \tag{4.20}$$

Let $e(x_1, x_2, x_3, t)$ be the internal energy per unit mass of the continuum then the total internal energy of the portion V is given by

$$E(t) = \int_V \rho(x_1, x_2, x_3, t) e d\tau. \tag{4.21}$$

We assume that the material or the continuum is a conductor of heat so that the material in V gets heat energy by flow it across the boundary S by conduction and also get heat by radiation due to internal heat sources. If $\vec{q}(x_1, x_2, x_3, t)$ be the heat flux vector per unit area across S per unit time then the rate at which heat entering the volume V across the surface S is equal to

$$-\int_S \vec{q} \cdot \vec{n} \, dS = -\int_S q_k n_k \, dS \tag{4.22}$$

where (n_1, n_2, n_3) are the direction cosines of the outward normal to an element dS of S. If there are internal sources of energy within V producing radiant heat energy h per unit mass in unit time then the total heat energy created by radiation is equal to

$$\int_V \rho(x_1, x_2, x_3, t) h \, d\tau. \tag{4.23}$$

At time $t = 0$, the continuum is at rest. When it is subjected to the action of external forces the continuum gets deform and hence the forces will do some work on the body. Let $\vec{F}(x_1, x_2, x_3, t)$ be the external body force per unit mass and $\overline{T^{(n)}}$ be the stress vector acting on unit area of S with normal $\vec{n}$ then if V be the velocity. The rate of displacement of a point is V. Hence the rate of doing work on matter within V by body forces F_k and surface forces $T_k^{(n)}$ equal to

$$\int_V \rho(x_1, x_2, x_3, t) F_k v_k \, d\tau + \int_S T_k^{(n)} v_k \, dS. \tag{4.24}$$

By the principle of conservation of energy, equating the time rate of (4.20) and (4.21) to the sum of (4.22), (4.23) and (4.24) we get

$$\frac{d}{dt}(T+E) = -\int_S q_k n_k \, dS + \int_V \rho h \, d\tau + \int_V \rho F_k v_k \, d\tau + \int_S T_k^{(n)} v_k \, dS$$

$$\frac{d}{dt}\left[\int_V \left(\frac{1}{2} v_k v_k + e\right) \rho \, d\tau\right] = -\int_S q_k n_k \, dS + \int_V \rho h \, d\tau + \int_V \rho F_k v_k \, d\tau + \int_S T_k^{(n)} v_k \, dS$$

$$\int_V [\rho \frac{d}{dt}(v_k v_k + e) - \rho F_k v_k - \rho h] d\tau = -\int_S q_k n_k \, dS + \int_S T_{kl} n_l v_k \, dS$$

by using Cauchy's formula. By using Gauss divergence theorem we get

$$\int_V \left[\rho \frac{d}{dt}(v_k v_k + e) - \rho F_k v_k - \rho h\right] d\tau = -\int_V \frac{\partial}{\partial x_l}(T_{kl} v_k - q_l) dV.$$

Since the volume V is arbitrary, therefore a point of continuity we have

$$\rho \frac{d}{dt}(v_k v_k + e) - \rho F_k v_k - \rho h = \frac{\partial}{\partial x_l}(T_{kl} v_k - q_l)$$

4.4 Thermodynamics of Continuous Media

or, $v_k \left\{ \rho \dfrac{dv_k}{dt} - \rho F_k - \dfrac{\partial T_{kl}}{\partial x_l} \right\} = \rho h - \rho \dfrac{de}{dt} + T_{kl} \dfrac{\partial v_k}{\partial x_l} - \dfrac{\partial q_l}{\partial x_l}.$

The term on the L.H.S. vanish by Cauchy's equation of motion and we get the equation

$$\rho \frac{de}{dt} = \rho h + T_{kl} \frac{\partial v_k}{\partial x_l} - \frac{\partial q_l}{\partial x_l} \tag{4.25}$$

$$\Rightarrow \rho \frac{de}{dt} = \rho h + T_{kl} \left\{ \frac{1}{2} \left[\frac{\partial v_k}{\partial x_l} + \frac{\partial v_l}{\partial x_k} \right] + \frac{1}{2} \left[\frac{\partial v_k}{\partial x_l} - \frac{\partial v_l}{\partial x_k} \right] \right\} - \frac{\partial q_l}{\partial x_l}$$

$$\Rightarrow \rho \frac{de}{dt} = \rho h + T_{kl} [d_{kl} + w_{kl}] - \frac{\partial q_l}{\partial x_l}. \tag{4.26}$$

where, d_{kl} = strain rate tensor which is second order and symmetric and w_{kl} = second order skew symmetric tensor which is known as spin tensor.

Since T_{kl} is symmetric and w_{kl} is antisymmetric, the product of T_{kl} and w_{kl} vanishes and the equation (4.25) becomes

$$\rho \frac{de}{dt} = \rho h + T_{kl} d_{kl} - \frac{\partial q_l}{\partial x_l}. \tag{4.27}$$

The equation (1.26) is the required energy equation for a thermodynamical continuum. The scalar quantity $T_{kl} d_{kl}$ is called stress power.

Deduction 4.2: If the heat flux $\vec{q}$ obeys the Fourier laws of heat conduction then

$$\vec{q} = -k \vec{\nabla} T; \quad i.e., q_i = -k \frac{\partial T}{\partial x_i}, \tag{4.28}$$

where k = thermal conductivity of the material and T = the absolute temperature. Using equation (4.28), the equation (4.27) becomes

$$\rho \frac{de}{dt} = \rho h + T_{ij} d_{ij} + \frac{\partial}{\partial x_i} \left(k \frac{\partial T}{\partial x_i} \right) = \rho h + T_{ij} d_{ij} + k \nabla^2 T; \nabla^2 = \frac{\partial^2}{\partial x_i^2}. \tag{4.29}$$

assuming k to be constant. For a perfect continuum $e = cT$, where denotes the specific heat of the continuum. Therefore, equation (4.29) becomes

$$\rho c \frac{dT}{dt} = \rho h + T_{ij} d_{ij} + k \nabla^2 T. \tag{4.30}$$

When the motion of the body consists of translation and rotation only then $d_{ij} = 0$ and in the absence of heat sources $h = 0$. In this case, equation (4.30) simplifies to

$$\frac{dT}{dt} = \frac{k}{\rho c} \nabla^2 T. \qquad (4.31)$$

Deduction 4.3: Equation of energy for mechanical continuum in the absence of heat conduction and radiation we have the heat flux $\vec{q} = \vec{0}, h = 0$. In this case the conservation of energy gives us

$$\frac{d}{dt}(K+E) = \int_V \rho F_k v_k d\tau + \int_S T_k^{(n)} v_k dS. \qquad (4.32)$$

If the external body forces $\vec{F}$ be conservative, time independent and derivable from potential function, then

$$F_k = -\frac{\partial U}{\partial x_k}; \frac{\partial U}{\partial t} = 0$$

In this case,

$$\int_V \rho F_k v_k d\tau = -\int_V \rho \frac{\partial U}{\partial x_k} v_k d\tau = -\iiint_V \rho \frac{\partial U}{\partial t} d\tau$$

$$= -\frac{d}{dt} \iiint_V \rho d\tau = -\frac{d\Omega}{dt}, \qquad (4.33)$$

where, $\Omega = \iiint_V \rho d\tau$ = total potential energy of the continuum in V. Using equation (4.33), the equation (4.32) becomes

$$\frac{d}{dt}(K+E) = -\frac{d\Omega}{dt} + \int_S T_k^{(n)} v_k dS$$

$$\Rightarrow \frac{d}{dt}(K+E+\Omega) = \int_S T_k^{(n)} v_k dS. \qquad (4.34)$$

If the surface be such that on S, $v_k = 0$, the stress vector is normal to velocity vector, the right hand side of equation (4.34) vanishes, i.e.,

$$\frac{d}{dt}(K+E+\Omega) = 0 \Rightarrow K+E+\Omega = \text{constant}. \qquad (4.35)$$

The equation (4.35) says that the total kinetic, potential and internal energy any portion of a mechanical continuum remain constant with the passage of time as it moves with medium.

4.4.2 Second Law of Thermodynamics

The complete characterization of a thermodynamic system (here, a continuum) is said to describe the state of the system. This description is

4.4 Thermodynamics of Continuous Media

specified, in general, by several thermodynamic and kinematics quantities called state variables. A change with time of the state variables characterizes a thermodynamic process. The state variables used to describe a given system are usually not all independent. Functional relationships exist among the state variables and these relationships are expressed by the so-called equations of state. Any state variable which may be expressed as a single-valued function of a set of other variables is known as a state function.

The second law of thermodynamics postulates the existence of two distinct state functions; T the absolute temperature and S the entropy, with certain following properties. T is a positive quatity which is a function of the empirical temperature θ only. The entropy is an extensive property; *i.e.*, the total entropy in the system is the sum of the entropies of its parts.

In continuum mechanics, the specific entropy (per unit mass), or entropy density is denoted by s so that the total entropy L is given by

$$L = \int_V \rho s \, dV.$$

The entropy of a system can change either by interactions that occur with the surroudings or by changes that take place within the system. Thus where, is the increase in specific entropy, $ds^{(e)}$ is the increase due to interaction with the exterior and $ds^{(i)}$ is the internal increase. The change $ds^{(i)}$ is never negative. It is zero for a reversible process and positive for an irreversible process. Therefore

$$ds^{(i)} > 0 \, (\text{irreversible process}); \quad ds^{(i)} = 0 (\text{reversible process}).$$

In the reversible process, if $dq_{(R)}$ denotes the heat supplied per unit mass to the system, the change $ds^{(e)}$ is given by

$$ds^{(e)} = \frac{1}{T} dq_{(R)}.$$

Exercise

Theoretical

1. During a thermomechanical process, mass density is produced at the rate of p per second. Find the local conservation of energy. What form would this equation take if, instead, there were a momentum creation?
2. Find the equation of the jump of energy across a discontinuity plane (a shock wave), moving with the velocity U in the *x*-direction when all physical quantities are functions of x only.

Numerical

1. Prove that possible liquid motion is possible when velocities at a point (x_1, x_2, x_3) are given by

$$v_1 = -\frac{2x_1 x_2 x_3}{(x_1^2 + x_2^2)^2}, v_2 = \frac{(x_1^2 - x_2^2)x_3}{(x_1^2 + x_2^2)^2}, v_3 = \frac{x_2}{x_1^2 + x_2^2}.$$

   Is the motion irrotational?

2. If the motion of a continuous medium is given by

$$x_1 = X_1 e^t - X_3(e^t - 1); x_2 = X_2 e^{-t} + X_3(1 - e^{-t}); x_3 = X_3$$

   determine the displacement field in both material and spatial descriptions.

   Answer:

$$u_1 = (X_1 - X_3)(e^t - 1) = (x_1 - x_3)(e^t - 1);$$
$$u_2 = (X_2 - X_3)(e^{-t} - 1)$$
$$= (x_2 - x_3)(1 - e^t); u_3 = 0$$

3. The displacement field in a body is given by

$$x_1 = (2AX_1 + B)^{1/2}; x_2 = CX_2; x_3 = DX_3$$

   where (X_1, X_2, X_3) and (x_1, x_2, x_3) are rectangular coordinates, and A, B, C and D are constants.

   (i) Determine the deformation tensors.
   (ii) Determine the lagrangian strain tensor and find the change in the square of arc length.
   (iii) The infinitesimal strains and rotations.

   If the coordinates (x_1, x_2, x_3) are taken to be cylindrical coordinates having the same origin as X_k. Find a geometrical meaning for the displacement field.

■ ■ ■

5

Linear Elasticity

A continuum is called elastic if the stress tensor is a continuous function of the strain tensor such that the stress tensor automatically vanishes when the strain tensor is zero and vice-versa. An elastic body recovers its original shape and size completely when the stresses are removed. Therefore, elastic behavior is characterized by the following two conditions:

   (*i*) where the stress in a material is a unique function of the strain, and
   (*ii*) where the material has the property for complete recovery to a "natural" shape upon removal of the applied forces.

If the behavior of a material is not elastic, we say that it is *inelastic*. Also, we acknowledge that elastic behavior may be *linear* or *non-linear*. For many engineering applications, especially those involving structural materials such as metals and concrete, the conditions for elastic behavior are realized, and for these cases the theory of elasticity offers a very useful and reliable model for design.

The property by which a continuum body recovers from strain is called elasticity.

5.1 Linearly Elastic Solid

We define linearly elastic body to be a continuum which undergoes very small change of shape when subjected to forces of moderate magnitude such that, every stress components is a linear function of all strain components. Metal, concrete, wood are the examples of linearly elastic solid.

If the forces are not too large, it is a natural shape to which it will return whenever all forces causing the deformation are removed. It is restricted to the case in which deformation and gradients are small. The linear theory is inadequate to describe the mechanical behavior of materials (*e.g.* rubber) which is capable of undergoing large deformation.

(a) Generalized Hooke's law: In a continuous medium, the state of stress is completely determined by the stress tensor and the state of deformation by the strain tensor. In general, it may be expressed as

$$T = f(e) = T_{ij} = f_{ij}(e_{kl}); i, j, k, l = 1, 2, 3$$

where, *f* is a symmetric tensor-valued function of various strain tensor *e*. This is the generalized Hooke's law, which states that at each point of a continuous medium at a fixed temperature each of the six components is a function of the six strain components. Expanding this equation by Taylors series theorem, we obtain

$$T_{ij} = f_{ij}(0,0,\ldots,0) + \left(\frac{\partial f_{ij}}{\partial e_{kl}}\right)_0 e_{kl} + \ldots$$

$$= b_{ij} + a_{ijkl} e_{kl} + \ldots$$

Within the context of the above assumptions, we write the constitutive equation for a linear elastic body as

$$T_{ij} = b_{ij} + a_{ijkl} e_{kl}. \tag{5.1}$$

Since, in the initial unstrained state, the body will be unstressed, $T_{ij} = 0$ when all $e_{ij} = 0$, we must have $f_{ij}(0,0,\ldots,0) = b_{ij} = 0$; for $i,j = 1,2,3$. Therefore, the constitutive equations for a linear elastic solid relate the stress and strain tensors through the expression

$$T = ae \Rightarrow T_{ij} = a_{ijkl} e_{kl}; \quad i, j, k, l = 1, 2, 3. \tag{5.2}$$

where the tensor of elastic coefficients a_{ijkl} has $3^4 = 81$ components. However, due to the symmetry of both the stress and strain tensors, it is clear that

$$a_{ijkl} = a_{jikl} = a_{ijlk}$$

which reduces the 81 possibilities to 36 distinct coefficients at most. The constitutive, linear law for the relation (5.2) between stress and strain is known as the generalized Hooke's law for linear elastic body. The coefficients a_{ijkl} are called the elastic constant or elastic moduli and they are describing the elastic properties of the material.

(*i*) In general, the coefficients a_{ijkl} may depend upon temperature, but here we assume adiabatic (no heat gain or loss) and isothermal (constant temperature) conditions.

(*ii*) We also ignore strain-rate effects and consider the components a_{ijkl} to be at most a function of position. If these constants vary from

5.2 Strain energy

point to point of the medium, then the body is called elastically non-homogeneous or inhomogeneous. Reinforced concrete is an example of elastically non-homogeneous or inhomogeneous.

(*iii*) But, if these elastic coefficients are constants, i.e., have the same value throughout the body, the body is called elastically homogeneous. Mild steel is an example of elastically homogeneous body.

(b) Newton's law of Viscocity: In a viscous fluid, the viscous stress tensor T_{ij} is a function of the rate of deformation tensor d_{ij}. In general, the functional relationship is non-linear, it may be expressed as

$$T_{ij} = f_{ij}(d_{kl})$$

The fluid is then called a Stokesian fluid. When the functional relationship is linear, it can be written as

$$T_{ij} = g_{ijkl} d_{kl}$$

where the constants a_{ijkl} are called viscosity coefficients. The fluid is then termed as Newtonian fluid.

5.2 Strain Energy

The work done in deforming a body by the surface forces (i.e. the stresses) is transform completely into potential energy which is stored in that body. This potential energy due to deformation or strain is called the strain energy or the stress potential of the elastic body. It may be derive as follows:

From the first law of thermodynamics, i.e., the principle of conservation of energy, we have

$$\rho \frac{de}{dt} = T_{ij} d_{ij} - q_{i,j} + \rho h \qquad (5.3)$$

where,

e = internal energy per unit mass
h = rate per unit mass at which the heat energy is produced by radiation.
q = influx of heat pet unit area in unit time by conduction.
d_{ij} = strain rate tensor
ρ = mass density of the body.

For linear elastic solid the heat conduction may be neglected and the heat energy is produced entirely by internal sources only, so that one can be written as

$$\rho \frac{de}{dt} = T_{ij} d_{ij} + \rho h \qquad (5.4)$$

For the small strain components we have $d_{ij} = e_{ij}$, where $e_{ij} = u_{i,j} + u_{j,i}$ and $v_i = \dfrac{\partial u_i}{\partial t}$, then equation (5.4) can be written as

$$\rho \frac{de}{dt} = T_{ij} \dot{e}_{ij} + \rho h \tag{5.5}$$

For small displacement gradient we may replace ρ by ρ_0, the density in the undeformed state. If Q_1 be the quantity of heat per unit mass produced by internal sources at time t, then $h = \dfrac{dQ_1}{dt}$. The equation (5.5) can be written as

$$T_{ij}\dot{e}_{ij} = \rho_v\left(h - \frac{de}{dt}\right) = \rho_v\left(\frac{de}{dt} - \frac{dQ_1}{dt}\right) = \rho_v\left(\dot{e} - \dot{Q}_1\right) = \dot{\mho} - \dot{Q} \tag{5.6}$$

where, $\mho = \rho_0 e$, and $Q = \rho_0 Q_1$. Here $\mho$ represents the internal energy per unit volume of the uncertain state and Q is the quantity of heat produce from internal sources per unit volume of the unstrain state. From the second law of thermodynamics

$$\dot{Q} = T\frac{dS}{dt} \tag{5.7}$$

where, S = the entropy per unit volume, T = absolute temperature per unit volume. If the deformation takes place adiabatically then $\dot{Q} = 0$ and the equation (5.6) becomes

$$T_{ij}\dot{e}_{ij} = \dot{\mho}, i.e., T_{ij}de_{ij} = d\mho \tag{5.8}$$

On the other hand if the body is strained very slowly then the temperature remain practically constant (isothermal process) so that, $\dot{T} = 0 = \dfrac{dT}{dt}$ in this case equation (5.7) can be written as

$$\dot{Q} = \frac{\partial}{\partial t}(TS)$$

and hence from the equation (5.6) we get

$$T_{ij}\dot{e}_{ij} = \frac{\partial}{\partial t}(\mho - TS) = \frac{\partial F}{\partial t} \tag{5.9}$$

where, $F = \mho - TS$ is known as Helmholtz's free energy per unit volume of a body. From equation (5.9) we get

$$T_{ij}de_{ij} = d\mho = dF \tag{5.10}$$

In either case, *i.e.*, whether deformation takes place isothermally or adiabatically the left side of equation (5.10) is exact differential, say dW, so that we may write

5.2 Strain energy

$$T_{ij}de_{ij} = dW \qquad (5.11)$$

The double summation of the left hand side of equation (5.11), can be simplified if we introduce the notation

$$T_1 = T_{11}, T_2 = T_{22}, T_3 = T_{33}$$

Similarly, $\quad T_4 = T_{23} = T_{32}, T_5 = T_{31} = T_{13}, T_6 = T_{12} = T_{21}.$

$$e_1 = e_{11}, e_2 = e_{22}, e_3 = e_{33}$$

$$e_4 = e_{23} = e_{32}, e_5 = e_{31} = e_{13}, e_6 = e_{12} = e_{21}.$$

Then we can write equation (5.11) as

$$T_i de_i = dW = \frac{\partial W}{\partial e_i} de_i \text{ i.e., } T_i = \frac{\partial W}{\partial e_i}; \ i = 1, 2, \ldots, 6 \qquad (5.12)$$

Now $T_{ij}de_{ij}$ represents the work done per unit volume at a point by all surface forces. Hence dW represents the work done per unit volume and w is called the stress potential or the strain energy function for an unit volume of the elastic body. For an unique value of the elastic body, the equation (5.12) states that the stress components are partial derivatives of strain energy with respect to the corresponding strain components.

It is worthwhile noting at this point that elastic behavior is sometimes defined on the basis of the existence of a strain energy function from which the stresses may be determined by the differentiation in Eq (5.12). A material defined in this way is called a *hyperelastic* material. The stress is still a unique function of strain so that this energy approach is compatible with our earlier definition of elastic behavior. Thus, in keeping with our basic restriction to infinitesimal deformations, we shall develop the linearized form of Eq (5.12).

Since W represents potential energy per unit volume stored up in the body by strain deformation alone, W must be a function of components of strain so that we can write

$$W = W(e_1, e_2, \ldots, e_6)$$

Expanding W in a power series about the origin ($e_i = 0$), we have

$$W = C_0 + C_i e_i + \frac{1}{2} C_{ij} e_i e_j + \ldots$$

where, $C_0 = W(0, 0, \ldots, 0) = $ a constant, $C_i = \left(\frac{\partial W}{\partial e_i}\right)_0, C_{ij} = \left(\frac{\partial^2 W}{\partial e_i \partial e_j}\right)_0$, etc.

Since the magnitude of derivatives of W does not depend on the order of differentiation,

$$\frac{\partial^2 W}{\partial e_i \partial e_j} = \frac{\partial^2 W}{\partial e_j \partial e_i} \Rightarrow C_{ij} = C_{ji}.$$

Since we are interested in the derivative of W, we can set $C_0 = 0$. For elastic body, $T_i = 0$ whenever $e_j = 0$, consequently, $C_i = 0$. Therefore, from equation (5.12) we get

$$T_i = \frac{\partial W}{\partial e_i} = C_i + \frac{1}{2}\frac{\partial}{\partial e_i}\left(C_{pq} e_p e_q\right) + \ldots$$

$$= \frac{1}{2} C_{pq}\left(e_p \frac{\partial e_q}{\partial e_i} + \frac{\partial e_p}{\partial e_i} e_q\right) + \ldots = \frac{1}{2} C_{pq}\left(e_p \delta_{qi} + \delta_{pi} e_q\right) + \ldots$$

$$= \frac{1}{2} C_{pi} e_p + \frac{1}{2} e_q C_{iq} + \ldots = \frac{1}{2}\left(C_{ji} + C_{ij}\right) e_j + \ldots$$

For linear elastic solid, stresses are linear functions of strain, we neglect all terms of order 2 and higher in strain. Therefore,

$$T_i = C_{ij} e_j; \quad \text{as} \quad C_{ij} = C_{ji} \tag{5.13}$$

Equation (5.13) is called generalized Hooke's law, in which the number of independent elastic constants are given as

(i) 6 different constants located at the diagonal

(ii) $\frac{36-6}{2} = 15$ among the remaining constants

so that they are 15 + 6 = 21 in numbers. Because of this symmetry on C_{ij}, the number of independent elastic constants is at most 21 if a strain energy function exists. Since the elastic properties of a Hookean solid are expressed through the coefficients C_{ij}, a general anisotropic body will have an elastic-constant matrix of the form

$$(C_{ij}) = \begin{pmatrix} C_{11} & C_{12} & C_{13} & C_{14} & C_{15} & C_{16} \\ C_{12} & C_{22} & C_{23} & C_{24} & C_{25} & C_{26} \\ C_{13} & C_{23} & C_{33} & C_{34} & C_{35} & C_{36} \\ C_{14} & C_{24} & C_{34} & C_{44} & C_{45} & C_{46} \\ C_{15} & C_{25} & C_{35} & C_{45} & C_{55} & C_{56} \\ C_{16} & C_{26} & C_{36} & C_{46} & C_{56} & C_{66} \end{pmatrix}$$

The material with 21 elastic constants as displayed above is called anisotropic linearly elastic material. The expression of strain-energy function is given by

$$W = \frac{1}{2}C_{ij}e_i e_j; C_{ij} = C_{ji} \tag{5.14}$$

which gives the stress potential in case of linearly elastic body. Also

$$W = \frac{1}{2}T_i e_i (i = 1, 2, \ldots, 6) = \frac{1}{2}T_{ij}\rho_{ij}; i = 1, 2, 3; j = 1, 2, 3. \tag{5.15}$$

which is the Clapeyron's formula. The equation (5.15) also written as by using equation (5.13) as

$$2W = T_i e_i = \frac{\partial W}{\partial e_i} e_i$$

i.e., W is a homogeneous function of second degree in (e_1, e_1, ..., e_2). An example of anistropic body is provided by crystal.

5.3 Elastic Symmetry

A plane of elastic symmetry exists at a point where the elastic constants have the same values for every pair of coordinate systems which are the reflected images of one another with respect to the plane. The axes of such coordinate systems are referred to as "equivalent elastic directions." It is obvious from the stress-strain relation for linearly elastic body

$$T_i = C_{ij}e_j \tag{5.16}$$

that the elastic coefficients c_{ij} in general depend on the reference frame as the stress components T_i and the strain components T_i also depend on the choice of coordinate systems. When the T_i remain invariant under a given transformation of coordinates for a medium, the medium is called elastically symmetric. Now elastic symmetry imposes restrictions on these coefficients, thus further reducing the number of independent constants. We proceed to discuss the various elastic symmetries.

(a) Monotropic Material

Consider a material elastically symmetry with respect to a plane. Let this plane be taken as x_1, x_1 plane. For elastic symmetry strain-energy W is invariant if the direction of Ox_3 axis is reserved (Fig. 5.1). The reversal of Ox_3 axis changes the sign x_3 and u_3 and hence signs of strain components

$$e_5 = 2e_{31} = \left(\frac{\partial u_3}{\partial x_1} + \frac{\partial u_1}{\partial x_3}\right); e_4 = 2e_{23} = \left(\frac{\partial u_2}{\partial x_3} + \frac{\partial u_3}{\partial x_2}\right).$$

250   Chapter 5 Linear Elasticity

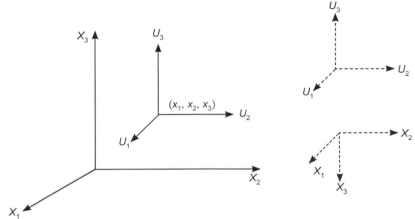

Fig. 5.1: Monotropic material

Now strain-energy function W being the potential energy due to deformation per unit volume has a physical meaning independent of the choice of coordinate axes. Hence the strain-energy function W of the form

$$W = \frac{1}{2}C_{ij}e_i e_j ; C_{ij} = C_{ji}$$

must be invariant, under the reversal of Ox_3 axis. Consequently, the first degree terms in e_4 and e_5 in the expression of strain-energy function vanish,

$$C_{14} = C_{15} = C_{24} = C_{25} = C_{34} = C_{35} = C_{46} = C_{56} = 0.$$

For a material with one plane of symmetry the matrix reduces to

$$(C_{ij}) = \begin{pmatrix} C_{11} & C_{12} & C_{13} & 0 & 0 & C_{16} \\ .. & C_{22} & C_{23} & 0 & 0 & C_{26} \\ .. & .. & C_{33} & 0 & 0 & C_{36} \\ .. & .. & .. & C_{44} & C_{45} & 0 \\ .. & .. & .. & .. & C_{55} & 0 \\ .. & .. & .. & .. & .. & C_{66} \end{pmatrix} \quad (5.17)$$

in which the number of nonzero independent elastic constants = 21 – 8 = 13. A material with 13 elastic coefficients as given in equation (5.17) is called monotropic material. For this material, the expression of strain-energy function is given by

$$W = \frac{1}{2}\left[C_{11}e_1^2 + C_{22}e_2^2 + C_{33}e_3^2 + C_{44}e_4^2 + C_{55}e_5^2 + C_{66}e_6^2 \right.$$
$$\left. + 2C_{12}e_1 e_2 + 2C_{13}e_1 e_3 + 2C_{16}e_1 e_6 + 2C_{23}e_2 e_3 \right.$$

5.3 Elastic Symmetry

$$+ 2C_{26}e_2e_6 + 2C_{36}e_3e_6 + 2C_{45}e_4e_5 \Big]. \tag{5.18}$$

(b) Orthotropic Material

Consider a material elastically symmetric with respect to two mutually perpendicular planes. Let these two planes be taken as x_1x_2 plane and x_2x_3 plane. Since the material is symmetric with respect to x_1x_2 plane, condition

$$C_{14} = C_{15} = C_{24} = C_{25} = C_{34} = C_{35} = C_{46} = C_{56} = 0$$

is satisfied and elastic matrix of the coefficients

$$(C_{ij}) = \begin{pmatrix} C_{11} & C_{12} & C_{13} & 0 & 0 & C_{16} \\ .. & C_{22} & C_{23} & 0 & 0 & C_{26} \\ .. & .. & C_{33} & 0 & 0 & C_{36} \\ .. & .. & .. & C_{44} & C_{45} & 0 \\ .. & .. & .. & .. & C_{55} & 0 \\ .. & .. & .. & .. & .. & C_{66} \end{pmatrix}$$

and the expression of strain-energy function is given by

$$W = \frac{1}{2}\Big[C_{11}e_1^2 + C_{22}e_2^2 + C_{33}e_3^2 + C_{44}e_4^2 + C_{55}e_5^2 + C_{66}e_6^2$$
$$+ 2C_{12}e_1e_2 + 2C_{13}e_1e_3 + 2C_{16}e_1e_6 + 2C_{23}e_2e_3$$
$$+ 2C_{26}e_2e_6 + 2C_{36}e_3e_6 + 2C_{45}e_4e_5 \Big].$$

Besides the symmetry with respect to x_1x_2 plane, the material is also symmetric with respect to x_2x_3 plane. Consequently, the expression of strain-energy function W is invariant if the direction of OX_1 axis is reserved (Fig. 5.2). It changes the sign x_1 and u_1 and signs of strain components.

$$e_5 = 2e_{31} = \left(\frac{\partial u_3}{\partial x_1} + \frac{\partial u_1}{\partial x_3} \right); e_4 = 2e_{12} = \left(\frac{\partial u_1}{\partial x_2} + \frac{\partial u_2}{\partial x_1} \right).$$

In order that the expression of strain-energy function W invariant under the reversal of OX_1 axis, we must set

$$C_{16} = C_{26} = C_{36} = C_{45} = 0. \tag{5.19}$$

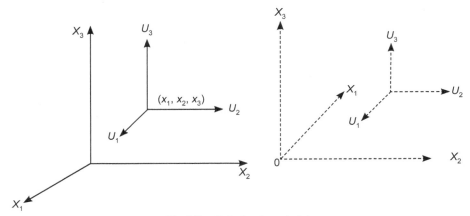

Fig 5.2: Orthotropic material

For a plane material with one plane of symmetry the matrix reduces to

$$(C_{ij}) = \begin{pmatrix} C_{11} & C_{12} & C_{13} & 0 & 0 & 0 \\ C_{12} & C_{22} & C_{23} & 0 & 0 & 0 \\ C_{13} & 0 & C_{33} & 0 & 0 & 0 \\ 0 & .. & .. & C_{44} & 0 & 0 \\ .. & .. & .. & 0 & C_{55} & 0 \\ .. & .. & .. & .. & .. & C_{66} \end{pmatrix} \qquad (5.20)$$

in which the number of nonzero independent elastic constants $= (21-8) - 4 = 13 - 4 = 9$. It is obvious from equations (5.12d) and (5.12e) that material is also symmetric with respect to the $x_1 x_2$ plane. Thus if there are two orthogonal planes of elastic symmetry in a material, then third orthogonal plane is automatically a plane of elastic symmetry. Such a material having three mutually perpendicular planes of elastic symmetry is said to be orthotropic (for example, wood). For such material, the expression of strain-energy function W is given by

$$W = \frac{1}{2}\left[C_{11}e_1^2 + C_{22}e_2^2 + C_{33}e_3^2 + C_{44}e_4^2 + C_{55}e_5^2 + C_{66}e_6^2 \right. \qquad (5.21)$$

$$\left. + 2C_{12}e_1e_2 + 2C_{13}e_1e_3 + 2C_{16}e_1e_6 + 2C_{23}e_2e_3 \right].$$

(c) Symmetry with respect to each of three orthogonal planes

If the body exhibits the same elastic properties with respect to each of three orthogonal planes of symmetry, the strain-energy density function is to remain invariant when axes OX_1, OX_2, OX_3 are permuted cyclically, i.e., when e_1, e_2, e_3 or e_4, e_5, e_6 are permuted cyclically. Consequently, in addition to condition

5.4 Isotropic Elastic Body

$C_{14} = C_{15} = C_{24} = C_{25} = C_{34} = C_{35} = C_{46} = C_{56} = 0$ and $C_{16} = C_{26} = C_{36} = C_{45} = 0$, the following condition must be fulfilled.

$$C_{11} = C_{22} = C_{33}; C_{44} = C_{55} = C_{66}; C_{12} = C_{13} = C_{23}. \qquad (5.22)$$

The matrix of the elastic constants are given by

$$(C_{ij}) = \begin{pmatrix} C_{11} & C_{12} & C_{12} & 0 & 0 & 0 \\ C_{12} & C_{11} & C_{12} & 0 & 0 & 0 \\ C_{12} & C_{12} & C_{11} & 0 & 0 & 0 \\ 0 & .. & .. & C_{44} & 0 & 0 \\ .. & .. & .. & 0 & C_{44} & 0 \\ .. & .. & .. & .. & .. & C_{44} \end{pmatrix} \qquad (5.23)$$

The number of independent elastic constants reduce to 3 and the expression of strain-energy function is given by

$$W = \frac{1}{2}\left[C_{11}\left(e_1^2 + e_2^2 + e_3^2\right) + C_{44}(e_4^2 + e_5^2 + e_6^2) + 2C_{12}\left(e_1 e_2 + e_1 e_3 + e_2 e_3\right) \right] \qquad (5.24)$$

5.4 Isotropic Elastic Body

If the behavior of a material is elastic under a given set of circumstances, it is customarily spoken of as an elastic material when discussing that situation even though under a different set of circumstances its behavior may not be elastic. Furthermore, if a body's elastic properties as described by the coefficients a_{ijkl} are the same in every set of reference axes at any point for a given situation, we call it an *isotropic elastic material*. For such materials, the constitutive equation has only two elastic constants. A material that is not isotropic is called *anisotropic*; we shall define some of these based upon the degree of elastic symmetry each possesses.

Isotropy requires the elastic tensor a_{ijkl} of Eq (5.2) to be a fourth-order isotropic tensor. In general, an isotropic tensor is defined as one whose components are unchanged by any orthogonal transformation from one set of Cartesian axes to another. Zero-order tensors of any order—and all zeroth-order tensors (scalars)—are isotropic, but there are no first-order isotropic tensors (vectors). The only nontrivial third-order isotropic tensor is the permutation symbol.

Therefore, a linearly elastic solid is said to be isotropic if it has the symmetry of elastic properties in all directions. Its mean that the strain energy function W is an invariant under all orthogonal transformation of material axis.

5.4.1 Hooke's Law for Linear Isotropic Solid

For a linear elastic solid we have from generalized Hooke's law

$$T_{ij} = a_{ijkl} e_{kl}; i, j, k, l = 1, 2, 3. \qquad (5.25)$$

We may write the elastic constant a_{ijkl} as

$$a_{ijkl} = \frac{1}{2}\left(a_{ijkl} - a_{ijlk}\right) + \frac{1}{2}\left(a_{ijkl} + a_{ijlk}\right) = b_{ijkl} + c_{ijkl} \qquad (5.26)$$

where, $b_{ijkl} = \frac{1}{2}\left(a_{ijkl} - a_{ijlk}\right)$ and $c_{ijkl} = \frac{1}{2}\left(a_{ijkl} + a_{ijlk}\right).$
Now

$$b_{ijkl} = \frac{1}{2}\left(a_{ijkl} - a_{ijlk}\right) = -b_{ijlk}$$

$$c_{ijkl} = \frac{1}{2}\left(a_{ijkl} + a_{ijlk}\right) = c_{ijlk}$$

so that b_{ijkl} is skew symmetric and c_{ijkl} is symmetric. Using equation (5.26), equation (5.25) gives us

$$b_{ijkl} e_{kl} = b_{ijlk} e_{lk} = -b_{ijkl} e_{kl}; i.e., b_{ijkl} e_{kl} = 0. \qquad (5.27)$$

Now,

$$T_{ij} = a_{ijkl} e_{kl} = \left(b_{ijkl} + c_{ijkl}\right) e_{kl} = b_{ijkl} e_{kl} + c_{ijkl} e_{kl}$$

Therefore,

$$T_{ij} = c_{ijkl} e_{kl}; \quad i, j, k, l = 1, 2, \qquad (5.28)$$

Since T_{ij} and e_{ij} are second order tensors, therefore c_{ijkl} must be a tensor of order 4. For isotropic elastic media the elastic constant c_{ijkl} remains the same under all orthogonal transformation of the coordinate axes. The most general fourth order tensor may be shown to have a form in terms of Kronecker deltas which we now introduce as a prototype for c_{ijkl}, namely

$$c_{ijkl} = \lambda \delta_{ij} \delta_{kl} + \mu \delta_{ik} \delta_{jl} + \gamma \delta_{il} \delta_{jk} + \beta(\delta_{ik} \delta_{jl} - \delta_{il} \delta_{jk})$$

where, λ, μ, γ and β are scalars. Now $c_{ijkl} = c_{jikl} = c_{ijlk}$. This implies that β must be zero for stated symmetries since by interchanging i and j in the expression

$$\beta\left(\delta_{ik}\delta_{jl} - \delta_{il}\delta_{jk}\right) = \beta\left(\delta_{jk}\delta_{il} - \delta_{jl}\delta_{ik}\right) \Rightarrow \beta = -\beta \Rightarrow \beta = 0$$

Therefore, c_{ijkl} must be an tensor of order 4 for a isotropic body, which can be written in terms of three arbitrary constants, λ, μ, β as

$$c_{ijkl} = \lambda \delta_{ij} \delta_{kl} + \mu \delta_{ik} \delta_{jl} + \gamma \delta_{il} \delta_{jk} \qquad (5.29)$$

5.4 Isotropic Elastic Body

and
$$c_{ijlk} = \lambda \delta_{ij}\delta_{lk} + \mu \delta_{il}\delta_{jk} + \gamma \delta_{ik}\delta_{jl}$$

Since δ_{ij} is symmetric therefore using the equation $c_{ijkl} = c_{ijlk}$, we have

$$\lambda \delta_{ij}\delta_{kl} + \mu \delta_{ik}\delta_{jl} + \gamma \delta_{il}\delta_{jk} = \lambda \delta_{ij}\delta_{lk} + \mu \delta_{il}\delta_{jk} + \gamma \delta_{ik}\delta_{jl}$$

$$\Rightarrow (\mu - \gamma)(\delta_{ik}\delta_{jl} - \delta_{il}\delta_{jk}) = 0 \qquad (5.30)$$

The relation (5.30) is true for all values of i, j, k, l. If we take $i = k = 1$ and $j = l = 2$, the equation (5.30) becomes

$$(\mu - \gamma)(\delta_{11}\delta_{22} - \delta_{12}\delta_{21}) = 0 \Rightarrow (\mu - \gamma)(1 - 0) = 0 \Rightarrow \mu = \gamma.$$

Thus we can write the relation (5.29) as

$$c_{ijkl} = \lambda \delta_{ij}\delta_{kl} + \mu(\delta_{il}\delta_{jk} + \delta_{ik}\delta_{jl}) \qquad (5.31)$$

Therefore, from equation (5.28) we get

$$T_{ij} = c_{ijkl}e_{kl} = \lambda \delta_{ij}\delta_{kl}e_{kl} + \mu\left(\delta_{il}\delta_{jk} + \delta_{ik}\delta_{jl}\right)e_{kl}$$

$$= \lambda \delta_{ij}e_{kk} + \mu\left(\delta_{il}e_{jl} + \delta_{ik}e_{kj}\right)$$

$$= \lambda \delta_{ij}e_{kk} + \mu\left(e_{ij} + e_{ji}\right)$$

$$= \lambda \theta \delta_{ij} + 2\mu e_{ij} \text{ as } e_{ij} = e_{ji} \qquad (5.32)$$

where, $\theta = e_{kk} = e_{11} + e_{22} + e_{33}$, which are constitutive equations or stress-strain law for linearly elastic isotropic body or *Hooke's law for isotropic elastic behavior*. As mentioned earlier, we see that for isotropic elastic behavior the 21 constants of the generalized law have been reduced to two, λ and μ, known as the *Lamé constants*. The matrix of elastic constants c_{ijkl} in relation $T_{ij} = c_{ijkl}e_{kl}$ is given by

$$(c_{ijkl}) = \begin{pmatrix} \lambda + 2\mu & \lambda & \lambda & 0 & 0 & 0 \\ \lambda & \lambda + 2\mu & \lambda & 0 & 0 & 0 \\ \lambda & \lambda & \lambda + 2\mu & 0 & 0 & 0 \\ 0 & 0 & 0 & \mu & 0 & 0 \\ 0 & 0 & 0 & 0 & \mu & 0 \\ 0 & 0 & 0 & 0 & 0 & \mu \end{pmatrix}.$$

In which the number of independent elastic constants reduce to 2. Note that for an isotropic elastic material $c_{ijkl} = c_{klij}$ i.e., an isotropic elastic material is necessarily hyperelastic.

Example 5.1: **Show that principal directions of strain at each point of a linearly elastic isotropic body are coincident with the principal directions of stress.**

Solution: Let us take the principal direction of strain at some point of the body as coordinates axes. Let e_{ij} be strain tensor and T_{ij} be stress tensor at that point. Then

$$e_{31} = 0, e_{12} = 0, e_{23} = 0.$$

Now constitutive equation for isotropic elastic body is given by the equation (5.20) is

$$T_{ij} = \lambda \theta \delta_{ij} + 2\mu e_{ij}.$$

Therefore,

$$T_{12} = \lambda \theta \delta_{12} + 2\mu e_{12} = 2\mu e_{12} = 0$$
$$T_{23} = \lambda \theta \delta_{23} + 2\mu e_{23} = 2\mu e_{23} = 0$$
$$T_{31} = \lambda \theta \delta_{31} + 2\mu e_{31} = 2\mu e_{31} = 0$$

Hence, coordinate axes must be along the principal directions of stress. For isotropic body no distinction is therefore made between principal direction of strain and those of stress. Both are referred to us as principal directions.

Example 5.2: **A displacement field is given by $u_1 = 3x_1 x_2^2, u_2 = 2x_1 x_3$, $u_3 = x_3^2 - x_1 x_2$. Determine the strain components and check whether they satisfy the compatibility equations. Also find the stress components.**

Solution: Let us take the principal direction of strain at some point of the body as coordinates axes. Let e_{ij} be strain tensor and T_{ij} be stress tensor at that point. Then

$$e_{11} = \frac{\partial u_1}{\partial x_1} = 3x_2^2, \quad e_{22} = \frac{\partial u_2}{\partial x_2} = 0, \quad e_{33} = \frac{\partial u_3}{\partial x_3} = 2x_3$$

$$e_{23} = \frac{1}{2}\left(\frac{\partial u_2}{\partial x_3} + \frac{\partial u_3}{\partial x_2}\right) = x_1; \quad e_{31} = \frac{1}{2}\left(\frac{\partial u_1}{\partial x_3} + \frac{\partial u_3}{\partial x_1}\right) = -x_2;$$

$$e_{12} = \frac{1}{2}\left(\frac{\partial u_1}{\partial x_2} + \frac{\partial u_2}{\partial x_1}\right) = 6x_1 x_2 + 2x_3.$$

Therefore, $\theta = e_{11} + e_{22} + e_{33} = 3x_2^2 + 2x_3$. The constitutive equation for isotropic elastic body given by the equation (5.32) is $T_{ij} = \lambda \theta \delta_{ij} + 2\mu e_{ij}$, where λ and μ are Lame's constant. Therefore, the stress components are

$$T_{11} = \lambda\theta\delta_{11} + 2\mu e_{11} = (3\lambda + 6\mu)x_2^2 + 2\lambda x_3$$
$$T_{22} = \lambda\theta\delta_{22} + 2\mu e_{22} = \lambda(3x_2^2 + 2x_3)$$
$$T_{33} = \lambda\theta\delta_{33} + 2\mu e_{33} = 3\lambda x_2^2 + (2\lambda + 4\mu)x_3$$
$$T_{12} = \lambda\theta\delta_{12} + 2\mu e_{12} = 2\mu e_{12} = 2\mu(6x_1 x_2 + 2x_3)$$
$$T_{23} = \lambda\theta\delta_{23} + 2\mu e_{23} = 2\mu e_{23} = 2\mu x_1$$
$$T_{31} = \lambda\theta\delta_{31} + 2\mu e_{31} = 2\mu e_{31} = -2\mu x_2.$$

5.5 Strains in Terms of Stresses

Hooke's law for isotropic bodies can be written in terms of elastic constants. In equation (5.32), if we put $j = i$ and sum over i, we get

$$T_{ii} = 3\lambda\theta + 2\mu e_{ii} = 3\lambda\ \theta + 2\mu\ \theta \text{ i.e., } \Theta = (3\lambda + 2\mu)\theta$$

where $\Theta = T_{kk} = T_{11} + T_{22} + T_{33}$ = first stress invariant.

Now,

$$2\mu e_{ij} = T_{ij} - \lambda\ \theta\delta_{ij};$$

$$e_{ij} = \frac{T_{ij}}{2\mu} - \frac{\lambda\ \theta}{2\mu}\delta_{ij} = \frac{T_{ij}}{2\mu} - \frac{\lambda}{2\mu}\frac{\Theta}{3\lambda + 2\mu}\delta_{ij} = \frac{T_{ij}}{2\mu} - \frac{\lambda\ \Theta}{2\mu(3\lambda + 2\mu)}\delta_{ij} \quad (5.33)$$

for $\mu \neq 0$ and $3\lambda + 2\mu \neq 0$. Equation (5.33) is the inversion of Hooke's law and give us the strain-stress relation.

5.5.1 Equation of Equilibrium and Motion in Terms of Displacement

Using the principle of balance of linear momentum, the stress equation of equilibrium of a continuum under external body forces per unit volume is given by

$$T_{ij,j} + F_i = 0;\ i, j = 1, 2, 3 \quad (5.34)$$

where, T_{ij} are the stress components. The stress-strain relation for a linear isotropic solid is given by

$$T_{ij} = \lambda\theta\delta_{ij} + 2\mu e_{ij}; \theta = e_{kk} \quad (5.35)$$

$$2e_{ij} = \left[\frac{\partial u_i}{\partial x_j} + \frac{\partial u_j}{\partial x_i}\right] = u_{i,j} + u_{j,i} \quad (5.36)$$

u_i being the displacement component. Putting (5.36) in equation (5.35), we get

$$T_{ij} = \lambda\theta\delta_{ij} + 2\mu e_{ij} = \lambda\theta\delta_{ij} + \mu(u_{i,j} + u_{j,i}) \tag{5.37}$$

Differentiating equation (5.37) with respect to x_j and summing over j we get

$$T_{ij,j} = \lambda\delta_{ij}\theta_{,j} + \mu\left(u_{i,jj} + u_{j,ij}\right) = \lambda\theta_{,i} + \mu\left\{\nabla^2 u_i + \left(u_{j,j}\right)_i\right\}$$

$$= \lambda\theta_{,i} + \mu\nabla^2 u_i + \mu\theta_{,i} = (\lambda+\mu)\theta_{,i} + \mu\nabla^2 u_i \tag{5.38}$$

Putting (5.38) in equation (5.34), we get

$$\mu\nabla^2 u_i + (\lambda+\mu)\theta_{,i} + F_i = 0; \theta_{,i} = \left(u_{1,1} + u_{1,1} + u_{1,1}\right)_i \tag{5.39}$$

These are the displacement equation of equilibrium. Putting $i = 1,2,3$, we get three such equations.

The equation of motion in terms of displacements can similarly obtained from the stress equation of motion namely

$$\rho\ddot{u}_i = \rho\frac{\partial^2 u}{\partial t^2} = T_{ij,j} + F_i = \mu\nabla^2 u_i + (\lambda+\mu)\theta_{,i} + F_i \tag{5.40}$$

In vector notation this equation (5.40) becomes

$$\rho\ddot{\vec{u}} = \mu\nabla^2\vec{u} + (\lambda+\mu)\,grad\left(div\vec{u}\right) + \vec{F} = \vec{0} \tag{5.41}$$

These are displacement equations of motion.

5.5.2 Physical Significance of Elastic Constants

For linearly isotropic bodies the constitutive equation is

$$T_{ij} = \lambda\theta\delta_{ij} + 2\mu e_{ij};\ \theta = e_{kk} = e_{11} + e_{22} + e_{33}. \tag{5.42}$$

This equation contain two elastic constants λ and μ. In order to find the physical significance of the elastic constants, we consider the behavior of elastic bodies subjected to

(*i*) Simple tension
(*ii*) Pure shear
(*iii*) Hydrostatic pressure.

(i) Simple Tension

We assume that a right elastic cylinder with its axis along the axis is subjected to a longitudinal force applied to ends of the cylinder of the applied force give rise to (produce) a uniform tension in every cross-section of the cylinder then

5.5 Strains in terms of stresses

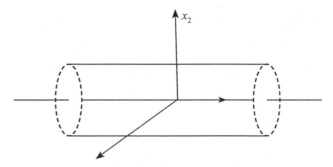

Fig. 5.3: Simple tension

$$T_{11} = T, T_{22} = 0 = T_{33}; T_{12} = T_{23} = T_{31} = 0 \tag{5.43}$$

Let $\vec{n} = (n_1, n_2, n_3)$ be the unit outward normal to the curved boundary, so $n_1 = 0, n_2^2 + n_3^2 = 1$. The body forces are absent in the interior of the cylinder, the stress given by equation (5.43) satisfies the equation of equilibrium, $T_{ij,j} = 0$

$$T_{ij,j} = 0 \tag{5.44}$$

within the body. At any point on the curved surface the components of the stress-vector is given by

$$T_1^{(n)} = T_{11} n_1 + T_{12} n_2 + T_{13} n_3 = 0$$
$$T_2^{(n)} = T_{21} n_1 + T_{22} n_2 + T_{23} n_3 = 0$$
$$T_3^{(n)} = T_{31} n_1 + T_{32} n_2 + T_{33} n_3 = 0$$

We see that the lateral surface is stress free. Also from the relation, we have

$$T_{ii} = 3\lambda\theta + 2\mu e_{ii} = 3\lambda\theta + 2\mu(e_{11} + e_{22} + e_{33}) = 3\lambda\theta + 2\mu,$$

i.e., $T = (3\lambda + 2\mu)\theta$, i.e., $\theta = \dfrac{T}{3\lambda + 2\mu}$

$$T_{11} = \lambda\theta + 2\mu e_{11} = T, \text{i.e.}, e_{11} = \frac{T}{2\mu} - \frac{\lambda}{2\mu}\frac{T}{3\lambda + 2\mu} = \frac{T(\lambda + \mu)}{\mu(3\lambda + 2\mu)} \tag{5.45}$$

$$T_{22} = \lambda\theta + 2\mu e_{22} = 0, \text{i.e.}, e_{22} = -\frac{T\lambda}{2\mu(3\lambda + 2\mu)} \tag{5.46}$$

$$T_{33} = \lambda\theta + 2\mu e_{33} = 0, \text{i.e.}, e_{33} = -\frac{T\lambda}{2\mu(3\lambda + 2\mu)} \tag{5.47}$$

and the other strain components $e_{12} = e_{23} = e_{31} = 0$. The strain system given by the expressions e_{11}, e_{22} and e_{33} are compatible. Hence, the stress and the strain systems given by equation (5.43) and equation (5.45)-(5.47) actually corresponds to those which exist in the above deformation. We now introduce two constants E and σ by

$$E = \frac{\mu(3\lambda + 2\mu)}{(\lambda + \mu)}, \sigma = \frac{\lambda}{2(\lambda + \mu)} \tag{5.48}$$

Then from equation (5.45) we have

$$\frac{T}{e_{11}} = \frac{\mu(3\lambda + 2\mu)}{(\lambda + \mu)} = E; \quad \frac{e_{22}}{e_{11}} = -\frac{\lambda}{2(\lambda + \mu)} = -\sigma \tag{5.49}$$

$$e_{22} = -\sigma e_{11}; e_{33} = -\sigma e_{11}; e_{12} = e_{23} = e_{31} = 0. \tag{5.50}$$

$T > 0$ represents the tension. The tensile strain will produce an extension in the direction of the axis of the cylinder and a contraction in the cross section. Accordingly, $e_{11} > 0$, $e_{22} < 0$, $e_{33} < 0$. Hence, two constants E and σ are both positive and we may write

$$E = \frac{T}{e_{11}} = \frac{\text{Tensile stress}}{\text{logitudinal extension}}.$$

$$\sigma = \left|\frac{e_{22}}{e_{11}}\right| = \left|\frac{e_{33}}{e_{11}}\right| = \text{ratio of lateral contraction to longitudinal extension.}$$

E is called Young's modulus or modulus of elasticity and σ is called the Poisson ratio. Now from equation (5.48) we have

$$1 + \sigma = 1 + \frac{\lambda}{2(\lambda + \mu)} = \frac{(3\lambda + 2\mu)}{2(\lambda + \mu)} = \frac{E}{2\mu}$$

$$1 - 2\sigma = 1 - \frac{\lambda}{\lambda + \mu} = \frac{\mu}{\lambda + \mu}$$

$$(1 + \sigma)(1 - 2\sigma) = \frac{E}{2\mu} \cdot \frac{\mu}{\lambda + \mu} = \frac{E}{2(\lambda + \mu)} = \frac{E\sigma}{\lambda}$$

$$\lambda = \frac{E\sigma}{(1+\sigma)(1-2\sigma)} \text{ and } \mu = \frac{E}{2(1+\sigma)} \tag{5.51}$$

Example 5.3: (Displacement under inform pressure): When a body is subjected to uniform pressure such that $T_{11} = T_{22} = T_{33} = -p$,

5.5 Strains in terms of stresses

$T_{23} = T_{31} = T_{12} = 0$, prove that $\dfrac{u_1}{x_1} = \dfrac{u_2}{x_2} = \dfrac{u_3}{x_3} = \dfrac{-p}{3k}$, assuming that displacement and rotation at $x_1 = 0, x_2 = 0, x_3 = 0$ are zero.

Solution: Clearly, the stress equations of equilibrium under no body forces are satisfied. Using Cauchy's formula $T_i^{(n)} = T_{ij} n_j; i, j = 1, 2, 3,$ we get

$$T_1^{(n)} = T_{11} n_1 + T_{12} n_2 + T_{13} n_3 = -p n_1$$
$$T_2^{(n)} = T_{21} n_1 + T_{22} n_2 + T_{23} n_3 = -p n_2$$
$$T_3^{(n)} = T_{31} n_1 + T_{32} n_2 + T_{33} n_3 = -p n_3$$

Therefore, $|\vec{T}^{(n)}| = p,$ on the boundary surface in the inward normal direction.

Hence, $\vec{T}^{(n)} = -p\vec{n}$. Thus there is a distributed normal pressure of magnitude p on the boundary. The displacement components,

$$\dfrac{\partial u_1}{\partial x_1} = e_{11} = \dfrac{1+\sigma}{E}\left[T_{11} - \dfrac{\sigma}{1+\sigma}\Theta\delta_{11}\right] = \dfrac{1+\sigma}{E}\left[T_{11} - \dfrac{\sigma}{1+\sigma}(T_{11} + T_{22} + T_{33})\right]$$

$$= \dfrac{1+\sigma}{E}\left[-p + \dfrac{3p\sigma}{1+\sigma}\right] = -\dfrac{1-2\sigma}{E} p$$

$$= -\dfrac{1 - \dfrac{\lambda}{\lambda+\mu}}{\mu(3\lambda+2\mu)} = -\dfrac{1}{3}\dfrac{1}{\lambda + \dfrac{2}{3}\mu} = \dfrac{-p}{3k}.$$

Similarly, $\dfrac{\partial u_2}{\partial x_2} = e_{22} = \dfrac{-p}{3k}$ and $\dfrac{\partial u_3}{\partial x_3} = e_{33} = \dfrac{-p}{3k}$, therefore

$$u_3(x_1, x_2, x_3) = \dfrac{-p}{3k} x_3 + u_3^0(x_1, x_2),$$

where, $u_3^0(x_1, x_2)$ is an arbitrary function. Now

$$e_{23} = \left(\dfrac{\partial u_2}{\partial x_3} + \dfrac{\partial u_3}{\partial x_2}\right) = \dfrac{1+\sigma}{E}\left[T_{23} - \dfrac{\sigma}{1+\sigma}\Theta\delta_{23}\right] = 0$$

$$e_{31} = \left(\dfrac{\partial u_3}{\partial x_1} + \dfrac{\partial u_1}{\partial x_3}\right) = \dfrac{1+\sigma}{E}\left[T_{13} - \dfrac{\sigma}{1+\sigma}\Theta\delta_{13}\right] = 0$$

$$e_{12} = \left(\dfrac{\partial u_1}{\partial x_2} + \dfrac{\partial u_2}{\partial x_1}\right) = \dfrac{1+\sigma}{E}\left[T_{12} - \dfrac{\sigma}{1+\sigma}\Theta\delta_{12}\right] = 0$$

262   Chapter 5   Linear Elasticity

Thus, from the equation of e_{31}, we get

$$\frac{\partial u_3}{\partial x_1} + \frac{\partial u_1}{\partial x_3} = 0 \Rightarrow \frac{\partial u_1}{\partial x_3} = -\frac{\partial u_3}{\partial x_1} = -\frac{\partial u_3^0}{\partial x_1}$$

$$\Rightarrow u_1(x_1, x_2, x_3) = -\frac{\partial u_3^0}{\partial x_1} x_3 + u_1^0(x_1, x_2),$$

where, $u_1^0(x_1, x_2)$ is an arbitrary function. Now,

$$\frac{\partial u_1}{\partial x_1} = \frac{-p}{3k} \Rightarrow -\frac{\partial^2 u_3^0}{\partial x_1^2} x_3 + \frac{\partial u_1^0}{\partial x_1} = \frac{-p}{3k}$$

$$\Rightarrow \frac{\partial^2 u_3^0}{\partial x_1^2} = 0 \text{ and } \frac{\partial u_1^0}{\partial x_1} = \frac{-p}{3k} \Rightarrow u_1^0 = \frac{-p}{3k} x_1 + f_1(x_2)$$

where, $f_1(x_2)$ is an arbitrary function. Thus,

$$u_1 = -\frac{\partial u_3^0}{\partial x_1} x_3 - \frac{p}{3k} x_1 + f_1(x_2).$$

Also, from the equation of e_{23}, we get

$$\frac{\partial u_2}{\partial x_3} + \frac{\partial u_3}{\partial x_2} = 0 \Rightarrow \frac{\partial u_2}{\partial x_3} = -\frac{\partial u_3}{\partial x_2} = -\frac{\partial u_3^0}{\partial x_2}$$

$$\Rightarrow u_2(x_1, x_2, x_3) = -\frac{\partial u_3^0}{\partial x_2} x_3 + u_2^0(x_1, x_2),$$

where, $u_2^0(x_1, x_2)$ is an arbitrary function. Now,

$$\frac{\partial u_2}{\partial x_2} = \frac{-p}{3k} \Rightarrow -\frac{\partial^2 u_3^0}{\partial x_2^2} x_3 + \frac{\partial u_2^0}{\partial x_2} = \frac{-p}{3k}$$

$$\Rightarrow \frac{\partial^2 u_3^0}{\partial x_2^2} = 0 \text{ and } \frac{\partial u_2^0}{\partial x_2} = \frac{-p}{3k} \Rightarrow u_2^0 = \frac{-p}{3k} x_2 + f_2(x_1)$$

where, $f_2(x_1)$ is an arbitrary function. Lastly, using the relation of e_{12}, we get

$$\frac{\partial u_2}{\partial x_1} + \frac{\partial u_1}{\partial x_2} = 0$$

$$\Rightarrow -\frac{\partial^2 u_3^0}{\partial x_1 \partial x_2} x_3 + f_2'(x_1) - \frac{\partial^2 u_3^0}{\partial x_2 \partial x_1} x_3 + f_1'(x_2) = 0$$

$$\Rightarrow -2\frac{\partial^2 u_3^0}{\partial x_1 \partial x_2} x_3 + f_2'(x_1) + f_1'(x_2) = 0$$

5.5 Strains in terms of stresses

$$\Rightarrow -2\frac{\partial^2 u_3^0}{\partial x_1 \partial x_2}x_3 = 0 \text{ and } f_2'(x_1) + f_1'(x_2) = 0$$

$\Rightarrow u_3^0 =$ a linear function in x_1 and x_2 and $f_2'(x_1) = -f_1'(x_2)$

Let $u_3^0 = \alpha x_1 + \beta x_2 + c$ and $f_2'(x_1) = -f_1'(x_2) =$ constant $= \gamma$, say. Thus

$$f_2(x_1) = \gamma x_1 + b \text{ and } f_1(x_2) = -\gamma x_2 + a$$

Therefore, the displacement components u_1, u_2 and u_3 are given by

$$u_1(x_1, x_2, x_3) = \beta x_3 - \frac{p}{3k}x_1 - \gamma x_2 + a = -\frac{p}{3k}x_1 - \gamma x_2 + \beta x_3 + a$$

$$u_2(x_1, x_2, x_3) = -\alpha x_3 - \frac{p}{3k}x_2 + \gamma x_1 + b = -\frac{p}{3k}x_2 + \gamma x_1 - \alpha x_3 + b$$

$$u_3(x_1, x_2, x_3) = \alpha x_2 - \frac{p}{3k}x_3 - \beta x_1 + c = -\frac{p}{3k}x_3 - \beta x_1 + \alpha x_2 + c$$

where, α, β, γ and a, b, c correspond to rigid body motion. Thus if the displacement and rotation at $x_1 = 0, x_2 = 0, x_3 = 0$ are zero, then

$$u_1(x_1, x_2, x_3) = -\frac{p}{3k}x_1, u_2(x_1, x_2, x_3) = -\frac{p}{3k}x_2, u_3(x_1, x_2, x_3) = -\frac{p}{3k}x_3$$

$$\Rightarrow \frac{u_1}{x_1} = \frac{u_2}{x_2} = \frac{u_3}{x_3} = \frac{-p}{3k}$$

Example 5.4: When a bar is stretched by a tension such that $T_{11} = T_{22} = T_{33} = 0, T_{23} = \mu\alpha x_1, T_{31} = -\mu\alpha x_2, T_{12} = 0$, evaluate $u_1(x_1, x_2, x_3)$, $u_2(x_1, x_2, x_3)$ and $u_3(x_1, x_2, x_3)$.

Solution: Given the stress system $T_{11} = T_{22} = T_{33} = 0, T_{23} = \mu\alpha x_1$, $T_{31} = -\mu\alpha x_2$, $T_{12} = 0$. The strain-stress constitutive relations are

$$e_{ij} = \frac{T_{ij}}{2\mu} - \frac{\lambda\Theta}{2\mu(3\lambda + 2\mu)}\delta_{ij} = \frac{1+\sigma}{E}T_{ij} - \frac{\sigma}{E}\Theta\delta_{ij} = \frac{1+\sigma}{E}\left[T_{ij} - \frac{\sigma}{1+\sigma}\Theta\delta_{ij}\right]$$

where, the two constants E and σ are related with the Lame's constant by

$$E = \frac{\mu(3\lambda + 2\mu)}{(\lambda + \mu)}, \quad \sigma = \frac{\lambda}{2(\lambda + \mu)}$$

Therefore, the strain components are given by

$$e_{11} = \frac{T_{11}}{2\mu} - \frac{\lambda\Theta}{2\mu(3\lambda+2\mu)}\delta_{11} = \frac{1}{2\mu}\left[T_{11} - \frac{\lambda}{3\lambda+2\mu}(T_{11}+T_{22}+T_{33})\right]$$

$$= 0 = \frac{\partial u_1}{\partial x_1}; \text{ as } T_{11} = T_{22} = T_{33} = 0$$

$$e_{22} = \frac{T_{22}}{2\mu} - \frac{\lambda\Theta}{2\mu(3\lambda+2\mu)}\delta_{22} = \frac{1}{2\mu}\left[T_{22} - \frac{\lambda}{3\lambda+2\mu}(T_{11}+T_{22}+T_{33})\right]$$

$$= 0 = \frac{\partial u_2}{\partial x_2}; \text{ as } T_{11} = T_{22} = T_{33} = 0$$

$$e_{33} = \frac{T_{33}}{2\mu} - \frac{\lambda\Theta}{2\mu(3\lambda+2\mu)}\delta_{33} = \frac{1}{2\mu}\left[T_{33} - \frac{\lambda}{3\lambda+2\mu}(T_{11}+T_{22}+T_{33})\right]$$

$$= 0 = \frac{\partial u_3}{\partial x_3}; \text{ as } T_{11} = T_{22} = T_{33} = 0$$

$$e_{23} = \frac{T_{23}}{2\mu} - \frac{\lambda\,\Theta}{2\mu(3\lambda+2\mu)}\delta_{23} = \frac{1}{2\mu}\left[T_{23} - \frac{\lambda}{3\lambda+2\mu}(T_{11}+T_{22}+T_{33}).0\right]$$

$$= \alpha x_1 = \frac{1}{2}\left(\frac{\partial u_2}{\partial x_3} + \frac{\partial u_3}{\partial x_2}\right); \text{ as } \delta_{23} = 0 \text{ and } T_{23} = \mu\alpha x_1$$

$$e_{31} = \frac{T_{31}}{2\mu} - \frac{\lambda\,\Theta}{2\mu(3\lambda+2\mu)}\delta_{31} = \frac{1}{2\mu}\left[T_{31} - \frac{\lambda}{3\lambda+2\mu}(T_{11}+T_{22}+T_{33}).0\right]$$

$$= -\alpha x_2 = \frac{1}{2}\left(\frac{\partial u_1}{\partial x_3} + \frac{\partial u_3}{\partial x_1}\right); \text{ as } \delta_{31} = 0 \text{ and } T_{31} = -\mu\alpha x_2$$

$$e_{12} = \frac{T_{12}}{2\mu} - \frac{\lambda\,\Theta}{2\mu(3\lambda+2\mu)}\delta_{12} = \frac{1}{2\mu}\left[T_{12} - \frac{\lambda}{3\lambda+2\mu}(T_{11}+T_{22}+T_{33}).0\right]$$

$$= 0 = \frac{1}{2}\left(\frac{\partial u_1}{\partial x_2} + \frac{\partial u_2}{\partial x_1}\right); \text{ as } \delta_{12} = 0 \text{ and } T_{12} = 0$$

Since $e_{33} = 0 = \dfrac{\partial u_3}{\partial x_3}$, the solution gives

$$u_3(x_1, x_2, x_3) = u_3^0(x_1, x_2),$$

5.5 Strains in terms of stresses

where, $u_3^0(x_1, x_2)$ is an arbitrary function.
Now,

$$e_{31} = -\alpha x_2 = \frac{1}{2}\left(\frac{\partial u_1}{\partial x_3} + \frac{\partial u_3}{\partial x_1}\right) \Rightarrow \frac{\partial u_1}{\partial x_3} = -\alpha x_2 - \frac{\partial u_3}{\partial x_1} = -\alpha x_2 - \frac{\partial u_3^0}{\partial x_1}$$

$$\Rightarrow u_1(x_1, x_2, x_3) = -\alpha x_2 x_3 - \frac{\partial u_3^0}{\partial x_1} x_3 + u_1^0(x_1, x_2),$$

where, $u_1^0(x_1, x_2)$ is an arbitrary function. Now,

$$e_{23} = \alpha x_1 = \frac{1}{2}\left(\frac{\partial u_2}{\partial x_3} + \frac{\partial u_3}{\partial x_2}\right) \Rightarrow \frac{\partial u_2}{\partial x_3} = \alpha x_1 - \frac{\partial u_3}{\partial x_2} = \alpha x_1 - \frac{\partial u_3^0}{\partial x_2}$$

$$\Rightarrow u_2(x_1, x_2, x_3) = \alpha x_1 x_3 - \frac{\partial u_3^0}{\partial x_2} x_3 + u_2^0(x_1, x_2),$$

where, $u_2^0(x_1, x_2)$ is an arbitrary function. Now,

$$e_{11} = 0 = \frac{\partial u_1}{\partial x_1} \Rightarrow \frac{\partial^2 u_3^0}{\partial x_1^2} x_3 + \frac{\partial u_1^0}{\partial x_1} = 0$$

$$\Rightarrow \frac{\partial^2 u_3^0}{\partial x_1^2} = 0 \text{ and } \frac{\partial u_1^0}{\partial x_1} = 0 \Rightarrow u_1^0 = f_1(x_2)$$

where, $f_1(x_2)$ is an arbitrary function. Thus,

$$u_1(x_1, x_2, x_3) = -\frac{\partial u_3^0}{\partial x_1} x_3 + f_1(x_2).$$

Also, from the equation of e_{22} and $u_2(x_1, x_2, x_3)$ we get

$$\frac{\partial u_2}{\partial x_2} = 0 \Rightarrow -\frac{\partial^2 u_3^0}{\partial x_2^2} x_3 + \frac{\partial u_2^0}{\partial x_2} = 0$$

$$\Rightarrow \frac{\partial^2 u_3^0}{\partial x_2^2} = 0 \text{ and } \frac{\partial u_2^0}{\partial x_2} = 0 \Rightarrow u_2^0 = f_2(x_1)$$

where, $f_2(x_1)$ is an arbitrary function. Thus,

$$u_2(x_1, x_2, x_3) = -\frac{\partial u_3^0}{\partial x_2} x_3 + f_2(x_1).$$

Lastly, using the relation of e_{12}, we get

$$e_{12} = \frac{1}{2}\left(\frac{\partial u_1}{\partial x_2} + \frac{\partial u_2}{\partial x_1}\right) = 0 \Rightarrow -\frac{\partial^2 u_3^0}{\partial x_1 \partial x_2}x_3 + f_2'(x_1) - \frac{\partial^2 u_3^0}{\partial x_2 \partial x_1}x_3 + f_1'(x_2) = 0$$

$$\Rightarrow -2\frac{\partial^2 u_3^0}{\partial x_1 \partial x_2}x_3 + f_2'(x_1) + f_1'(x_2) = 0$$

$$\Rightarrow -2\frac{\partial^2 u_3^0}{\partial x_1 \partial x_2}x_3 = 0 \text{ and } f_2'(x_1) + f_1'(x_2) = 0$$

$\Rightarrow u_3^0 =$ a linear function in x_1 and x_2 and $f_2'(x_1) = -f_1'(x_2)$

Let $u_3^0 = \alpha_1 x_1 + \beta_1 x_2 + c$ and $f_2'(x_1) = -f_1'(x_2) =$ constant $= \gamma$, say. Thus

$$f_2(x_1) = \gamma x_1 + b \text{ and } f_1(x_2) = -\gamma x_2 + a$$

Therefore, the displacement components u_1, u_2 and u_3 are given by

$$u_1(x_1, x_2, x_3) = -\alpha x_2 x_3 - \alpha_1 x_3 - \gamma x_2 + a$$

$$u_2(x_1, x_2, x_3) = -\frac{\partial u_3^0}{\partial x_2}x_3 + f_2(x_1) = \alpha x_1 x_3 - \beta_1 x_3 + \gamma x_1 + b$$

$$u_3(x_1, x_2, x_3) = u_3^0(x_1, x_2), = \alpha_1 x_1 + \beta_1 x_2 + c$$

where, the parts of u_1, u_2, u_3 containing α_1, β_1, γ and a, b, c correspond to rigid body motion. Omitting the linear part of the displacements which correspond to the rigid body motion, we have

$$u_1(x_1, x_2, x_3) = -\alpha x_2 x_3, \ u_2(x_1, x_2, x_3) = \alpha x_1 x_3, \ u_3(x_1, x_2, x_3) = 0.$$

(ii) Pure Shear

We consider the state of stress characterized by the stress components given by

$$T_{23} = S = \text{a constant.}$$

$$T_{11} = T_{22} = T_{33} = T_{12} = T_{31} = 0 \tag{5.52}$$

The state of stress is possible in the deformation of a long rectangular parallelepiped of a square across section *OABC* which is sheared in the plane containing *OA* and *OC* by a shearing stress of magnitude *S*, acting per unit area on the side *CB*. The stress *S* will tend to slide the plane of the material originally perpendicular to *OC*. The axes x_2 and x_3 are taken along *OA* and *OC* respectively, x_1 axis being perpendicular to *OAC* plane by the traction *S*. The right angle ∠*AOC* will be diminished by an angle Φ (say). It is easily seen that the above stress system satisfies the equation of equilibrium under

5.5 Strains in terms of stresses

no body forces at each interior point and with the boundary condition on the surface.

Now, from the stress-strain relation, we have

$$T_{23} = 2\mu e_{23} = S; \quad T_{12} = T_{31} = 0. \tag{5.53}$$

Also, $0 = T_{11} = \lambda\theta + 2\mu e_{11}; \quad 0 = T_{22} = \lambda\theta + 2\mu e_{22}; \quad 0 = T_{33} = \lambda\theta + 2\mu e_{33}$

Adding these results, we get

$$3\lambda\theta + 2\mu\theta = 0, \quad i.e., \quad \theta = e_{11} + e_{22} + e_{33} = 0,$$

assuming $3\lambda + 2\mu \neq 0$. Therefore, the strains satisfy the compatibility equations. From the above equation, we get

$$e_{11} = e_{22} = e_{33} = 0; \quad e_{23} = \frac{S}{2\mu}, \quad e_{31} = e_{12} = 0 \tag{5.54}$$

Clearly, the strain system (5.44) is compatible and the state of stress given by the equation (5.53) actually corresponds to equation (5.32), $T_{ij} = \lambda\theta\delta_{ij} + 2\mu e_{ij}$ that can exists in a deformed parallelepiped. Also from the definition, we have

$$2e_{23} = \Phi_{23} = \text{change in angle}$$

Therefore,

$$\text{modulus of rigidity} = \frac{S}{\Phi_{23}} = \frac{S}{2e_{23}} = \mu \tag{5.55}$$

where, Φ_{23} is the change in the right angle between two linear elements parallel to x_2 and x_1 axis due to deformation. Thus the modulus of rigidity is identical with Lame's constant. μ = ratio of the shearing stress to the corresponding change in the angle Φ_{23} produced by the shearing stress. For this reason, μ is called shear modulus.

Example 5.5: **Displacement due to tension parallel to the axis of an elastic prism: When a bar is stretched by a tension such that** $T_{11} = T, T_{22} = T_{33} = T_{23} = T_{31} = T_{12} = 0$, **prove that**

$$\frac{u_1}{x_1} = \frac{(\lambda+\mu)T}{\mu(3\lambda+2\mu)}; \frac{u_2}{x_2} = \frac{u_3}{x_3} = -\frac{\lambda T}{2\mu(3\lambda+2\mu)}$$

assuming that displacement and rotation as $x_1 = 0, x_2 = 0, x_3 = 0$ **are zero.**

Solution: Given the stress system $T_{11} = T =$ a constant, and $T_{22} = T_{33} = T_{23} = T_{31} = T_{12} = 0$. The strain-stress constitutive relations are

$$e_{ij} = \frac{T_{ij}}{2\mu} - \frac{\lambda\Theta}{2\mu(3\lambda+2\mu)}\delta_{ij} = \frac{1+\sigma}{E}T_{ij} - \frac{\sigma}{E}\Theta\delta_{ij} = \frac{1+\sigma}{E}\left[T_{ij} - \frac{\sigma}{1+\sigma}\Theta\delta_{ij}\right]$$

where, the two constants E and σ are related with the Lame's constant by

$$E = \frac{\mu(3\lambda + 2\mu)}{(\lambda + \mu)}, \quad \sigma = \frac{\lambda}{2(\lambda + \mu)}$$

Therefore, the strain components are given by

$$e_{11} = \frac{T_{11}}{2\mu} - \frac{\lambda\Theta}{2\mu(3\lambda + 2\mu)}\delta_{11} = \frac{1}{2\mu}\left[T_{11} - \frac{\lambda}{3\lambda + 2\mu}(T_{11} + T_{22} + T_{33})\right]$$

$$= \frac{2(\lambda + \mu)}{2\mu(3\lambda + 2\mu)}T = \frac{T}{E} = \frac{\partial u_1}{\partial x_1}; \text{ as } T_{11} = T \text{ and } T_{22} = T_{33} = 0$$

$$e_{22} = \frac{T_{22}}{2\mu} - \frac{\lambda\Theta}{2\mu(3\lambda + 2\mu)}\delta_{22} = \frac{1}{2\mu}\left[T_{22} - \frac{\lambda}{3\lambda + 2\mu}(T_{11} + T_{22} + T_{33})\right]$$

$$= -\frac{\lambda}{2\mu(3\lambda + 2\mu)}T = -\frac{\sigma T}{E} = \frac{\partial u_2}{\partial x_2}; \text{ as } T_{11} = T \text{ and } T_{22} = T_{33} = 0$$

$$e_{33} = \frac{T_{33}}{2\mu} - \frac{\lambda\Theta}{2\mu(3\lambda + 2\mu)}\delta_{33} = \frac{1}{2\mu}\left[T_{33} - \frac{\lambda}{3\lambda + 2\mu}(T_{11} + T_{22} + T_{33})\right]$$

$$= -\frac{\lambda}{2\mu(3\lambda + 2\mu)}T = -\frac{\sigma T}{E} = \frac{\partial u_3}{\partial x_3}; \text{ as } T_{11} = T \text{ and } T_{22} = T_{33} = 0$$

$$e_{23} = \frac{T_{23}}{2\mu} - \frac{\lambda\,\Theta}{2\mu(3\lambda + 2\mu)}\delta_{23} = \frac{1}{2\mu}\left[T_{23} - \frac{\lambda}{3\lambda + 2\mu}(T_{11} + T_{22} + T_{33}).0\right]$$

$$= 0 = \frac{1}{2}\left(\frac{\partial u_2}{\partial x_3} + \frac{\partial u_3}{\partial x_2}\right); \text{ as } \delta_{23} = 0 \text{ and } T_{23} = 0$$

$$e_{31} = \frac{T_{31}}{2\mu} - \frac{\lambda\,\Theta}{2\mu(3\lambda + 2\mu)}\delta_{31} = \frac{1}{2\mu}\left[T_{31} - \frac{\lambda}{3\lambda + 2\mu}(T_{11} + T_{22} + T_{33}).0\right]$$

$$= 0 = \frac{1}{2}\left(\frac{\partial u_1}{\partial x_3} + \frac{\partial u_3}{\partial x_1}\right); \text{ as } \delta_{31} = 0 \text{ and } T_{31} = 0$$

$$e_{12} = \frac{T_{12}}{2\mu} - \frac{\lambda\,\Theta}{2\mu(3\lambda + 2\mu)}\delta_{12} = \frac{1}{2\mu}\left[T_{12} - \frac{\lambda}{3\lambda + 2\mu}(T_{11} + T_{22} + T_{33}).0\right]$$

$$= 0 = \frac{1}{2}\left(\frac{\partial u_1}{\partial x_2} + \frac{\partial u_2}{\partial x_1}\right); \text{ as } \delta_{12} = 0 \text{ and } T_{12} = 0$$

5.5 Strains in terms of stresses

Since $e_{33} = -\dfrac{\sigma T}{E} = \dfrac{\partial u_3}{\partial x_3}$, the solution gives

$$u_3(x_1, x_2, x_3) = -\dfrac{\sigma T}{E} x_3 + u_3^0(x_1, x_2),$$

where, $u_3^0(x_1, x_2)$ is an arbitrary function. Now,

$$e_{31} = 0 = \dfrac{1}{2}\left(\dfrac{\partial u_1}{\partial x_3} + \dfrac{\partial u_3}{\partial x_1}\right) \Rightarrow \dfrac{\partial u_1}{\partial x_3} = -\dfrac{\partial u_3}{\partial x_1} = -\dfrac{\partial u_3^0}{\partial x_1}$$

$$\Rightarrow u_1(x_1, x_2, x_3) = -\dfrac{\partial u_3^0}{\partial x_1} x_3 + u_1^0(x_1, x_2),$$

where, $u_1^0(x_1, x_2)$ is an arbitrary function. Now,

$$e_{23} = 0 = \dfrac{1}{2}\left(\dfrac{\partial u_2}{\partial x_3} + \dfrac{\partial u_3}{\partial x_2}\right) \Rightarrow \dfrac{\partial u_2}{\partial x_3} = -\dfrac{\partial u_3}{\partial x_2} = -\dfrac{\partial u_3^0}{\partial x_2}$$

$$\Rightarrow u_2(x_1, x_2, x_3) = -\dfrac{\partial u_3^0}{\partial x_2} x_3 + u_2^0(x_1, x_2),$$

where, $u_2^0(x_1, x_2)$ is an arbitrary function. Now,

$$e_{11} = \dfrac{T}{E} = \dfrac{\partial u_1}{\partial x_1} \Rightarrow \dfrac{\partial^2 u_3^0}{\partial x_1^2} x_3 + \dfrac{\partial u_1^0}{\partial x_1} = \dfrac{T}{E}$$

$$\Rightarrow \dfrac{\partial^2 u_3^0}{\partial x_1^2} = 0 \text{ and } \dfrac{\partial u_1^0}{\partial x_1} = \dfrac{T}{E} \Rightarrow u_1^0 = \dfrac{T}{E} x_1 + f_1(x_2)$$

where, $f_1(x_2)$ is an arbitrary function. Thus,

$$u_1(x_1, x_2, x_3) = -\dfrac{\partial u_3^0}{\partial x_1} x_3 + \dfrac{T}{E} x_1 + f_1(x_2).$$

Also, from the equation of e_{22} and $u_2(x_1, x_2, x_3)$ we get

$$\dfrac{\partial u_2}{\partial x_2} = -\dfrac{\sigma T}{E} \Rightarrow -\dfrac{\partial^2 u_3^0}{\partial x_2^2} x_3 + \dfrac{\partial u_2^0}{\partial x_2} = -\dfrac{\sigma T}{E}$$

$$\Rightarrow \dfrac{\partial^2 u_3^0}{\partial x_2^2} = 0 \text{ and } \dfrac{\partial u_2^0}{\partial x_2} = -\dfrac{\sigma T}{E} \Rightarrow u_2^0 = -\dfrac{\sigma T}{E} x_2 + f_2(x_1)$$

where, $f_2(x_1)$ is an arbitrary function. Thus,

$$u_2(x_1,x_2,x_3) = -\frac{\partial u_3^0}{\partial x_2}x_3 - \frac{\sigma T}{E}x_2 + f_2(x_1).$$

Lastly, using the relation of e_{12}, we get

$$e_{12} = \frac{1}{2}\left(\frac{\partial u_1}{\partial x_2} + \frac{\partial u_2}{\partial x_1}\right) = 0 \Rightarrow -\frac{\partial^2 u_3^0}{\partial x_1 \partial x_2}x_3 + f_2'(x_1) - \frac{\partial^2 u_3^0}{\partial x_2 \partial x_1}x_3 + f_1'(x_2) = 0$$

$$\Rightarrow -2\frac{\partial^2 u_3^0}{\partial x_1 \partial x_2}x_3 + f_2'(x_1) + f_1'(x_2) = 0$$

$$\Rightarrow -2\frac{\partial^2 u_3^0}{\partial x_1 \partial x_2}x_3 = 0 \text{ and } f_2'(x_1) + f_1'(x_2) = 0$$

$\Rightarrow u_3^0$ = a linear function in x_1 and x_2 and $f_2'(x_1) = -f_1'(x_2)$

Let $u_3^0 = -\alpha x_1 + \beta x_2 + c$ and $f_2'(x_1) = -f_1'(x_2) = $ constant $= \gamma$, say. Thus

$$f_2(x_1) = \gamma x_1 + b \text{ and } f_1(x_2) = -\gamma x_2 + a$$

Therefore, the displacement components u_1, u_2 and u_3 are given by

$$u_1(x_1,x_2,x_3) = \beta x_3 - \frac{T}{E}x_1 - \gamma x_2 + a = \frac{T}{E}x_1 - \gamma x_2 + \beta x_3 + a$$

$$u_2(x_1,x_2,x_3) = -\alpha x_3 - \frac{\sigma T}{E}x_2 + \gamma x_1 + b = \gamma x_1 - \frac{\sigma T}{E}x_2 - \alpha x_3 + b$$

$$u_3(x_1,x_2,x_3) = \alpha x_2 - \frac{\sigma T}{E}x_3 - \beta x_1 + c = -\frac{\sigma T}{E}x_3 - \beta x_1 + \alpha x_2 + c$$

where, the parts of u_1, u_2, u_3 containing α, β, γ and a, b, c correspond to rigid body motion. Thus for pure deformation, we must have

$$u_1(x_1,x_2,x_3) = \frac{T}{E}x_1; \ u_2(x_1,x_2,x_3) = -\frac{\sigma T}{E}x_2; \ u_3(x_1,x_2,x_3) = -\frac{\sigma T}{E}x_3$$

$$\Rightarrow \frac{u_1}{x_1} = \frac{(\lambda+\mu)T}{\mu(3\lambda+2\mu)}; \ \frac{u_2}{x_2} = \frac{u_3}{x_3} = -\frac{\lambda T}{2\mu(3\lambda+2\mu)}$$

(iii) Hydrostatic Pressure

We consider an elastic body of arbitrary shape subjected to a hydrostatic stress diminishes the volume of the body. It turns out that the ratio of the hydrostatic

5.5 Strains in terms of stresses

stress to the decrease in volume per unit volume is a constant of the material and is known as Bulk modulus or modulus of compression. It is denoted by K. To obtain K in terms of λ and μ we take the state of stress possible in such a deformed body as

$$T_{11} = T_{22} = T_{33} = -p = \text{constant}, \quad T_{23} = T_{31} = T_{12} = 0. \tag{5.56}$$

This state of stress satisfies equilibrium equation in the interior of the body. If $T_i^{(n)}$ be stress vector acting on the surface with normal n_i, then

$$T_i^{(n)} = T_{ij}n_j = -pn_i \text{ at every point on the surface.} \tag{5.57}$$

It is obvious that state of stress in equation (5.56) also satisfies the boundary condition on the surface. Substituting (5.56) into equation (5.33),

$$e_{11} = e_{22} = e_{33} = -\frac{p}{(3\lambda + 2\mu)} = \text{constant}, \quad e_{23} = e_{31} = e_{12} = 0. \tag{5.58}$$

Strains given by equation (5.58) clearly satisfies the equation of compatibility and state of stress given by the equation (5.56) actually corresponds to the one that can exist in a body. If θ be the cubical dilation, then the decrease in volume per unit volume $= -\theta$. Therefore

$$K = \frac{p}{-\theta} = \frac{p}{-e_{ii}} = \frac{p(3\lambda + 2\mu)}{3p} = \frac{3\lambda + 2\mu}{3} = \lambda + \frac{2\mu}{3} \tag{5.59}$$

For isotropic elastic materials, any two elastic constants completely define the material's response. In addition to that, any elastic constant can be determined in terms of any two other constants.

(a) We can express the Bulk modulus K in terms of E and σ as

$$K = \lambda + \frac{2\mu}{3} = \frac{E\sigma}{(1+\sigma)(1-2\sigma)} + \frac{2}{3} \cdot \frac{E}{2(1+\sigma)} = \frac{E}{3(1-2\sigma)} \tag{5.60}$$

For, $K > 0$, $E > 0$ we have from equation (5.60) $\sigma < \frac{1}{2}$, and $\sigma > 0$ and so $0 < \sigma < \frac{1}{2}$. Since $0 < \sigma < \frac{1}{2}$, and $E > 0$, we get from equation (5.51)

$$\lambda = \frac{E\sigma}{(1+\sigma)(1-2\sigma)} > 0 \text{ and } \mu = \frac{E}{2(1+\sigma)} > 0$$

(b) If the material is incompressible

$$\theta = 0, K \to \infty, \sigma = \frac{1}{2}, \mu = \frac{E}{3}.$$

(c) For many solids and rocks $\lambda = \mu$, this leads to

$$K = \frac{5\mu}{3}, \quad E = \frac{5\mu}{2}$$

(d) We can obtain simpler form of stress-strain law. From equation (5.51), we get

$$\lambda = \frac{E\sigma}{(1+\sigma)(1-2\sigma)}, \quad \mu = \frac{E}{2(1+\sigma)}$$

$$\Rightarrow \sigma = \frac{\lambda}{2(\lambda + \mu)}, \quad E = \frac{\mu(3\lambda + 2\mu)}{\lambda + \mu}$$

$$\Rightarrow \frac{1+\sigma}{E} = \frac{1}{2\mu}, \quad \frac{\sigma}{E} = \frac{\lambda}{\mu(3\lambda + 2\mu)}$$

Therefore, equation (5.33) becomes

$$e_{ij} = \frac{T_{ij}}{2\mu} - \frac{\lambda \Theta}{2\mu(3\lambda + 2\mu)}\delta_{ij} = \frac{1+\sigma}{E}T_{ij} - \frac{\sigma}{E}\Theta\delta_{ij} \qquad (5.61)$$

Equation (5.61) is the simpler form of stress-strain relation.

(e) Now,

$$\theta = e_{ii} = \frac{1+\sigma}{E}T_{ii} - \frac{\sigma}{E}\Theta\delta_{ii} = \frac{1+\sigma}{E}\Theta - \frac{\sigma}{E}\Theta 3 = \frac{1-2\sigma}{E}\Theta = \frac{\Theta}{3K} \qquad (5.62)$$

Therefore, $K\theta = \frac{\Theta}{3}$ and thus dilation depends on average normal stress.

(f) We can also express strain-energy function in terms of stresses only. Now from the Clapeyron's formula

$$W = \frac{1}{2}T_{ij}e_{ij} = \frac{1}{2}T_{ij}\left[\frac{1+\sigma}{E}T_{ij} - \frac{\sigma}{E}\Theta\delta_{ij}\right] = \frac{1+\sigma}{2E}T_{ij}T_{ij} - \frac{\sigma\Theta^2}{2E}$$

$$= \frac{1+\sigma}{2E}\left[T_{11}^2 + T_{22}^2 + T_{33}^2 + 2T_{23}^2 + 2T_{31}^2 + 2T_{12}^2\right] - \frac{\sigma}{2E}\left[T_{11} + T_{22} + T_{33}\right]^2 \qquad (5.63)$$

(g) **Relationship among elastic constants in an isotropic material:** Young's modulus and Poisson's ratio are the most common properties used to characterize elastic solids, but other measures are also used. For example, we define the shear modulus, bulk modulus and Lame's modulus, of an elastic solid as follows:

Bulk modulus $K = \dfrac{E}{3(1-2\gamma)}$, Shear modulus $\mu = \dfrac{E}{2(1+\gamma)}$,

5.5 Strains in terms of stresses

$$\text{Lame modulus } \lambda = \frac{\gamma E}{(1+\gamma)(1-2\gamma)}$$

The table relating to all the possible combinations of moduli to all other possible combinations is given below:

Table 5.1: Relationship among elastic constants in an isotropic material

	Bulk Modulus K	Young's Modulus E	Lame's Modulus λ	Poisson Ratio γ	Shear Modulus μ
λ, μ	$\lambda + \dfrac{2\mu}{3}$	$\mu\dfrac{3\lambda + 2\mu}{\lambda + \mu}$	—	$\dfrac{\lambda}{2(\lambda + \mu)}$	—
λ, E	Irrotational			Irrotational	Irrotational
λ, K	—	$9K\dfrac{K - \lambda}{3K - \lambda}$	—	$\dfrac{\lambda}{3K - \lambda}$	$3\dfrac{K - \lambda}{2}$
μ, K	—	$\dfrac{9K\mu}{3K + \mu}$	$K - \dfrac{2\mu}{3}$	$\dfrac{3K - 2\mu}{2(3K + \mu)}$	—
μ, E	$\dfrac{E\mu}{3(3\mu - E)}$	—	$\mu\dfrac{E - 2\mu}{3\mu - E}$	$\dfrac{E}{2\mu} - 1$	—
E, K			$3K\dfrac{3K - E}{9K - E}$	$\dfrac{3K - E}{6K}$	$\dfrac{3KE}{9K - E}$
μ, γ	$\dfrac{2\mu(1+\gamma)}{3(1-2\gamma)}$	$2\mu(1+\gamma)$	$\dfrac{2\mu\gamma}{1-2\gamma}$		
E, γ	$\dfrac{E}{3(1-2\gamma)}$		$\dfrac{\gamma E}{(1+\gamma)(1-2\gamma)}$		$\dfrac{E}{2(1+\gamma)}$
K, γ		$3K(1-2\gamma)$	$\dfrac{3K\gamma}{(1+\gamma)}$		$\dfrac{3K(1-2\gamma)}{2(1+\gamma)}$
λ, γ	$\lambda\dfrac{1+\gamma}{3\gamma}$	$\lambda\dfrac{(1+\gamma)(1-2\gamma)}{3\gamma}$			$\lambda\dfrac{1-2\gamma}{2\gamma}$

Example 5.6: Test whether the following system of strain components is possible in an elastic body:

$$e_{xx} = k(x^2 + y^2); \quad e_{yy} = k(y^2 + z^2); \quad e_{xy} = kxyz; \quad e_{yz} = e_{zx} = e_{xy} = e_{zz} = 0; \quad k \neq 0.$$

Solution:

$$\frac{\partial e_{xx}}{\partial y} = 2ky; \quad \frac{\partial^2 e_{xx}}{\partial y^2} = 2k; \quad \frac{\partial e_{yy}}{\partial x} = 0; \quad \frac{\partial^2 e_{yy}}{\partial x^2} = 0$$

$$\frac{\partial^2 e_{xy}}{\partial x \partial y} = \frac{\partial}{\partial x}\left(\frac{\partial e_{xy}}{\partial y}\right) = \frac{\partial}{\partial x}(kxz) = kz$$

$$\frac{\partial^2 e_{xx}}{\partial y^2} + \frac{\partial^2 e_{yy}}{\partial x^2} \neq \frac{\partial^2 e_{xy}}{\partial x \partial y}$$

Therefore, the compatibility equations are not all satisfied and hence, the system of strain components is not possible in an elastic body.

5.6 Fundamental Boundary Value Problem in Elastostatics

In elastostatics, when linearly elastic solid is in equilibrium the shape of the body and the distribution of external body forces throughout the material are given, the problem is to find 15 unknowns: 6 stresses T_{ij}, 6 strains e_{ij} and 3 displacement functions u_i which satisfy the basic 15 appropriate field equations:

(*i*) Three equations of equilibrium:

$$T_{ij} + \rho F_i = 0; \quad i = 1, 2, 3,$$

(*ii*) Six stress-strain constitutive relations:

$$T_{ij} = \lambda \Theta \delta_{ij} + 2\mu e_{ij}; \quad \Theta = e_{kk}; \quad i = 1, 2, 3, \quad j = 1, 2, 3$$

or six equivalent strain-stress constitutive relations

$$e_{ij} = \frac{1+\sigma}{E} T_{ij} - \frac{\sigma}{E} \Theta \delta_{ij} = \frac{1+\sigma}{E}\left[T_{ij} - \frac{\sigma}{1+\sigma} \Theta \delta_{ij}\right].$$

(*iii*) and six strain-displacement kinematic relations:

$$e_{ij} = \frac{1}{2}\left[u_{i,j} + u_{j,i}\right]$$

at all interior points of linearly elastic body. The solution of these 15 partial differential equations will contain arbitrary functions. Also, prescribed conditions on stress and/or displacements must be satisfied on the bounding surface of the body. The boundary value problems of elasticity are usually classified according to boundary conditions into problems for which

(*i*) displacements are prescribed everywhere on the boundary,
(*ii*) stresses (surface tractions) are prescribed everywhere on the boundary,

5.6 Fundamental boundary value problem in elastostatics

(*iii*) displacements are prescribed over a portion of the boundary, stresses are prescribed over the remaining part.

For all three categories the body forces are assumed to be given throughout the continuum. In order to determine unique solution we have to use these arbitrary functions by using a set of boundary conditions. On the bounding surface of the body either stress vectors or displacements are prescribed everywhere in the shape of boundary conditions.

(a) The stress vector is given at each point of the boundary: When the functions $f_i(x_1,x_2,x_3)$, are prescribed on the boundary surface of the body representing the stress vector acting on surface element with normal n_i the stresses T_{ij}, in addition, must satisfy 3 boundary conditions

$$T_{ij} n_j = f_i(x_1,x_2,x_3); \quad i=1,2,3. \tag{5.64}$$

Thus the problem of obtaining the displacements, strains and stresses in linearly elastic isotropic solid body in equilibrium which satisfy basic equations

(*i*) Three equations of equilibrium: $T_{ij} + \rho F_i = 0; \quad i=1,2,3$,

(*ii*) Six strain-stress constitutive relations

$$e_{ij} = \frac{1+\sigma}{E} T_{ij} - \frac{\sigma}{E} \Theta \delta_{ij} = \frac{1+\sigma}{E}\left[T_{ij} - \frac{\sigma}{1+\sigma}\Theta \delta_{ij}\right]$$

(*iii*) and six strain-displacement kinematic relations: $e_{ij} = \frac{1}{2}\left[u_{i,j} + u_{j,i}\right]$
in addition the boundary condition (5.64) is known as the first fundamental boundary value problem in elastostatics.

(b) The displacement is prescribed at each point of the boundary:
When the functions $g_i(x_1,x_2,x_3)$ are prescribed on the boundary of the body representing the displacements, the displacements u_i, in addition, must satisfy 3 boundary conditions

$$u_i = g_i(x_1,x_2,x_3); i=1,2,3 \tag{5.65}$$

The problem of obtaining the displacements strains and stresses which satisfy the basic equations

(*i*) Three equations of equilibrium: $T_{ij} + \rho F_i = 0; i=1,2,3$,

(*ii*) Six strain-stress constitutive relations

$$e_{ij} = \frac{1+\sigma}{E} T_{ij} - \frac{\sigma}{E} \Theta \delta_{ij} = \frac{1+\sigma}{E}\left[T_{ij} - \frac{\sigma}{1+\sigma}\Theta \delta_{ij}\right]$$

(*iii*) and six strain-displacement kinematic relations: $e_{ij} = \frac{1}{2}\left[u_{i,j} + u_{j,i}\right]$
and in addition the boundary condition (5.65) is known as the second fundamental boundary value problem in elastostatics.

(c) Beltrami-Michell compatibility equations for stress components: Let us consider those problems in which surface tractions are prescribed everywhere on the boundary. The equations of compatibility for strains

$$e_{ij,kl} + e_{kl,ij} - e_{ik,jl} - e_{jl,ik} = 0 \tag{5.66}$$

Putting $l = k$ and summing over k, we get

$$e_{ij,kk} + e_{kk,ij} - e_{ik,jk} - e_{jk,ik} = 0 \tag{5.67}$$

From the stress-strain relation for isotropic elastic body

$$e_{ij} = \frac{1+\sigma}{E} T_{ij} - \frac{\sigma}{E} \Theta \delta_{ij} = \frac{1+\sigma}{E}\left[T_{ij} - \frac{\sigma}{1+\sigma} \Theta \delta_{ij} \right]; \tag{5.68}$$

where, $\sigma = \dfrac{\lambda}{2(\lambda+\mu)}$, $E = \dfrac{\mu(3\lambda+2\mu)}{\lambda+\mu}$. Using the results $T_{ij,kk} = \dfrac{\partial^2 T_{ij}}{\partial x_k^2} = \nabla^2 T_{ij}$, $T_{kk,ij} = \dfrac{\partial T_{kk}}{\partial x_i \partial x_j} = \Theta_{,ij}$, from equation (5.68), we get

$$e_{ij,kk} = \frac{1+\sigma}{E}\left[T_{ij,kk} - \frac{\sigma}{1+\sigma} \Theta_{,kk} \delta_{ij} \right] = \frac{1+\sigma}{E}\left[\nabla^2 T_{ij} - \frac{\sigma}{1+\sigma} \Theta_{,kk} \delta_{ij} \right]$$

$$e_{kk,ij} = \frac{1+\sigma}{E}\left[T_{kk,ij} - \frac{\sigma}{1+\sigma} \Theta_{,ij} \delta_{kk} \right] = \frac{1+\sigma}{E}\left[\Theta_{,ij} - \frac{3\sigma}{1+\sigma} \Theta_{,ij} \right]$$

$$e_{ik,jk} = \frac{1+\sigma}{E}\left[T_{ik,jk} - \frac{\sigma}{1+\sigma} \Theta_{,jk} \delta_{ik} \right]; \; e_{jk,ik} = \frac{1+\sigma}{E}\left[T_{jk,ik} - \frac{\sigma}{1+\sigma} \Theta_{,ik} \delta_{jk} \right]$$

Using these relations, the equation (5.67) reduces to

$$\frac{1+\sigma}{E}\left[\nabla^2 T_{ij} - \frac{\sigma}{1+\sigma} \Theta_{,kk} \delta_{ij} + \Theta_{,ij} - \frac{3\sigma}{1+\sigma} \Theta_{,ij} - T_{ik,jk} + \frac{\sigma}{1+\sigma} \Theta_{,jk} \delta_{ik} - T_{jk,ik} \right.$$
$$\left. + \frac{\sigma}{1+\sigma} \Theta_{,ik} \delta_{jk} \right] = 0$$

$$\Rightarrow \nabla^2 T_{ij} + \Theta_{,ij} - T_{ik,jk} - T_{jk,ik} = \frac{\sigma}{1+\sigma}[\Theta_{,kk}\delta_{ij} + 3\Theta_{,ij} - \Theta_{,jk}\delta_{ik} - \Theta_{,ik}\delta_{jk}]$$

$$\Rightarrow \nabla^2 T_{ij} + \Theta_{,ij} - T_{ik,jk} - T_{jk,ik} = \frac{\sigma}{1+\sigma}[\Theta_{,kk}\delta_{ij} + 3\Theta_{,ij} - \Theta_{,ji} - \Theta_{,ij}]$$

$$\Rightarrow \nabla^2 T_{ij} + \left[1 - \frac{3\sigma}{1+\sigma} + \frac{2\sigma}{1+\sigma}\right]\Theta_{,ij} - \frac{\sigma}{1+\sigma}\nabla^2\Theta\delta_{ij} = T_{ik,jk} + T_{jk,ik}$$

$$\Rightarrow \nabla^2 T_{ij} + \frac{1}{1+\sigma}\Theta_{,ij} - \frac{\sigma}{1+\sigma}\delta_{ij}\nabla^2\Theta = T_{ik,jk} + T_{jk,ik}. \tag{5.69}$$

5.6 Fundamental boundary value problem in elastostatics

The solution for this type of problem is given by specifying the stress tensor which satisfies (5.69) throughout the continuum and fulfills traction condition prescribed everywhere on the boundary. Now from the stress equation of equilibrium we have, $T_{ik,k} + \rho F_i = 0$, where, F_i's are the body force components per unit mass, differentiating this with respect to x_j, we get

$$T_{ik,jk} + \rho F_{i,j} = 0, \text{ i.e., } T_{ik,jk} = -\rho F_{i,j}$$

Similarly, differentiating this with respect to x_i, we get $T_{jk,ik} = -\rho F_{j,i}$. Putting these values in equation (5.69), we get

$$\nabla^2 T_{ij} + \frac{1}{1+\sigma}\Theta_{,ij} - \frac{\sigma}{1+\sigma}\delta_{ij}\nabla^2\Theta = -\rho\left[F_{i,j} + F_{j,i}\right] \qquad (5.70)$$

Putting $j = i$ and summing over i we get

$$\nabla^2 T_{ii} + \frac{1}{1+\sigma}\Theta_{,ii} - \frac{\sigma}{1+\sigma}\delta_{ii}\nabla^2\Theta = -\rho\left[F_{i,i} + F_{i,i}\right]$$

$$\nabla^2\Theta + \frac{1}{1+\sigma}\nabla^2\Theta - \frac{3\sigma}{1+\sigma}\nabla^2\Theta = -2\rho F_{i,i}$$

$$\nabla^2\Theta = -\frac{1+\sigma}{1-\sigma}\rho\, div\, \vec{F} \qquad (5.71)$$

Using equation (5.71) in equation (5.70) we get

$$\nabla^2 T_{ij} + \frac{1}{1+\sigma}\Theta_{,ij} = -\frac{\sigma}{1-\sigma}\rho\delta_{ij} div\,\vec{F} - \rho\left[F_{i,j} + F_{j,i}\right] \qquad (5.72)$$

The equation (5.72) contains 6 independent equations. They are known as the Beltrami-Michell compatibility equation for stresses. These equations for stresses are suitable only for isotropic elastic body whereas Saint Venant's compatibility equations for strain $T_{ik,k} + \rho F_i = 0$ are suitable for any body.

Thus, we have seen that the first boundary value problem is reduced to the determination of six stresses which must satisfy

(i) the three equations of equilibrium

$$T_{ik,k} + \rho F_i = 0; \quad i = 1, 2, 3$$

(ii) six compatibility equations

$$\nabla^2 T_{ij} + \frac{1}{1+\sigma}\Theta_{,ij} = -\frac{\sigma}{1-\sigma}\rho\delta_{ij}\vec{F} - \rho\left[F_{i,j} + F_{j,i}\right]$$

instead of 15 subject to boundary condition $T_{ij}n_j = f_i(x_1, x_2, x_3)$.

Once stresses T_{ij} are known, the strains c_{ij} may be determined from strain-displacement kinematic relations and then the displacements u_i can be obtained by integrating: $e_{ij} = \frac{1}{2}\left[u_{i,j} + u_{j,i}\right]$ and they must be single valued.

In combination with the equilibrium equations, these equations comprise a system for the solution of the stress components, but it is not an especially easy system to solve. As was the case with the infinitesimal strain equation of compatibility, the body must be simply connected.

Result 5.1 If the body forces are constants, the invariants Θ and θ are harmonic functions and the stress components T_{ij} and strain components e_{ij} are biharmonic functions.

Proof: The Beltrami-Michell compatibility equation for stresses are given by

$$\nabla^2 T_{ij} + \frac{1}{1+\sigma}\Theta_{,ij} = -\frac{\sigma}{1-\sigma}\rho\delta_{ij}\text{div}\vec{F} - \rho\left[F_{i,j} + F_{j,i}\right]$$

If $\vec{F}$ be a constant vector, then $\text{div}\vec{F} = \vec{0}$, $F_{i,j} = 0 = F_{j,i}$. Therefore, the above equation becomes

$$\nabla^2 T_{ij} + \frac{1}{1+\sigma}\Theta_{,ij} = 0$$

Putting $j = i$ and summing over i we get

$$\nabla^2 T_{ii} + \frac{1}{1+\sigma}\Theta_{,ii} = 0 \Rightarrow \nabla^2\Theta + \frac{1}{1+\sigma}\nabla^2\Theta = 0$$

$$\Rightarrow \frac{2+\sigma}{1+\sigma}\nabla^2\Theta = 0 \Rightarrow \nabla^2\Theta = 0$$

which shows that Θ is harmonic. Since, $\Theta = (3\lambda + 2\mu)\theta$ we have

$$\nabla^2\Theta = \nabla^2(3\lambda + 2\mu)\theta = 0 \Rightarrow \nabla^2\theta = 0$$

which shows that θ is harmonic also. Operating on both sides by ∇^2, on the Beltrami-Michell compatibility equation for stresses, we get

$$\nabla^4 T_{ij} + \frac{1}{1+\sigma}(\nabla^2\Theta)_{,ij} = 0 \Rightarrow \nabla^4 T_{ij} = 0,$$

which shows that the stress component T_{ij} is biharmonic. Using the stress-strain constitutive relations $T_{ij} = \lambda\theta\delta_{ij} + 2\mu e_{ij}$, for isotropic elastic body, we get, operating on both sides by ∇^2,

$$\nabla^2 T_{ij} = \lambda\nabla^2\theta\delta_{ij} + 2\mu\nabla^2 e_{ij} = 2\mu\nabla^2 e_{ij}; \text{ as } \nabla^2\theta = 0$$

5.6 Fundamental boundary value problem in elastostatics

Again operating on both sides by ∇^2,

$$\nabla^4 T_{ij} = 2\mu \nabla^4 e_{ij} = 0; \text{ as } \nabla^4 T_{ij} = 0$$

Hence

$$2\mu \nabla^4 e_{ij} = 0 \Rightarrow \nabla^4 e_{ij} = 0,$$

which shows that the strain components e_{ij} is biharmonic.

Example 5.7 Show that the following stress components are not solutions of the problem in elasticity, even though they satisfy the equations of equilibrium with zero body forces:

$$T_{11} = \alpha \left[x_2^2 + \sigma \left(x_1^2 - x_2^2 \right) \right]; \quad T_{12} = -2\alpha \, \sigma x_1 x_2; \quad T_{13} = 0$$

$$T_{22} = \alpha \left[x_1^2 + \sigma \left(x_2^2 - x_1^2 \right) \right]; \quad T_{23} = 0; \quad T_{33} = \alpha \, \sigma \left(x_1^2 + x_2^2 \right).$$

Solution: The quantity $\Theta = T_{kk} = T_{11} + T_{22} + T_{33}$ is given by

$$\Theta = T_{kk} = \alpha \left[x_2^2 + \alpha \left(x_1^2 - x_2^2 \right) \right] + \alpha \left[x_1^2 + \sigma \left(x_2^2 - x_1^2 \right) \right] + \alpha \, \sigma \left(x_1^2 + x_2^2 \right)$$

$$= \alpha \left(x_1^2 + x_2^2 \right) + \alpha \, \sigma \left(x_1^2 + x_2^2 \right) = \alpha (1+\sigma) \left(x_1^2 + x_2^2 \right)$$

In absence of body forces, Beltrami-Michell compatibility equation (5.72), containing 6 independent equations for stresses are

$$\nabla^2 T_{ij} + \frac{1}{1+\sigma} \Theta_{,ij} = -\frac{\sigma}{1-\sigma} \rho \delta_{ij} \mathrm{div}\vec{F} - \rho \left[F_{i,j} + F_{j,i} \right] = 0.$$

For $i = 1,2,3; \ j = 1,2,3$ this equation becomes

$$\nabla^2 T_{11} + \frac{1}{1+\sigma} \Theta_{,11} = \frac{\partial^2 T_{11}}{\partial x_1^2} + \frac{\partial^2 T_{11}}{\partial x_2^2} + \frac{\partial^2 T_{11}}{\partial x_3^2} + \frac{1}{1+\sigma} \frac{\partial^2 \Theta}{\partial x_1^2}$$

$$= 2\alpha\sigma + (2\alpha - 2\alpha\sigma) + 0 + \frac{1}{1+\sigma} 2\alpha(1+\sigma) = 4\alpha \neq 0.$$

$$\nabla^2 T_{22} + \frac{1}{1+\sigma} \Theta_{,22} = \frac{\partial^2 T_{22}}{\partial x_1^2} + \frac{\partial^2 T_{22}}{\partial x_2^2} + \frac{\partial^2 T_{22}}{\partial x_3^2} + \frac{1}{1+\sigma} \frac{\partial^2 \Theta}{\partial x_2^2}$$

$$= (2\alpha - 2\alpha\sigma) + 2\alpha\sigma + 0 + \frac{1}{1+\sigma} 2\alpha(1+\sigma) = 4\alpha \neq 0.$$

$$\nabla^2 T_{33} + \frac{1}{1+\sigma}\Theta_{,33} = \frac{\partial^2 T_{33}}{\partial x_1^2} + \frac{\partial^2 T_{33}}{\partial x_2^2} + \frac{\partial^2 T_{33}}{\partial x_3^2} + \frac{1}{1+\sigma}\frac{\partial^2 \Theta}{\partial x_3^2}$$

$$= 2\alpha\sigma + 2\alpha\sigma + 0 + \frac{1}{1+\sigma}.0 = 4\alpha\sigma \neq 0.$$

$$\nabla^2 T_{12} + \frac{1}{1+\sigma}\Theta_{,12} = \frac{\partial^2 T_{12}}{\partial x_1^2} + \frac{\partial^2 T_{12}}{\partial x_2^2} + \frac{\partial^2 T_{12}}{\partial x_3^2} + \frac{1}{1+\sigma}\frac{\partial^2 \Theta}{\partial x_1 \partial x_2}$$

$$= 0 + 0 + 0 + \frac{1}{1+\sigma}2\alpha(1+\sigma).0 = 0.$$

$$\nabla^2 T_{13} + \frac{1}{1+\sigma}\Theta_{,13} = \frac{\partial^2 T_{13}}{\partial x_1^2} + \frac{\partial^2 T_{13}}{\partial x_2^2} + \frac{\partial^2 T_{13}}{\partial x_3^2} + \frac{1}{1+\sigma}\frac{\partial^2 \Theta}{\partial x_1 \partial x_3}$$

$$= 0 + 0 + 0 + \frac{1}{1+\sigma}2\alpha(1+\sigma).0 = 0.$$

$$\nabla^2 T_{23} + \frac{1}{1+\sigma}\Theta_{,13} = \frac{\partial^2 T_{23}}{\partial x_1^2} + \frac{\partial^2 T_{23}}{\partial x_2^2} + \frac{\partial^2 T_{23}}{\partial x_3^2} + \frac{1}{1+\sigma}\frac{\partial^2 \Theta}{\partial x_2 \partial x_3}$$

$$= 0 + 0 + 0 + \frac{1}{1+\sigma}2\alpha(1+\sigma).0 = 0.$$

Thus, the Beltrami-Michell compatibility equations are not all satisfied. Hence, the given stress components are not solutions of the problem in elasticity. The stress equations of equilibrium for a continuous medium is given by

$$T_{ij,j} + \rho F_i = 0; \quad i, j = 1, 2, 3.$$

In absence of body forces, the stress equation of equilibrium becomes $T_{ij,j}=0$; $i, j = 1, 2, 3$,

$$T_{1j,j} = \frac{\partial T_{11}}{\partial x_1} + \frac{\partial T_{12}}{\partial x_2} + \frac{\partial T_{13}}{\partial x_3} = 2\alpha\,\sigma x_1 - 2\alpha\,\sigma x_1 + 0 = 0$$

$$T_{2j,j} = \frac{\partial T_{21}}{\partial x_1} + \frac{\partial T_{22}}{\partial x_2} + \frac{\partial T_{23}}{\partial x_3} = -2\alpha\sigma x_2 + 2\alpha\sigma x_2 + 0 = 0$$

$$T_{3j,j} = \frac{\partial T_{31}}{\partial x_1} + \frac{\partial T_{32}}{\partial x_2} + \frac{\partial T_{33}}{\partial x_3} = 0 + 0 + 0 = 0.$$

5.6 Fundamental boundary value problem in elastostatics

Therefore, in absence of body forces, the given stress components satisfy the equations of equilibrium.

(ii) Navier's equation of equilibrium: Here we are to reduce the basic equations in the second boundary value problem. The formulation of second boundary value problem of elasto-statics suggests the desirability of expressing the basic equations entirely in terms of displacements and we take the displacements as the basic unknowns.

To obtain the equations entirely in terms of displacements substitute the values of strain components given by the equations of strain-displacement kinematic relations as

$$e_{ij} = \frac{1}{2}\left[u_{i,j} + u_{j,i}\right]$$

into the stress-strain constitutive relations given by

$$T_{ij} = \lambda\theta\delta_{ij} + 2\mu e_{ij}; \theta = e_{kk}$$

to obtain

$$T_{ij} = \lambda\theta\delta_{ij} + 2\mu e_{ij} = \lambda e_{kk}\delta_{ij} + \mu\left[u_{i,j} + u_{j,i}\right] = \lambda\delta_{ij}u_{k,k} + \mu\left[u_{i,j} + u_{j,i}\right].$$

Therefore,

$$T_{ij,j} = \lambda\delta_{ij}(u_{k,k})_{,j} + \mu\left[u_{i,jj} + u_{j,ij}\right] = \lambda(u_{k,k})_{,i} + \mu\left(\nabla^2 u_i\right) + \mu(u_{j,j})_{,i}$$

$$= \lambda\theta_{,i} + \mu\nabla^2 u_i + \mu\theta_{,i} = (\lambda+\mu)\theta_{,i} + \mu\nabla^2 u_i$$

Substituting these values of stresses in equilibrium equations $T_{ij,j} + \rho F_i = 0$, we have

$$(\lambda+\mu)\theta_{,i} + \mu\nabla^2 u_i + \rho F_i = 0; , = e_{kk} \qquad (5.73)$$

The equation (5.73) contains 3 independent equations. They are known as the Navier's equation of equilibrium. In vector notation, the above equation becomes

$$(\lambda+\mu)\operatorname{grad}\theta + \mu\nabla^2 \vec{u} + \rho\vec{F} = \vec{0}.$$

Using the relation, *curl curl $\equiv$ grad div $-\nabla^2$* the Navier's equation of equilibrium equation (5.73) can be written in the vector notation as

$$(\lambda+\mu)\operatorname{grad}\operatorname{div}\vec{u} + \mu\nabla^2 \vec{u} + \rho\vec{F} = \vec{0}$$

$$(\lambda+\mu)\operatorname{grad}\operatorname{div}\vec{u} + \mu[\operatorname{grad}\operatorname{div}\vec{u} - \operatorname{curl}\operatorname{curl}\vec{u}] + \rho\vec{F} = \vec{0}$$

$$(\lambda + 2\mu)\operatorname{grad}\operatorname{div}\vec{u} - \mu\operatorname{curl}\vec{R} + \rho\vec{F} = \vec{0} \tag{5.74}$$

where, $\vec{R} = \operatorname{curl}\vec{u} = \operatorname{rot}\vec{u} =$ rotation vector. Thus we have seen that the second boundary value problem is reduced to the determination of three displacement components which must satisfy the strain-displacement kinematic relations as

$$e_{ij} = \frac{1}{2}\left[u_{i,j} + u_{j,i}\right]$$

instead of 15 subject to boundary condition $u_i = g_i(x_1, x_2, x_3)$. Once obtain the displacements components u_i, strains and stresses can be determined from

(i) six strain-displacement kinematic relations: $e_{ij} = \frac{1}{2}\left[u_{i,j} + u_{j,i}\right]$
(ii) Six stress-strain constitutive relations:

$$T_{ij} = \lambda\theta\delta_{ij} + 2\mu e_{ij}; \quad , = e_{kk}; \quad i = 1,2,3, \quad j = 1,2,3.$$

Result 5.2 If an elastic body is in equilibrium under no body forces, the invariants and rotation tensor are harmonic functions, while T_{ij} and e_{ij} are biharmonic functions.

Proof: From the displacement equation of equilibrium

$$(\lambda + \mu)\operatorname{grad}\theta + \mu\nabla^2\vec{u} = \vec{0}; \text{ as } \vec{F} = \vec{0}.$$

Taking the divergence of both sides

$$(\lambda + \mu)\operatorname{div}(\operatorname{grad}\theta) + \mu\nabla^2(\operatorname{div}\vec{u}) = 0$$

$$\Rightarrow (\lambda + \mu)\nabla^2\theta + \mu\nabla^2\theta = 0; \text{ as } \theta = \operatorname{div}\vec{u}$$

$$\Rightarrow (\lambda + 2\mu)\nabla^2\theta = 0 \Rightarrow \nabla^2\theta = 0$$

which shows that, θ is harmonic. Since, $\Theta = (3\lambda + 2\mu)\theta$ we have

$$\nabla^2\Theta = \nabla^2(3\lambda + 2\mu)\theta \Rightarrow \nabla^2\Theta = 0; \text{ as } \nabla^2\theta = 0$$

which shows that Θ is harmonic also. Differentiating partially the relation $(\lambda + \mu)\theta_{,i} + \mu\nabla^2 u_i = 0;\ i = 1,2,3$ with respect to x_i and x_j we get

$$(\lambda + \mu)\theta_{,ji} + \mu\nabla^2 u_{j,i} = 0; \quad (\lambda + \mu)\theta_{,ij} + \mu\nabla^2 u_{i,j} = 0$$

Adding the above results and multiplying by $\frac{1}{2}$, we get

$$(\lambda + \mu)\frac{1}{2}\left[\theta_{,ji} + \theta_{,ij}\right] + \mu\nabla^2\frac{1}{2}\left[u_{i,j} + u_{j,i}\right] = 0$$

5.6 Fundamental boundary value problem in elastostatics

or, $\quad (\lambda + \mu)\theta_{,ij} + \mu\nabla^2 e_{ij} = 0.$

Operating on both sides by ∇^2, we get

$$(\lambda + \mu)(\nabla^2\theta)_{,ij} + \mu\nabla^4 e_{ij} = 0 \Rightarrow \nabla^4 e_{ij} = 0; \text{ as } \nabla^2\theta = 0$$

which shows that the strain components e_{ij} is biharmonic. Subtracting the above results and multiplying by $\dfrac{1}{2}$, we get

$$(\lambda + \mu)\frac{1}{2}\left[\theta_{,ji} - \theta_{,ij}\right] + \mu\nabla^2 \frac{1}{2}\left[u_{i,j} - u_{j,i}\right] = 0$$

$$\Rightarrow \mu\nabla^2 \omega_{ij} = 0 \Rightarrow \nabla^2 \omega_{ij} = 0$$

showing that, the rotation tensor is harmonic. Again, operating on both sides of the relation $(\lambda + \mu)\theta_{,i} + \mu\nabla^2 u_i = 0$ by ∇^2, we get

$$(\lambda + \mu)\nabla^2 \theta_{,i} + \mu\nabla^4 u_i = 0$$

$$\Rightarrow \mu\nabla^4 u_i = 0 \Rightarrow \nabla^4 u_i = 0; \text{ as } \nabla^2 \theta_{,i} = 0$$

which shows that the displacement components u_{ij} are biharmonic. Using the stress-strain constitutive relations $T_{ij} = \lambda\theta\delta_{ij} + 2\mu e_{ij}$, for isotropic elastic body, we get, operating on both sides by ∇^2,

$$\nabla^2 T_{ij} = \lambda\nabla^2 \theta\delta_{ij} + 2\mu\nabla^2 e_{ij} = 2\mu\nabla^2 e_{ij}; \text{ as } \nabla^2\theta = 0$$

Again operating on both sides by ∇^2,

$$\nabla^4 T_{ij} = 2\mu\nabla^4 e_{ij} = 0; \text{ as } \nabla^4 e_{ij} = 0 \Rightarrow \nabla^4 T_{ij} = 0,$$

which shows that the stress components T_{ij} is biharmonic.

Theorem 5.1: **(Uniqueness of solutions of fundamental boundary value problems in elastostatics cases): If a body is in equilibrium under a given system of external body forces and surface forces, then the work done by the external forces of the equilibrium state in deforming the body from unstressed state to the state of equilibrium is equal to twice the strain energy of deformation.**

Proof: The uniqueness of a solution to the general elastostatic problem of elasticity may be established by use of the superposition principle, together with the law of conservation of energy. Consider a linearly elastic body in a deformed state of rest under the action of body force F_i per unit mass and the surface force $\bar{T}^{(n)}$ per unit area. The work done by the above forces during the displacement u_i is

$$W = \iiint_V \rho F_i u_i d\tau + \iint_S T_i^{(n)} u_i dS.$$

Using Gauss divergence theorem, we get

$$\iint_S T_i^{(n)} u_i dS = \iint_S T_{ij} n_j u_i dS = \iiint_V \left(T_{ij} u_i\right)_{,j} d\tau$$

$$= \iiint_V (T_{ij,j} u_i + T_{ij} u_{i,j}) d\tau$$

$$= \iiint_V T_{ij,j} u_i d\tau + \iiint_V T_{ij} e_{ij} d\tau + \iiint_V T_{ij} r_{ij} d\tau, \qquad (5.75)$$

where, $e_{ij} = \dfrac{1}{2}\left[\dfrac{\partial u_i}{\partial x_j} + \dfrac{\partial u_j}{\partial x_i}\right] =$ Symmetric tensor of order 2 $= e_{ji}$ and

$r_{ij} = \dfrac{1}{2}\left[\dfrac{\partial u_i}{\partial x_j} - \dfrac{\partial u_j}{\partial x_i}\right] =$ Skew-symmetric tensor of order 2 $= -r_{ji}$. Also, $T_{ij} = T_{ji}$, therefore

$$T_{ij} r_{ij} = T_{ji} r_{ji} = -T_{ji} r_{ij} = -T_{ij} T_{ij} \text{ and } 2T_{ij} r_{ij} = 0.$$

Therefore, Eq. (5.75) reduces to

$$\iint_S T_i^{(n)} u_i dS = \iiint_V T_{ij,j} u_i d\tau + \iiint_V T_{ij} e_{ij} d\tau.$$

If the deformation process takes place adiabatically or isothermally, we have by Clapeyron's formula

$$\text{Strain energy per unit volume} = W = \dfrac{1}{2} T_{ij} e_{ij}$$

Therefore, the expression becomes

$$W = \iiint_V (\rho F_i + T_{ij,j}) u_i d\tau + \iiint_V T_{ij} e_{ij} d\tau$$

$$= \iiint_V T_{ij} e_{ij} d\tau = \iiint_V 2W d\tau$$

As $\rho F_i + T_{ij,j} = 0$ for equilibrium.

Uniqueness: To prove the uniqueness of solutions consider an elastic body in a state of rest subjected to a specific given body force F_i. In addition to body forces either surface force $T_i^{(n)}$ or surface displacements are prescribed on the boundary. Let us assume that it is possible to obtain two sets of solutions

5.6 Fundamental boundary value problem in elastostatics

$$u'_i, T'_i \text{ and } u''_i, T''_i$$

which satisfy 15 basic equations of elasticity and boundary conditions.

Let us define

$$u_i = u'_i - u''_i, T_{ij} = T'_{ij} - T''_{ij}, e_{ij} = e'_{ij} - e''_{ij}, T_i^{(n)} = T_i'^{(n)} - T_i''^{(n)}$$

For the first state of stress, we have

$$\rho F_i + T_{ij,j}' = 0$$

as well as the following boundary condition

$$T_{ij}' n_j = f_i(x_1, x_2, x_3); i = 1, 2, 3$$

if surface forces are prescribed, or

$$u_i' = g_i(x_1, x_2, x_3); i = 1, 2, 3$$

if the boundary displacements are prescribed. Similarly, for the second state of stress

$$\rho F_i + T''_{ij,j} = 0, T''_{ij} n_j = f_i(x_1, x_2, x_3), u_i'' = g_i(x_1, x_2, x_3)$$

on boundary. Subtracting, we have

$$T'_{ij,j} - T''_{ij,j} = 0$$

and either $T'_{ij} n_j - T''_{ij} n_j = 0$ or $u'_i - u''_i = 0$ on the boundary. In other words $T_{ij,j} = 0$ at every interior point and either $T_i^{(n)} = T_{ij} n_j = 0$ or $u_i = 0$ on the boundary. Thus we have a new state of stress in which body forces are absent and either surface forces or surface displacements vanish. On the surface of the body, boundary conditions are either $T_i^{(n)} = 0$ or $u_i = 0$. In either case

$$T_i^{(n)} u_i = 0$$

at every point on the boundary. By Clapeyron's theorem

$$\iiint_V 2W d\tau = \iiint_V \rho F_i u_i d\tau + \iint_S T_i^{(n)} u_i dS.$$

For the new state of stress $F_i = 0$ in V and $T_i^{(n)} u_i = 0$ on S. Hence

$$\iiint_V W d\tau = 0.$$

But W is a positive definite quadratic form in components of strain. Hence, the integral can vanish only when $W = 0$ that is when $e_{ij} = 0$. Using Hooke's law, we get $T_{ij} = \lambda \theta \delta_{ij} + 2\mu e_{ij}$ that $T_{ij} = 0$. Therefore

$$e'_{ij} = e''_{ij} \text{ and } T'_{ij} = T''_{ij}$$

Consequently, components of strain tensor and components of stress tensor are identical. As regards the uniqueness of displacements, we recall that they, as the solutions of equations

$$\frac{\partial u_i}{\partial x_j} + \frac{\partial u_j}{\partial x_i} = 2e_{ij} = 0$$

are determined to within quantities representing rigid body displacement which has no effect on the state of stress or state of strain in the body.

5.7 Fundamental Boundary Value Problem in Elasto-dynamics

In elastodynamics, the equilibrium equations must be replaced by the equations of motion in the system of basic field equations. Therefore, all field quantities are now considered functions of time as well as of the coordinates, so that a solution for the displacement field appears in the form $u_i = u_i(x,t)$. In elastodynamics, when linearly elastic solid is in motion the shape of the body and the distribution of external body forces throughout the material are given, the problem is to find 15 unknowns: 6 stresses T_{ij}, 6 strains e_{ij} and 3 displacement functions u_i which satisfy the basic 15 equations:

(*i*) Three equations of motion:

$$T_{ij} + \rho F_i = \rho \ddot{u}_i; \quad i = 1, 2, 3,$$

(*ii*) Six stress-strain constitutive relations:

$$T_{ij} = \lambda \theta \delta_{ij} + 2\mu e_{ij}; \quad \theta = e_{kk}; \quad i = 1,2,3, \quad j = 1,2,3$$

(*iii*) and six strain-displacement kinematic relations:

$$e_{ij} = \frac{1}{2}\left[u_{i,j} + u_{j,i}\right]$$

at all interior points of linearly elastic body. The solution of these 15 partial differential equations will be regarded as functions of x_1, x_2, x_3, t. In order to determine unique solution we have to use these arbitrary functions by using a set of boundary conditions. On the bounding surface of the body either stress vectors or displacements are prescribed everywhere in the shape of boundary conditions.

(a) The stress vector is given at each point of the boundary: When the surface traction $f_i(x_1, x_2, x_3)$, are prescribed on the boundary surface of the body at time t, representing the stress vector acting on surface element with normal n_i the stresses T_{ij}, in addition, must satisfy 3 boundary conditions

5.7 Fundamental boundary value problem in elasto-dynamics

$$T_{ij}n_j = f_i(x_1, x_2, x_3); i = 1, 2, 3 \qquad (5.76)$$

To these equations it is necessary to adjoin the initial conditions specifying displacement and velocity of a point of the body at initial time $t = 0$ i.e.,

$$u_i(x_1, x_2, x_3, 0) = F_i(x_1, x_2, x_3); \frac{\partial u_i}{\partial t}(x_1, x_2, x_3, 0) = G_i(x_1, x_2, x_3) \qquad (5.77)$$

throughout the volume. Thus the problem of obtaining the displacements, strains and stresses in linearly elastic isotropic solid body in equilibrium which satisfy basic equations

(*i*) Three equations of motion: $T_{ij} + \rho F_i = \rho \ddot{u}_i$; $i = 1, 2, 3$,

(*ii*) Six stress-strain constitutive relations:

$$T_{ij} = \lambda \theta \delta_{ij} + 2\mu e_{ij}; \; \theta = e_{kk}; \; i = 1, 2, 3, \; j = 1, 2, 3$$

(*iii*) and six strain-displacement kinematic relations: $e_{ij} = \frac{1}{2}[u_{i,j} + u_{j,i}]$

in addition the boundary condition (5.76) together with the initial conditions (5.77) is known as the first fundamental boundary value problem in elastodynamics.

(b) The displacement is prescribed at each point of the boundary:
On the other hand when the functions $g_i(x_1, x_2, x_3)$ are prescribed on the boundary of the body representing the displacements u_i, the displacements, the displacements u_i, in addition, must satisfy 3 boundary conditions

$$u_i = g_i(x_1, x_2, x_3); \; i = 1, 2, 3 \qquad (5.78)$$

The problem of obtaining the displacements, strains and stresses, which satisfy the basic equations

(*i*) Three equations of motion: $T_{ij} + \rho F_i = \rho \ddot{u}_i$; $i = 1, 2, 3$,

(*ii*) Six stress-strain constitutive relations:

$$T_{ij} = \lambda \theta \delta_{ij} + 2\mu e_{ij}; \theta = e_{kk}; i = 1, 2, 3, j = 1, 2, 3$$

(*iii*) and six strain-displacement kinematic relations: $e_{ij} = \frac{1}{2}[u_{i,j} + u_{j,i}]$

and in addition the boundary condition (5.78) together with the initial conditions (5.77) is known as the second fundamental boundary value problem in elastodynamics.

5.7.1 Navier's Equation of Motion:

When the displacement are prescribed on the surface, the formulation of boundary value problems in elastodynamics suggests that we express basic equations entirely in terms of displacements. To obtain the equations entirely in terms of displacements substitute the values of strain components given by the equations of strain-displacement kinematic relations as

$$e_{ij} = \frac{1}{2}\left[u_{i,j} + u_{j,i}\right]$$

into the stress-strain constitutive relations give by

$$T_{ij} = \lambda\theta\delta_{ij} + 2\mu e_{ij}; \theta = e_{kk}$$

to obtain

$$T_{ij} = \lambda\theta\delta_{ij} + 2\mu e_{ij} = \lambda e_{kk}\delta_{ij} + \mu\left[u_{i,j} + u_{j,i}\right] = \lambda\delta_{ij}u_{k,k} + \mu\left[u_{i,j} + u_{j,i}\right].$$

Therefore

$$T_{ij,j} = \lambda\delta_{ij}(u_{k,k})_{,j} + \mu\left[u_{i,jj} + u_{j,ij}\right] = \lambda(u_{k,k})_{,i} + \mu\left(\nabla^2 u_i\right) + \mu(u_{j,j})_{,i}$$

$$= \lambda\theta_{,i} + \mu\nabla^2 u_i + \mu\theta_{,i} = (\lambda + \mu)\theta_{,i} + \mu\nabla^2 u_i.$$

Substituting these values of stresses in equilibrium equations $T_{ij,j} + \rho F_i = \rho \ddot{u}_i$, we have

$$(\lambda + \mu)\theta_{,i} + \mu\nabla^2 u_i + \rho F_i = \rho \ddot{u}_i; , = e_{kk} \quad (5.79)$$

The equation (5.79) contains 3 independent equations. They are known as the Navier's equation of motion.

(i) When body forces are absent, $F_i = 0$, Navier's equation of motion (5.79) reduces to

$$(\lambda + \mu)\theta_{,i} + \mu\nabla^2 u_i = \rho \ddot{u}_i \quad (5.80)$$

(ii) Propagation of waves in an infinite region: When the motion is irrotational, there is a potential function Φ such that $u_i = \Phi_{,i}$ and therefore,

$$\theta = u_{i,i} = \Phi_{,ii} = \nabla^2\Phi \Rightarrow \theta_{,i} = (\nabla^2\Phi)_{,i} = \nabla^2\left(\Phi_{,i}\right) = \nabla^2 u_i$$

Therefore, the equation (5.80) reduces to

$$(\lambda + \mu)\theta_{,i} + \mu\nabla^2 u_i = \rho\ddot{u}_i \Rightarrow (\lambda + \mu)\nabla^2 u_i + \mu\nabla^2 u_i = \rho\ddot{u}_i$$

$$(\lambda + 2\mu)\nabla^2 u_i = \rho\ddot{u}_i \Rightarrow \frac{\partial^2 u_i}{\partial t^2} = c_1^2\nabla^2 u_i; \text{ where } c_1 = \sqrt{\frac{\lambda + 2\mu}{\rho}} \quad (5.81)$$

Equation (5.81) represents wave equation for displacement called irrotational wave propagating with the velocity $c_1 = \sqrt{\dfrac{\lambda + 2\mu}{\rho}}$.

5.7 Fundamental boundary value problem in elasto-dynamics

(iii) When the motion is isochoric, equivolumnal $\theta = 0$, equation (5.80) reduces to

$$\mu \nabla^2 u_i = \rho \ddot{u}_i \Rightarrow \frac{\partial^2 u_i}{\partial t^2} = c_2^2 \nabla^2 u_i; \text{ where } c_2 = \sqrt{\frac{\mu}{\rho}} \tag{5.82}$$

Equation (5.82) represents a wave equation for displacement called equivolumnal waves, propagating with velocity $c_2 = \sqrt{\frac{\mu}{\rho}} < c_1$. These waves are possible in an infinite region which is so large that the effects of the boundaries can be disregarded.

Theorem 5.2: (Uniqueness of solutions of fundamental boundary value problems in elastodynamics cases): **The time rate of change of work done by the external forces in altering the configuration of the natural state of an elastic body to the current state is equal to the sum of time rate of change of kinetic energy and time rate of change of strain energy.**

Proof: Suppose a body is acted on by a surface force $T_i^{(n)}$ per unit area and a body force F_i per unit mass. Let us determine the rate at which work is done by these forces in altering the configuration from initial moment $t = 0$ corresponding to the natural state to the moment t corresponding to the current state. Let u_i be the displacement of the point at time t. The displacement of the point during the time interval is $\frac{\partial u_i}{\partial t} dt$. The work done dW by the external forces during the time interval dt is

$$W = \iiint_V \rho F_i \dot{u}_i dt \, d\tau + \iint_S T_i^{(n)} \dot{u}_i dt dS$$

$$\Rightarrow \frac{dW}{dt} = \iiint_V \rho F_i \dot{u}_i d\tau + \iint_S T_i^{(n)} \dot{u}_i dS. \tag{5.83}$$

Now, using Gauss divergence theorem, we get

$$\iint_S T_i^{(n)} \dot{u}_i dS = \iint_S T_{ij} n_j \dot{u}_i dS = \iiint_V \left(T_{ij} \dot{u}_i \right)_{,j} d\tau$$

$$= \iiint_V (T_{ij,j} \dot{u}_i + T_{ij} \dot{u}_{i,j}) d\tau$$

$$= \iiint_V T_{ij,j} \dot{u}_i d\tau + \iiint_V T_{ij} d_{ij} d\tau + \iiint_V T_{ij} w_{ij} d\tau$$

$$= \iiint_V T_{ij,j} \dot{u}_i d\tau + \iiint_V T_{ij} d_{ij} d\tau; \text{ as } T_{ij} w_{ij} = 0$$

where, $d_{ij} = \frac{1}{2} \left(\frac{\partial v_i}{\partial x_j} + \frac{\partial v_j}{\partial x_i} \right) = \frac{1}{2} (\dot{u}_{i,j} + \dot{u}_{j,i}) = \dot{e}_{ij}$

and $w_{ij} = \frac{1}{2}\left(\frac{\partial v_i}{\partial x_j} - \frac{\partial v_j}{\partial x_i}\right) = \frac{1}{2}(\dot{u}_{i,j} - \dot{u}_{j,i})$, and so

$$T_{ij}w_{ij} = T_{ji}w_{ji} = T_{ji}w_{ji} = -T_{ij}w_{ij} \Rightarrow 2T_{ij}w_{ij} = 0.$$

Therefore, Eq. (5.83) reduces to

$$\frac{dW}{dt} = \iiint_V \rho F_i \dot{u}_i d\tau + \iiint_V T_{ij,j} \dot{u}_i d\tau + \iiint_V T_{ij} d_{ij} d\tau$$

$$= \iiint_V (\rho F_i + T_{ij,j}) \dot{u}_i d\tau + \iiint_V T_{ij} d_{ij} d\tau$$

$$= \iiint_V \rho \ddot{u}_i \dot{u}_i d\tau + \iiint_V T_{ij} d_{ij} d\tau; \text{ as } \rho F_i + T_{ij,j} = \rho \ddot{u}_i$$

$$= \frac{dK}{dt} + \iiint_V T_{ij} d_{ij} d\tau,$$

where, K = kinetic energy of the body = $\iiint_V \rho \ddot{u}_i \dot{u}_i d\tau$. Therefore,

$$\frac{dW}{dt} = \frac{dK}{dt} + \iiint_V T_{ij} \frac{\partial e_{ij}}{\partial t} d\tau; \text{ as } d_{ij} = \dot{e}_{ij} = \frac{\partial e_{ij}}{\partial t}$$

$$= \frac{dK}{dt} + \iiint_V \frac{\partial W}{\partial e_{ij}} \frac{\partial e_{ij}}{\partial t} d\tau; \text{ as } T_{ij} = \frac{\partial W}{\partial e_{ij}}$$

$$= \frac{dK}{dt} + \frac{d}{dt} \iiint_V W d\tau$$

where, $\iiint_V W d\tau$ is strain energy and W is strain energy per unit volume.

Uniqueness: To prove the uniqueness of solutions, consider an elastic body in motion subjected to a specific given body force F_i. In addition to body forces, either surface force $T_i^{(n)}$ or surface displacements are prescribed on the boundary. Let us assume that it is possible to obtain two sets of solutions

$$u_i', e_{ij}', T_{ij}' \text{ and } u_i'', e_{ij}'', T_{ij}''$$

which satisfy 15 basic equations of elasticity and boundary conditions. Let us define

$$u_i = u_i' - u_i'', T_{ij} = T_{ij}' - T_{ij}'', e_{ij} = e_{ij}' - e_{ij}'',$$

For the first state of stress, we have

$$\rho F_i + T_{ij,j}' = \rho \ddot{u}_i'$$

5.7 Fundamental boundary value problem in elasto-dynamics

as well as the following boundary condition

$$T'_{ij} n_j = f_i(x_1, x_2, x_3,); i = 1, 2, 3$$

if surface forces are prescribed, or

$$u'_i = g_i(x_1, x_2, x_3, t); i = 1, 2, 3$$

if the boundary displacements are prescribed. Similarly, for the second state of stress

$$\rho F_i + T''_{ij,j} = \rho \ddot{u}''_i, T''_{ij} n_j = f_i(x_1, x_2, x_3, t), u''_i = g_i(x_1, x_2, x_3, t)$$

on boundary. Subtracting, we have

$$T'_{ij,j} - T''_{ij,j} = \rho(\ddot{u}'_i - \ddot{u}''_i)$$

and either $T'_{ij}n_j - T''_{ij}n_j = 0$ or $u'_i - u''_i = 0$ on the boundary. In other words $T'_{ij,j} = \rho \ddot{u}_i$ at every interior point and either on the boundary.

$$T_i^{(n)} = T_{ij} n_j = 0 \text{ or } u_i = 0; \text{ for } t \geq 0$$

Thus we have a new state in which body forces are absent and surface forces or surface displacements vanish. On the surface of the body, boundary conditions are either $u_i = 0$ for $t \geq 0$, we must have $\dfrac{\partial u_i}{\partial t} = 0$ on the surface for $t \geq 0$. On the surface of the body, boundary conditions are

$$T_i^{(n)} = 0; \quad \frac{\partial u_i}{\partial t} = 0 \text{ for } t \geq 0$$

In either case, $T_i^{(n)} \dfrac{\partial u_i}{\partial t} = 0$ on the surface for $t \geq 0$. Since both the solutions of the problem must satisfy the same initial condition, we have

$$u_i = 0; \quad \frac{\partial u_i}{\partial t} = \dot{u}_i = 0 \text{ for } t = 0.$$

Now, we know that

$$\frac{dW}{dt} = \frac{dK}{dt} + \frac{d}{dt} \iiint_V W d\tau = \iiint_V \rho F_i u_i \, d\tau + \iint_S T_i^{(n)} u_i \, dS.$$

Since for the new state body forces are absent, $F_i = 0$, and $T_i^{(n)} \dfrac{\partial u_i}{\partial t} = 0$ on the surface for $t \geq 0$, so $\dfrac{dW}{dt} = 0$,

$$\Rightarrow \frac{dK}{dt} + \frac{d}{dt} \iiint_V W d\tau = 0 \Rightarrow K + \iiint_V W d\tau = \text{constant}.$$

Since, $u_i = 0; \dot{u}_i = 0$, for $t = 0$, constant of integration must be zero. Hence

$$K + \iiint_V W d\tau = 0.$$

Since both kinetic energy and are essentially positive definite, we have

$$K = 0 \text{ and } W = 0; \text{ for all } t \geq 0$$

$$\Rightarrow \frac{\partial u_i}{\partial t} = \dot{u}_i = 0 \text{ and } e_{ij} = 0; \text{ for all } t \geq 0$$

$$\Rightarrow u_i = \text{independent of time}; \frac{1}{2}\left[u_{i,j} + u_{j,i}\right] = 0$$

i.e., solution can represent only rigid body displacement of the body. But the displacements $u_i = 0$ at $t = 0$. Hence this rigid body displacement must be zero at all points of the body and at all time. Hence two solutions are completely identical.

Exercise

Short Answer Type Questions

1. State generalized Hooke's law.
2. Write down Hooke's law for isotropic elastic material and explain the different terms involved there.
3. Deduce the displacement equations of equilibrium and motion in an isotropic elastic medium.
4. Write down the compatibility equations for strain components.
5. Write down the expressions for strains in terms of stresses.
6. Show that the Lame' constants λ, μ are both positive
7. State the fundamental boundary value problems of elasto-statics and elasto-dynamics.

Long Answer Type Questions

1. What is energy function? Write down the form of strain energy function for an isotropic medium and obtain its expression in terms of strain invariants.
2. What is strain energy function? Write down the form of strain energy function for an isotropic medium and obtain its expression in terms of strain invariants.

$$u_1 = -\alpha x_2 x_3, u_2 = \alpha x_1 x_3, u_3 = \alpha\ \varphi(x_1, x_2)$$

5.7 Fundamental boundary value problem in elasto-dynamics

3. An isotropic solid is subjected to one-dimensional deformation given by $\bar{u}_1 = (u_1(x_1,t), 0, 0)$. Simplify the displacement equation of motion. Interpret the type of the equation so obtained. Interpret also the different terms.

4. Test whether the following strain components are possible in an elastic body:

$$e_{11} = k(x^2{}_1 + x^2{}_2); e_{12} = k(x^2{}_2 + x^2{}_3) e_{13}$$
$$= kx^2{}_1 x^2{}_2 x^2{}_3; e_{23} = e_{31} = e_{33} = 0 (k \neq 0)$$

5. What are "elastic modulii"? Define Young's E modulus and Poisson's ratio σ. Show that $E = \dfrac{\mu(3\lambda + 2\mu)}{\lambda + \mu}, \sigma = \dfrac{\lambda}{2(\lambda + \mu)}$ where λ, μ are Lame constants.

6. Deduce Beltrami-Michell compatibility equations for stresses.

7. Show that the following stress components satisfy the equations of equilibrium of an isotropic elastic solid with zero body force. Test whether they can be a solution of any problem of elasticity.

$$T_{11} = \alpha\left[x_2^2 + \sigma(x_1^2 - x_2^2)\right]; T_{12} = -2\alpha \, \sigma x_1 x_2; T_{13} = 0$$

$$T_{22} = \alpha\left[x_1^2 + \sigma(x_2^2 - x_1^2)\right]; T_{23} = 0; T_{33} = \alpha \, \sigma(x_1^2 + x_2^2); \alpha \neq 0.$$

8. Express the strain energy density function W in the form

$$W = -\dfrac{\sigma}{2E}\Theta^2 + \dfrac{1+\sigma}{2E} T_{ij} T_{ij}; \ i,j = 1,2,3,$$

where $\Theta = T_{ii}$ = sum of the normal stresses, T_{ij} = stress tensor.

9. An isotropic elastic solid is subjected to the following stress system under no body forces and is in equilibrium: $T_{11} = T_{22} = T_{33} = -p$, $T_{23} = T_{31} = T_{12} = 0$, where $p > 0$ is a constant. Find the displacement components.

10. An isotropic elastic prism whose lateral surface is free of stress is subjected to a uniform tension parallel to its axis in absence of body forces as follows: $T_{11} = T, T_{22} = T_{33} = T_{23} = T_{31} = T_{12} = 0$. Show that the displacements are given by

$$\dfrac{u_1}{x_1} = \dfrac{T}{E}; \dfrac{u_2}{x_2} = \dfrac{u_3}{x_3} = -\dfrac{\sigma T}{E}$$

11. The displacement field (u_1, u_2, u_3) in a isotropic deformable body is given by

$$u_1 = -\dfrac{\sigma p}{E} x_1, \ u_2 = -\dfrac{\sigma p}{E} x_2, \ u_1 = \dfrac{p}{E} x_3$$

where, $p > 0$ is a constant and Young's modulus E and Poisson's ratio σ. Find the strain and stress components producing the deformation.

12. Consider the following stress system: $T_{11} = T_{22} = T_{33} = 0, T_{23} = \mu\alpha x_1, T_3$ $T_{31} = -\mu\alpha x_2, T_{12} = 0$, where, α is a constant and μ is the modulus of rigidity. Find the displacement components.

13. An isotropic elastic solid in absence of body forces and subjected to the follow stress system is in equilibrium: $T_{11} = T_{22} = T_{33} = 0, T_{23} = \mu\alpha x_1$, $T_{31} = -\mu\alpha x_2, T_{12} = 0$, where, α is a constant and μ is the modulus of rigidity. Find the displacement components. Find also the expression for the strain energy density function.

14. Find the condition for which the following are the possible strain components: $T_{11} = ax_2^2, T_{22} = ax_1^2, T_{12} = bx_1x_2 ; T_{33} = T_{31} = T_{23} = 0$, where a and b are constants.

15. If the body forces are constants, prove that the invariants Θ and θ are harmonic functions and the stress components T_{ij} and strain components e_{ij} are biharmonic functions.

16. If an elastic body is in equilibrium under no body forces, prove that the invariants and rotation tensor are harmonic functions, while T_{ij} and e_{ij} are biharmonic functions.

17. Prove that the Navier's equation of equilibrium can be written in vector notation as

$$(\lambda + \mu)\operatorname{grad}\theta + \mu\nabla^2\vec{u} + \rho\vec{F} = \vec{0}.$$

18. Show that two types of body waves can propagate in an infinite isotropic elastic solid.

Bibliography

[1] William Prager (2004), Introduction to Mechanics of Continua, Dover Publications.
[2] Rabindranath Chatterjee (1999), Mathematical Theory of Continuum Mechanics, Narosa Publishing House.
[3] Fridtjov Irgens (2008), Continuum Mechanics, Springer.
[4] J.N. Raddy, An introduction to continuum mechanics, Cambridge University Press.
[5] A. J.M. Spancer, Continuum Mechanics, Longman, 1980.
[6] A.C.Eringen, Mechanics of continua, Wiley, 1967.
[7] W.Pager, Introduction to Mechanics of Continua, lexiton, Mass, Ginn, 1961
[8] R.George Mase, Schaum's outline of theory and problems of continuum mechanics, McGraw-Hill, 1970.
[9] P.K. Nayak, Textbook of Tensor Calculus and Differential Geometry. PHI, 2009.
[10] P.K. Nayak, Vector Algebra and Analysis with Applications, 2017, Universities Press.